Lars Bergström   Ariel Goobar

Cosmology and Particle Astrophysics

Second Edition

 Springer

Published in association with
**Praxis Publishing**
Chichester, UK

Professor Lars Bergström and Dr Ariel Goobar
Department of Physics
Stockholm University
Stockholm
Sweden

April 22, 23 and 24 UT. North is up, and East to the left; the field of view is 194 × 194 arcseconds. The supernova SN1998aq was still in its rising phase when the exposures were taken, and is visible as the bright blue object to the upper right fo the galaxy nucleus. Back cover inset: Composite image of the galzxy NGC 5965 taken on the early evening of 2001 August 19. North is up, and East to the left; the field of view is 260 × 260 arcseconds. The supernova SN2001cm in this galaxy is visible as the blue star just below the central dust land, above and to the left of the galaxy nucleus. Observation and image processing for both cover images by Hakon Dahle, which are reproduced here with his permission.

SPRINGER–PRAXIS BOOKS IN ASTRONOMY AND PLANETARY SCIENCE
SUBJECT *ADVISORY EDITORS*: Dr. Philippe Blondel, C.Geol., F.G.S., Ph.D., M.Sc., Senior Scientist, Department of Physics, University of Bath, Bath, UK; John Mason B.Sc., M.Sc., Ph.D.

ISBN 3-540-32924-2 Springer-Verlag Berlin Heidelberg New York

Springer-Verlag is a part of Springer Science + Business Media (springeronline.com)

Library of Congress Control Number: 2003067259

Bibliographic information published by Die Deutsche Bibliothek

Die Deutsche Bibliothek lists this publication in the Deutsche Nationalbibliografie; detailed bibliographic data are available from the Internet at http://dnb.ddb.de

Reprinted and issued in paperback, 2006

Typeset in LaTex by the authors
Cover design: Jim Wilkie

Printed on acid-free paper

Preface to the Second Edition

Astroparticle physics is an arena where things are rapidly evolving. When we, four years after the first edition, were editing the text we found many places where the first edition was already obsolete. This is in particular true in Chapter 4 which treats cosmological models. Since the first edition, the presence of 'dark energy' has become much more established, so we include a new section on that topic in the new edition. In Chapter 4 we have also included a section on the distance scale of the Universe.

We have also rearranged the sections slightly, moving the chapter on phase transitions (Chapter 10) to become Chapter 7. In that way, we assemble the particle physics material together in Chapters 6 and 7. We have also included a summary page for Chapter 6, which was missing in the previous version.

We have included a discussion of a topical field, that of weak gravitational lensing, in Chapter 5.

In Chapter 10, we have put together some of the material concerning the presently accelerating Universe with that of primordial inflation, and have added a section about dark energy and quintessence models.

Also in the neutrino sector much development has taken place, with two experiments SNO and KamLAND presenting new results that pin down the solar neutrino sector quite accurately. We have added some comments about these experimental results in Chapter 14, and also introduced a section on the Mikheev-Smirnov-Wolfenstein (MSW) mechanism and defined the MSW large angle solution as indicated by the results.

The subject where the most activity is taking place at present is that of the cosmic microwave background radiation (CMBR). During the last few years there has been a flurry of balloon- and ground-based experiments which have measured the angular dependence of the CMBR and used its dependence on cosmological parameters to limit a large number of such parameters. In February, 2003 the WMAP satellite released its first set of data. We have added a section which discusses these new developments in Chapter 11. We have also included a new Appendix, which deals with the problem of primordial structure formation in terms of quantum fluctuations of a scalar field, such as predicted in theories of inflation, and which is supported by the WMAP data. We think that this is one of the greatest intellectual triumphs of modern cosmology. The less sophisticated reader may, however, prefer to skip the details.

For the convenience of the reader, we have expanded the index considerably.

Finally, we have tried to correct all printing and other errors that existed in the book. Of course, new ones will most probably be found, so it is always wise to check on the book's homepage www.physto.se/~lbe/cosm_book.html.

For finding the errors so far and for discussions that have improved the contents of the book we wish to thank R. Amanullah, J. Edsjö, D. Enström, M. Eriksson, M. Gålfalk, U. Goerlach, Anne Green, C. Gunnarsson, M. Gustafsson, P.O. Hulth, L. Liljestad, and E. Mörtsell. We wish to thank John Mason for a careful reading of the manuscript, suggesting many improvements.

Stockholm, October 2003   Lars Bergström & Ariel Goobar

Preface to the First Edition

The fields of cosmology and particle astrophysics (sometimes collectively named astroparticle physics) are currently experiencing an era which will most probably be remembered as 'the golden age'. The developments during the last few years have been truly astonishing, and the planning and building of new detectors, telescopes and other experimental facilities will guarantee an interesting decade to follow.

When attempting to convey to students our enthusiasm for these exciting developments, and when trying to teach some of the material to undergraduate and beginning graduate students, we found that a textbook of the appropriate level and scope was simply missing. It is our hope that this book will be found to successfully fill the gap. In addition, we think the book will also be very helpful for researchers in these areas and especially for those from the many related fields of science.

It is true that there exist many excellent textbooks both in particle physics and cosmology (many of them mentioned at the end of the first chapter), but none which brings the student rapidly to the fields where the most exciting developments are taking place today. We think this may be especially problematical for astronomy students, who will hardly have the time and energy to take advanced field theory courses just to acquire some knowledge, for instance, about the meaning of the cosmological constant. Neither will he or she be likely to master the full gauge theory machinery of the Standard Model of particle physics to be able to compute the cross-section for solar neutrino scattering. Still, these are two examples of subjects of relevance to the present-day astronomer and astrophysicist where a university education should not leave them completely without knowledge.

A major problem we encountered when giving a course of this kind is the very diverse background of students. To grasp the material in this book the student should have some knowledge of advanced quantum mechanics and classical field theory. However, not all students have this – especially not astronomy students. To solve this without having to load the first chapters with material that would be repetitious for many readers, we decided to make fairly extended appendices, with summaries at the end, which provide the required background. It is easily possible to use the book for an introductory course in relativistic quantum mechanics by just using Chapter 1 and appendices B, C and D.

We have aimed at making the book self-contained, which means that most of the phenomena discussed in the later chapters of the book may be understood from 'first principles'. Of course this means that we have had to narrow down the scope of the book to those areas where we think the recent developments and future prospects are especially exciting. This includes neutrino astrophysics, structure formation and the microwave background, gamma-ray astronomy, gravitational lensing, determination of cosmological parameters, and gravitational waves. In addition, more 'traditional' topics like Big Bang cosmology, thermodynamics, nucleosynthesis, dark matter and inflation are treated. A chapter on phase transitions is also provided, which explains in elementary terms how 'exotic' objects like cosmic strings and textures may have been produced.

We have tried to be as up to date as possible. Among other things we treat the technique that uses distant supernovae to determine the energy density components of the Universe. We also explain the principles behind the atmospheric neutrino oscillation detection by the Super-Kamiokande collaboration, which seems to indicate that neutrinos are not massless. We present the ideas behind the large neutrino telescopes, perhaps reaching sizes of cubic kilometres, which are currently being planned.

Of course, the risk we take by including very new material is that the book may age more rapidly than if we had only included standard material. On the other hand, if the book becomes successful, the chances are high that we will update it in the not too distant future. For small changes, additions and other comments, please check our internet homepage for the book, at the internet web address http://www.physto.se/~lbe/cosm_book.html.

According to our experience, the book can serve as a textbook for a one-semester course. If the students have little previous experience with relativity, one should devote at least two weeks each to Chapters 2 and 3. As the material in the later chapters is quite extensive, the lecturer probably has to make a decision as to which topics to include. The chapter on phase transition (Chapter 7) is fairly advanced and can be omitted without affecting the understanding of the later chapters. It gives, however, a flavour of the exciting links between cosmology and condensed matter physics which at present are growing stronger. Chapters 12 – 15 are also quite independent and may be included or omitted in a course according to preference.

The authors are grateful to several colleagues, including J. Bahcall, P. Carlson, J. Edsjö, T.H. Hansson, P.O. Hulth, E. Mörtsell, H. Rubinstein, G. Smoot, H. Snellman, M. Tegmark and P. Ullio for many useful comments and suggestions, and to many students taking our course at Stockholm University, especially E. Dalberg, C. Gunnarsson, M. Kaufmann and L. Samuelsson, for a careful reading of the manuscript. Special thanks go to R.A. Marriott for many useful suggestions on the style of presentation.

Stockholm, November 1998   Lars Bergström & Ariel Goobar

Table of Contents

1 The Observable Universe

1.1 Introduction

One of the most impressive achievements of science is the development of a quite detailed understanding of the physical properties of the Universe, even at its earliest stages. Thanks to the fruitful interplay between theoretical analysis, astronomical observations and laboratory experiments, we have today very successful 'Standard Models' of both particle physics and cosmology. The Standard Model of particle physics involves matter particles: quarks which always form bound states such as neutrons and protons, and leptons such as the electron which is charged and therefore can make up neutral matter when bound to nuclei formed by neutrons and protons. There are also neutral leptons, neutrinos, which do not form bound states but which play a very important role in cosmology and particle astrophysics as we shall see throughout this book. The other important ingredients in the Standard Model of particle physics are the particles which mediate the fundamental forces: the photon, the gluons and the W and Z bosons.

The Standard Model of cosmology is the Hot Big Bang model, which states that the Universe is not infinitely old but rather came into existence some 13 to 14 billion years ago. It started out in a state which after a small fraction of a second was enormously compressed and therefore very hot. No bound states could exist because of the intense heat which caused immediate dissociation even of protons and neutrons into quarks if they, by chance, were formed in the so-called quark-gluon plasma. Subsequently, the Universe expanded and cooled making possible the formation of a sequence of ever more complex objects: protons and neutrons, nuclei, atoms, molecules, clouds, stars, planets,.... As we shall see, the observational support for the Big Bang model is overwhelming. The key observations are:

- the present expansion of the Universe,
- the existence of the cosmic microwave background radiation, CMBR, the relic radiation from the hot stage of the early Universe,
- the relative abundance of light elements in the Universe, which agrees accurately with what would be synthesized in an initially hot, expanding Universe.

Also, the fact that the oldest objects found in the Universe – globular clusters of stars and some radioactive isotopes – do not seem to exceed an age around 13 billion years gives strong evidence for a Universe with a finite age, as predicted by the Big Bang model.

Although there are still many puzzles and interesting details to fill in, both in the Standard Model of particle physics and in the Big Bang model, they do an amazingly good job at describing the majority of all phenomena we can observe in nature. Combined, they allow us to follow the history of our Universe back to only about 10^{-10} seconds after the Big Bang using established physical laws. Extending the models, there are scenarios that describe the evolution back to 10^{-43} seconds after the Big Bang! Behind this remarkable success are the theories of *General Relativity* and *Quantum Field Theory*, which we shall explain thoroughly in this book. However, many fundamental aspects of the laws of nature remain uncertain and are the subject of modern research. The key problem at present is to find a valid description of quantized gravity, something which is needed to push our limit of knowledge even closer to (and maybe eventually explaining) the Big Bang itself.

In this chapter we shall review some of the most striking observational facts about our Universe. Many observations and underlying physical phenomena will be explained in detail in later chapters.

1.2 Baryonic Matter

Ordinary matter, made of protons and neutrons, is generically called *baryonic matter*. The particle physics definition of baryonic matter also includes other shortlived particles (see Chapter 6) but those are not stable over cosmological time scales. In normal matter, electrons are of course also present in equal numbers as protons. However, being almost 2000 times lighter, they do not contribute much to the present mass density of the Universe. Baryons are found in a variety of forms: in gaseous clouds, either of neutral atoms or molecules, or in ionized plasma, in frozen condensations in comets and in dense, hot, environments such as planets, stars and stellar remnants such as white dwarfs, neutron stars and presumably in black holes. Except for planets and stellar remnants, baryonic condensations consist mainly of hydrogen and helium. One of the most striking confirmations of the Standard Model of cosmology is the observed relative mass abundances of the light elements: helium ^{4}He (24 per cent), deuterium ^{2}H ($3 \cdot 10^{-5}$), and lithium ^{7}Li (10^{-9}).

The synthesis of the light elements, nucleosynthesis, from neutrons and protons, which according to the Standard Model took place through nuclear reactions in the first hundred seconds after the Big Bang, cannot be explained by any other known astrophysical model. The observed abundance of ^{4}He, for example, is significantly higher than could have possibly been produced in stars. Heavier nuclei, on the other hand, were produced much later in astrophysical environments,[1] and in fact only make up a very small fraction of the total baryonic content of a typical galaxy (in that respect, the Earth is a highly unusual place in the Universe!). The combined set of abundances: helium, deuterium and lithium, agree nicely with observations provided that there were many more photons than baryons during nucleosynthesis, $\eta_B = \frac{n_B}{n_\gamma} \approx 10^{-9}$, where n_B and n_γ are the respective number densities.

The agreement between the data and the theory of nucleosynthesis also has implications for particle physics. The rate of reactions involved in the production of the light elements is sensitive to the ratio of the neutron and proton densities (n/p) in the early Universe. As (n/p) depends on the number of neutrino species, it is possible to constrain the existence of additional neutrino types. The details will be explained in Chapter 9.

Example 1.2.1 Estimate how much helium could have been formed by stars in our Galaxy assuming that the age of the Milky Way is 10^{10} years and that is has been radiating all the time at its current power, $L_* = 4 \cdot 10^{36}$ W. The conversion of one kilogram of hydrogen to helium yields an energy production of $6 \cdot 10^{14}$ J.

Answer: The total produced mass of helium through hydrogen burning[2] in stars of the Milky Way becomes:

$$M_{He} = \frac{4 \cdot 10^{36} \times 10^{10} \times 3 \cdot 10^7}{6 \cdot 10^{14}} = 2 \cdot 10^{39} \text{ kg}$$

The total mass of our Galaxy is believed to be about $3 \cdot 10^{41}$ kg, thus hydrogen burning in stars can account for less than 1 per cent helium abundance, much below the observed 24 per cent. In addition, there is no known mechanism that can expel large quantities of helium from hydrogen burning stars without also (like in supernovae) generating more mass in the form of heavy elements than is observed.

1.3 Antimatter

Astronomical observations indicate that there is not much antimatter (that is antiprotons, antineutrons, positrons, etc.) in the Universe. For example, if sizeable amounts of antimatter were present in our Galaxy, it would be disclosed by powerful explosions as a result of matter-antimatter annihilation. Even more strikingly, if matter and antimatter had existed in equal amounts at the time, say, of nucleosynthesis, they would have rapidly annihilated each

---

[1] As a poetic side-remark, you may notice that this means that most of the atoms in your body have resided in stars – you are made from star-stuff!

[2] *Hydrogen burning* refers to the nuclear reactions in stars that convert hydrogen into helium

other due to the high density, and stable matter would never have formed. However, there seems to have been a small excess of matter compared with antimatter. In fact, the small baryon to photon ratio $\eta_B \approx 10^{-9}$, is roughly a measure of the baryon antibaryon asymmetry that must somehow have been created in the early evolution of the Universe. The origin of this asymmetry has been a major puzzle for cosmologists. Today, there are particle physics models that explain this type of depletion of antimatter in the Universe. A key element of these models is that there are differences in the interactions of particles and antiparticles (charge-parity or CP violation), a fact that was discovered in the laboratory by experimental particle physicists. The detailed mechanism at work in the early Universe is not known, however, but is the topic of intense research at present.

1.4 The Expansion of the Universe

The crucial observational fact of modern cosmology is that the Universe is expanding. This revolutionary idea was first proposed by Alexander Alexandrovich Friedmann, a young Russian mathematician and meteorologist. Friedmann in 1922 and, independently in 1927, Georges Lemaître (a Belgian priest and scholar) found that the solutions of the equations of general relativity of gravity in an isotropic and homogeneous Universe were not static. Thus, the Universe should be either contracting or expanding. This prediction was confirmed when the Universe was in fact observed to be expanding – distant galaxies were found to be moving away from us (and from each other) with velocities proportional to their relative distances, $v = H_0 \cdot d$. This is known as the *Hubble law* after its discoverer, Edwin Hubble, who published this result in 1929. The so-called *Hubble constant*, H_0, is one of the most fundamental parameters of modern cosmology.[3] The dynamics of the expansion in the Friedmann solution is governed by two additional parameters: the mass density of the Universe, Ω_M, and the energy density associated with the so-called *cosmological constant*, Ω_Λ. The latter has an interesting history. As we shall see in Chapter 4, it was introduced by Einstein in order to make a static Universe possible. When the Hubble expansion was established, Einstein is said to have regarded the introduction of this constant as a big mistake. It has not been easy to remove it from the theory, however. There is nothing that really prevents it from existing, and during several epochs of cosmology it has been fashionable to introduce it and analyse the consequences. In fact, as we shall se, there are today several observations which indicate that the cosmological constant is indeed present and non-zero.

---

[3] Note that the modern cosmological models are homogeneous and isotropic on large scales. This means that all observers in all galaxies will find themselves to be at the 'centre' of the expansion, with all other distant galaxies moving away. A better way to express it is that there is no centre and no preferred position. We will return to this later.

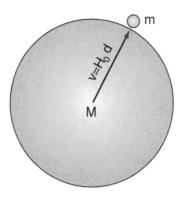

Fig. 1.1. A test particle on the surface of spherical piece of space escapes falling back into the gravitational field if the mass density of space is smaller than ρ_c.

While the presence of a mass density slows down the expansion due to the mutual gravitational attraction of massive bodies, the cosmological constant $\Omega_\Lambda > 0$ has the reverse effect. Both Ω_M and Ω_Λ are dimensionless parameters defined as $\Omega_M = \rho_M/\rho_c$ and $\Omega_\Lambda = \Lambda/3H_0^2$, where ρ_c is the critical density dividing an ever expanding Universe from one bound to collapse in the future for $\Lambda = 0$. The focus of modern observational cosmology is to measure H_0, Ω_M and Ω_Λ with ever better precision. Today, as we will see throughout this book, the parameters are constrained to be $H_0 = 60 - 80$ km s^{-1}Mpc^{-1}, $0.2 < \Omega_M < 0.6$ and $\Omega_\Lambda \geq 0$. (One megaparsec, Mpc, is $3.26 \cdot 10^6$ light years; one light year is $9.46 \cdot 10^{15}$ m, so in *SI* units H_0 is given in units of $3.2 \cdot 10^{-20}$ s^{-1}.)

Note that the energy density components of the Universe change with time as the Universe expands. We use capital indices, e.g. Ω_M, Ω_Λ whenever we refer to the density at the *present* time. Sometimes the index 0 is used to denote current values of a parameter, e.g. H_0. Lower case indices $(\Omega_m, \Omega_\lambda)$ denote energy densities at arbitrary times.

A 'concordance model', which is currently used to compare with observation very successfully, has $\Omega_M = 0.3$ and $\Omega_\Lambda = 0.7$.

A number of ongoing and future measurements will likely reduce the uncertainties of all three parameters to just a few per cent. The measurements giving $\Omega_\Lambda > 0$, if correct, imply that the expansion rate of the Universe is in fact accelerating.[4] This will be investigated in detail in Chapter 4.

Example 1.4.1 Use ordinary Newtonian gravity to show that a massive particle outside a spherical piece of the Universe expanding with a velocity $v = H_0 d$ (as shown in Fig. 1.1) will escape the gravitational attraction if the density of space is $\rho_M \leq \rho_c = \frac{3H_0^2}{8\pi G}$. (This result, as we will see later, is

---

[4] The condition for positive acceleration is $\Omega_\Lambda > \frac{\Omega_M}{2}$.

the same as in a full analysis based on Einstein's general relativity theory. However, as Einstein noticed, there may appear a cosmological constant in the Universe, the presence of which could change the result.)

Answer: For a test particle with exactly the escape velocity conservation of energy implies that:

$$\frac{mv^2}{2} - \frac{GMm}{d} = 0 \qquad (1.1)$$

where the first term is the kinetic energy and the second is the gravitational potential energy due to the matter within the sphere of radius d. That is, as the test particle reaches infinity it has lost all its kinetic and gravitational energy.

Inserting the Hubble law, $v = H_0 d$ and $M = \frac{4\pi \rho_c d^3}{3}$ we find the expression for the critical mass density:

$$\rho_c = \frac{3H_0^2}{8\pi G} \qquad (1.2)$$

Numerically, 1.2 gives the present value of the critical density (it is, as we will see later, time-dependent, and was much larger during earlier epochs of the Universe)

$$\rho_{crit}^0 = 1.9 \cdot 10^{-29} h^2 \text{ g cm}^{-3} \qquad (1.3)$$

where it is very convenient (we will do so throughout the book) to put the observational uncertainty of the present Hubble expansion rate into the dimensionless parameter h, defined as

$$h = \frac{H_0}{100 \text{ km s}^{-1} \text{ Mpc}^{-1}} \qquad (1.4)$$

where observations of H_0 presently restrict h to be in the range between 0.6 and 0.8.

1.5 Dark Matter

The success of the theory of primordial nucleosynthesis makes it also possible to put limits on the density of matter in the Universe in the form of baryons, $\Omega_B = \rho_B/\rho_c$. Using h defined as in (1.4), or equivalently $H_0 = h \cdot 100 \text{ km s}^{-1} \text{Mpc}^{-1}$, the obtained limits from the comparison of the abundance data to the theory can be written as: $0.014 \le \Omega_B h^2 \le 0.026$ [11]. Actually, the analysis of new data which we will describe in Chapter 11, gives the result $\Omega_B h^2 = 0.0224 \pm 0.009$.

Using the value of h between 0.6 and 0.8 we find that the mass density of baryons thus appears to be significantly lower (by a factor $5 - 8$) than the total matter density, $\Omega_M \sim 0.3$, which is our first contact with the *dark matter* problem. There is a large amount of astronomical evidence indicating that

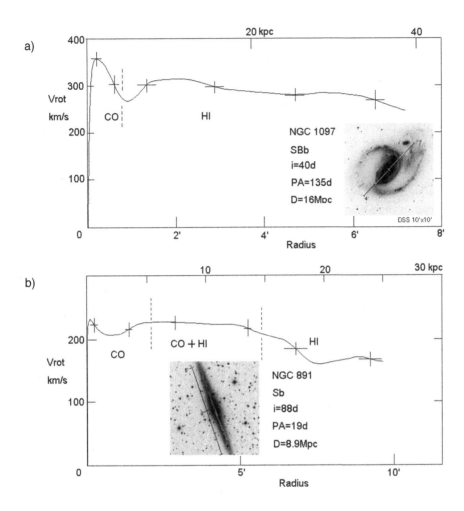

Fig. 1.2. Rotation curves of spiral galaxies. a) Optical image of the galaxy NGC1097 where the 4 arcmin distance from the core is shown in the right-hand side. (One arcminute is 1/60 of one degree; one arcsec is 1/60 of 1 arcmin.) To the left, a composite rotation curve made of radio observations of the Doppler shifts in CO line emission and the HI (21 cm) line. b) Similarly for the edge-on NGC891 galaxy, with a scale reaching 5 arcminutes from the galaxy center. Courtesy of Y. Sofue, Institute of Astronomy, Tokyo [45].

there is more matter than what can be associated with the luminous parts of the galaxies. For example, the orbital velocity of stars and gas clouds as a function of their distance from the galaxy core in spiral galaxies contradicts Kepler's third law unless the distribution of mass extends far beyond the visible galaxy core. For distances beyond the visible galaxy disk, the orbital velocities should decrease with the distance as $v_{orb} \propto r^{-\frac{1}{2}}$. Instead, the observations indicate that the rotation curves flatten out, that is $v_{orb} \approx$ const, as shown in Fig. 1.2, based on CO and neutral hydrogen (so-called HI) line emission outside the disk region. Then, unless classical Newtonian mechanics breaks down at scales of the size of spiral galaxies, a new massive source of gravity that does not emit any radiation, therefore called *dark matter*, has to be introduced. The rotation curves can be made compatible[5] with expectations if the galaxy cores are surrounded by a dark 'halo' with $\rho_{DM} \propto r^{-2}$ over some intermediate range of r. The mass density in luminous matter in galaxies is found to be only $\Omega_{LUM} \approx 0.01$. The dark matter halo is at least ten times as massive, $\Omega_{HALO} \approx 0.1 - 0.3$.

Example 1.5.1 Use Kepler's third law to show that orbital velocities of stars far from a massive core should fall as $v_{orb} \propto r^{-\frac{1}{2}}$, where r is the distance to the core.

Answer: According to Kepler's third law:

$$GM(r) = v_{orb}^2 \cdot r \tag{1.5}$$

Thus, for radial distances far from the galaxy core with mass M one expects to find:

$$v_{orb} = \sqrt{\frac{GM}{r}} \tag{1.6}$$

In fact, the amount of dark matter needed to explain the rotation curves may be in excess of what is consistent with the limits on baryonic matter from primordial nucleosynthesis. Although some fraction of the 'invisible' matter is certainly baryonic – for example, faint stars and black holes – other dark matter candidates have been proposed, for example exotic putative particles produced in connection with the Big Bang such as axions, massive neutrinos, neutralinos, monopoles or primordial black holes.

Dynamical tests of mass densities at even larger scales – for example, studies of the motions of galaxies in galaxy clusters – indicate that the mass density is possibly larger, $\Omega_M \geq 0.2$. Through the study of gravitational lensing and X-ray emission from the hot gas in such clusters it can be concluded that most of this matter is dark.

It should be noted that these arguments for a lower density of baryonic matter than what seems to be inferred for Ω_M from observations are somewhat indirect, depending on the observed luminosity density and then using estimates for the mass to light ratio of certain objects. On the largest scales, observations of the cosmic microwave background and the study of the growth of large-scale structure in the Universe also point to the existence of large amounts of dark, non-baryonic, matter.

---

[5] Modifications of the theory have been proposed to explain the rotation curves without changing the mass distribution of galaxies. These modifications are, however, incompatible with other types of astronomical observations.

Currently, the most accurate global measurements of the mass density and vacuum energy density of the Universe stem from the combined studies of brightness of high-redshift supernovae as a function of redshift and the anisotropies of the cosmic microwave background radiation. The principles and results of these methods are described in detail in Chapters 4 and 11. These methods, in particular the analysis of the microwave background, are based on fundamental physics to larger extent than other methods to determine mass and energy densities, but are on the other hand more dependent on a proper understanding of cosmology on the largest scales in the Universe.

The recent exciting progress in optical astronomy is mainly due to the development of large CCD cameras and telescopes such as the 2.4-metre Hubble Space Telescope (HST), the 10-metre ground-based Keck telescopes and the Very Large Telescope (VLT) at the European Southern Observatory (ESO) site in Chile, an array of four 8-metre telescopes which can work together forming the largest optical/IR telescope in the world. The imaging power of HST can be appreciated in colour Plate 1.[6] A next generation space telescope (the James Webb Space Telescope, JWST) is being designed and it will eventually replace NASA's successful Hubble Space Telescope.

1.6 The Age of the Universe

The equations of the Standard Model of cosmology due to Friedmann allow us, as we will see, to calculate the time t_U since the beginning of the expansion of the Universe. Of course, t_U as such is not a directly measurable number. Moreover, it depends on the entire history of the Universe, parts of which are unknown, in particular for the earliest history close to the Big Bang. However, in most models this only affects the estimated value of t_U by a completely negligible amount like a fraction of a second or so. We can thus compare calculations of t_U using simple cosmological models with estimates (or, rather, lower limits) of the age of the Universe, such as the ages of the oldest stars in the Milky Way, or the age of radioactive nuclei in cosmic rays.

The numerical answer to the age of the Universe for a given cosmology requires the knowledge of the three observable cosmological parameters H_0, Ω_M and Ω_Λ: $t_U = H_0^{-1} \cdot F(\Omega_M, \Omega_\Lambda)$. The functional dependence is such that the age of the Universe *increases* with Ω_Λ and *decreases* with Ω_M. With the present observational uncertainties, the age of the Universe is in the range 13–14 billion years. The range coincides roughly with the estimates of the ages of the oldest stars in our Galaxy. Stellar ages are measured by comparing models of stellar evolution to ensembles of stars in star clusters, the oldest of which are the *globular clusters*. Globular clusters are believed to be among the oldest gravitationally bound systems in the Universe, formed within one or two billion years after the Big Bang. The age of the solar system has been

---

[6] The colour plate section is positioned in the middle of the book.

determined through radioactive dating of terrestrial rocks and meteorites to be about 4.5 billion years (or giga-years, Gyr). The radioactive dating method can be extended to extract the age of the Galaxy provided one has a model of the time history of the production of the heavy long-lived isotopes of uranium, thorium and lead. The derived age of the Galaxy from such models also ranges between 10 and 20 Gyr. The consistency between the different dating methods and the estimate of the age of the Universe based on the present expansion rate is again an impressive example of the success of the Big Bang model.

1.7 The Left-Overs from the Big Bang

The discovery of the cosmic microwave background radiation (CMBR) in 1964[7] was crucial for the Standard Model of cosmology. Along with the discovery of the expansion it marked the beginning of a new era also for observational cosmology. The radiation, a relic from the hot Big Bang, has been cooling along with the expansion of the Universe. Its existence had been predicted by George Gamow and collaborators in the 1940s, but had been largely forgotten. Its measured Planck spectrum reveals an astonishing homogeneity at a temperature of 2.7 Kelvin. The angular variations in temperature are only $\Delta T/T \sim 10^{-5}$ and are of great importance for the understanding of structure formation and for the determination of the cosmological parameters (see Chapter 11).

About four hundred thousand years after the Big Bang, the expanding Universe became transparent to radiation, and the thermal photons which existed then as an effect of the hot early epochs could start to move freely. Today, there are around 400 such 'left-over' photons per cubic centimetre and the uniformity in the spectrum at all angles reflects the high level of homogeneity in the early Universe. The CMBR has been studied with ground-based and balloon-borne experiments for many years, recently with radically increased precision. However, the need for large sky coverage and good signal-to-noise ratio has also focused the observational efforts towards satellite experiments. NASA's first satellite dedicated for cosmology, COBE, was launched in 1989 and its successor, WMAP, presented its first set of data in 2003. Along with the planned European Planck mission, these are likely to become major milestones in the history of observational cosmology.

1.8 New Windows to Cosmology and Particle Physics

Besides electromagnetic radiation, a relic density of neutrinos and, possibly, other weakly interacting massive particles (WIMPs) are expected to populate

---

[7] The results were published in 1965.

the Universe, explaining the dark matter. The physics behind such hypothetical relics is easy to understand. In the earliest Universe, the temperature was so high that thermal kinetic energies of the constituents of the 'cosmic soup' were high enough to pair-produce these particles, if they exist. Once created, they could also annihilate each other to produce ordinary particles. However, as the Universe expanded the number density was diluted, eventually making it improbable that two particles would meet and annihilate. If these WIMPs are stable (or at least have a lifetime which is not much shorter than the age of the Universe) a population of such particles should thus have survived until today. It can be shown (Chapter 9) that massive particles with weak interactions indeed would contribute an amount to Ω_M which is close to what seems to be needed according to observations.

As these WIMPs do not interact with radiation they do not emit or absorb light and are therefore excellent dark matter candidates. Probing their existence is clearly of interdisciplinary interest. While they might solve the dark matter puzzle and could be of critical importance for understanding the growth of large structures in the Universe, they are potentially equally fundamental in the subatomic world.

The Standard Model of particle physics has been extremely successful in explaining most of the experimental results from non-gravitational interactions of particle beams with energies up to about 1 TeV (10^{12} eV, that is 1000 times the rest energy of the proton). However, the model fails to explain a number of fundamental facts such as why there are three fundamental interactions (gravity, electroweak, and strong interaction). The total number of particles in nature as well as their masses and electric charge are also important features that are not predictable by the theory. The Standard Model is therefore believed to be a low energy manifestation of a more fundamental theory, a 'Theory of Everything'. Attempts at formulating such theories reveal that they require the existence of more particles in the Universe than the ones yet observed and are therefore consistent with the astrophysical observation of the missing mass.

What experiments can prove (or disprove) their existence? The European particle physics laboratory CERN has launched an ambitious programme for this quest. The Large Electron Positron Collider (LEP), which was operating from 1989 until the year 2001 managed to explore the mass range up to around 115 GeV/c^2, with no new particles found. (As the beams collided their kinetic energy was available for the creation of new particles.) The same accelerator complex outside Geneva will be upgraded to accelerate protons to 8 TeV. This Large Hadron Collider (LHC), planned to be operational by 2007, has two interaction points were the beams from opposite directions interact. The centre-of-mass energy at the LHC collision points correspond to the temperature of the Universe just 10^{-11} seconds after the Big Bang!

If there are stable WIMPs of cosmological origin in the solar system they could be detected by two types of experiment. The 'direct' detection method is to look for nuclear recoils or ionization in a sensitive detector stored in an underground site (to shield it from cosmic rays which penetrate the atmosphere). The detector can then be excited by the weak elastic collisions caused by the passage of neutral WIMPs. One of the 'indirect' methods relies on the fact that since WIMPs lose energy through collisions inside the Sun and Earth, some will be gravitationally trapped. Neutrinos produced in WIMP annihilations in the centre of the Sun and the Earth may one day be detected with the large neutrino telescopes being constructed, as discussed in Chapter 14. Current results, based on experiments with a much smaller detection volumes, are shown in Fig. 1.3. The marked regions indicate the parameters for putative WIMPs, predicted in so-called supersymmetric models, presently excluded by the experiments.

WIMP annihilations in the galactic halo can also be searched for. A possible signature is a non-negligible flux of low-energy antiprotons or positrons. As we have remarked, there seems to be very little antimatter present in the Universe. A small amount of antiprotons and positrons has been detected in the cosmic rays, but this can be explained by pair-production when high-energy cosmic rays (mainly protons and nuclei) interact with gas in the interstellar medium. There will be several space experiments in the near future which will search for such a signal [3].

A striking experimental signature would be a monoenergetic gamma-ray line from $\chi\chi \to \gamma\gamma$, i.e. WIMP annihilations into two photons (each with $E_\gamma = M_\chi c^2$) in the Galaxy halo. Such a signal can be searched for with gamma-ray telescopes with good energy resolution, such as the future GLAST satellite [15] or the planned large arrays of ground-based gamma-ray telescopes.

1.9 Problems

1.1 Use Kepler's Third Law to estimate the orbital speed of the solar system at about 8 kpc from the centre of the Milky Way.

1.2 Use Newtonian gravity to calculate the escape velocity for a particle at the surface of the Earth and the Sun.

1.3 Show that a galaxy mass distribution $\rho \propto r^{-2}$ is consistent with flat rotation curves.

1.4 The estimated mass energy density of dark matter in the solar neighbourhood is around 0.3 GeV/cm^3. Suppose that it is made of WIMPs of rest energy (mass energy) 100 GeV.

(a) How many WIMPs are roughly inside your body at any particular time?

(b) What is the flux, i.e. the number of particles per cm^2 per second if they move with typical galactic velocities, $v \sim 200$ km/s?

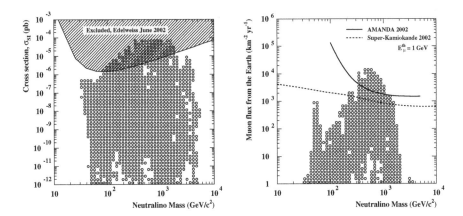

Fig. 1.3. Predicted rates and limits from direct and indirect WIMP detection. Each point represents one set of supersymmetric parameters. The figure on the left shows the predicted scattering cross-section on protons or neutrons in picobarns (1 pb = 10^{-36} cm^2) for supersymmetric WIMPs, neutralinos, as a function of neutralino mass. The shaded area is the region explored by current detectors which are sensitive to the 'wind' of WIMPs which the Earth traverses when the whole solar system moves through the Milky Way. The right-hand figure shows the muon flux from the direction of the centre of the Earth stemming from neutrinos generated in annihilations of accumulated WIMPs. Models above the lines, from current experiments, are excluded by upper limits on such a muon flux. Figures produced by L. Bergström, J. Edsjö, P. Gondolo and P. Ullio.

2 Special Relativity

2.1 Introduction

Much of the excitement in present-day particle astrophysics and cosmology has to do with energetic processes, where particles of high energy are interacting with each other and with ambient matter and radiation. We shall thus need to treat a number of processes where particles move with velocities very close to the speed of light. For those processes classical physics cannot provide reliable answers. Therefore we need to use the framework of special relativity. The theory of special relativity, developed by Einstein in the early 20th century, is essential for formulating a correct theory of elementary particles and their interactions (the other necessary ingredient is quantum mechanics). The marriage between special relativity and quantum mechanics was achieved in the 1940s and resulted in quantum field theory, the prime example being quantum electrodynamics (QED), which has been extremely successful in describing all electromagnetic phenomena. In the 1970s and 80s, the strong and weak interactions of elementary particles were also successfully formulated as relativistic quantum field theories, which we will discuss in Chapter 6.

In fact, there is one familiar interaction for which neither special relativity nor quantum mechanics is applicable, namely gravitation. In order to treat classical (that is non-quantum mechanical) gravity, Einstein succeeded in generalizing special relativity to a framework that is known as general relativity, which is a cornerstone of modern cosmology and which we shall return to in Chapter 3. The problem of constructing a quantum theory of gravitation remains unsolved, however – even if there is now hope that the so-called string theories or 'M theory' may finally provide a solution.

In this chapter, we give a brief review of some aspects of special relativity needed for this book. For a more thorough treatment, see standard textbooks [23, 39].

2.2 Frames, Coordinates and Metric

An important concept in special relativity is that of an *inertial frame*. This is a reference system where a body that is not subject to any forces remains at rest or in steady, rectilinear motion. Of course, such a system is strictly an idealization since there exist long-range forces in nature that cannot be screened. Also, for example, the rotation of the Earth around its axis and around the Sun means that there are inertial forces, so an object at rest with respect to the Earth's surface is not in an inertial frame. However, for practical purposes, this is rarely an important effect, and a frame at rest or in steady motion with respect to distant stars is an even better inertial frame.

Postulates of Special Relativity

The theory of special relativity rests on two innocent-looking postulates:

- The laws of physics take the same form in all inertial frames.
- The velocity of light in vacuum, c, is a universal constant, which has the same value ($\sim 3 \cdot 10^8$ m/s) in all inertial frames.

2.2.1 Coordinates

To develop the consequences of these postulates it is convenient to introduce four coordinates that parametrize space-time, $x^0 = ct$, $x^1 = x$, $x^2 = y$, $x^3 = z$, or

$$x^\mu = (ct, \mathbf{r}) = (ct, x^i) \tag{2.1}$$

where the Greek index μ runs over the values $0, 1, 2, 3$ and the latin index i takes the values $1, 2, 3$ (we will use this convention throughout the book for Greek and Latin indices, respectively).

The second postulate means that if we follow one and the same light-ray in two inertial coordinate systems, with sets of coordinates x^μ and x'^μ, and look at the time difference Δt, $\Delta t'$ of the passage of the light-ray through a distance $|\Delta \mathbf{r}|$, $|\Delta \mathbf{r}'|$, the velocity of light has to be the same, that is,

$$c = \frac{|\Delta \mathbf{r}|}{\Delta t} = \frac{|\Delta \mathbf{r}'|}{\Delta t'} \tag{2.2}$$

Or, put in another way, when we transform between two inertial systems the combination (the so-called line element)

$$ds^2 = c^2 dt^2 - |\mathbf{r}|^2 = (dx^0)^2 - (dx^1)^2 - (dx^2)^2 - (dx^3)^2 \tag{2.3}$$

has the same value in both systems. In the case of a light-ray, this value is $ds^2 = 0$, and the separation in space and time (space-time, for short) is said to be *light-like*. We characterize a *space-time event* by a value of time at a point in space. This could be an instant flash from a point source, or the passage of a particle at a particular point in space at a particular time. It is easy to see that ds^2 calculated between two nearby space-time events can, in contrast to the positive-definite three-dimensional Euclidean case, either be positive or negative:

$$ds^2 > 0 \;\; \text{time} - \text{like separations}$$
$$ds^2 < 0 \;\; \text{space} - \text{like separations}$$
$$ds^2 = 0 \;\; \text{light} - \text{like separations}$$

$$(2.4)$$

2.2.2 Metric and Transformations

With the use of the line element (2.3) we have a way to quantify the 'distance' between nearby space-time events. The equation for the line element has certain interesting properties when we change our coordinate system, our reference frame. We will soon discuss how to transform between two different inertial frames. Before doing this, it is useful to compare with the more elementary example of rotations in three-dimensional Euclidean space. The distance squared between two neighbouring points in space is

$$ds^2 = (dx^1)^2 + (dx^2)^2 + (dx^3)^2 =$$
$$\begin{pmatrix} dx^1 \\ dx^2 \\ dx^3 \end{pmatrix}^T \begin{pmatrix} 1 & 0 & 0 \\ 0 & 1 & 0 \\ 0 & 0 & 1 \end{pmatrix} \begin{pmatrix} dx^1 \\ dx^2 \\ dx^3 \end{pmatrix} = (d\mathbf{r})^T \mathbf{G} (d\mathbf{r}) \qquad (2.5)$$

Here we have introduced a 3×3 matrix $\mathbf{G}$ (in this case, equal to the unit matrix $\mathbf{I}$), which tells us how to construct the quadratic form in the dx^i which corresponds to (the square of) the distance. This matrix is called the *metric tensor*. If we did not use orthogonal axes, or if the basis vectors were not unit vectors, the matrix would be more complicated (but still symmetric). A more compact way of writing this equation is

$$ds^2 = \sum_{i,j=1}^{3} G_{ij} dx^i dx^j \equiv G_{ij} dx^i dx^j \qquad (2.6)$$

where the last form utilizes the *Einstein summation convention*: any index that is repeated twice is summed over. In cartesian coordinates, the elements of the $\mathbf{G}$ matrix are constant (that is independent of x^i), and are in fact equal to

$$G_{ij} = \delta_{ij} \qquad (2.7)$$

where δ_{ij} is the Kronecker delta function:

$$\delta_{ij} = \begin{cases} 1, & \text{if } i = j; \\ 0 & \text{otherwise} \end{cases} \qquad (2.8)$$

We know, however, that we can equally well use spherical coordinates r, θ and ϕ to parametrize Euclidean space, and then $\mathbf{G}$ is a function of the coordinates. Since in spherical coordinates

$$ds^2 = dr^2 + r^2 d\theta^2 + r^2 \sin^2 \theta d\phi^2 \qquad (2.9)$$

we see that $G_{rr} = 1$, $G_{\theta\theta} = r^2$, and $G_{\phi\phi} = r^2 sin^2\theta$. These coordinates are still orthogonal, which means that we obtain a volume element dV by in turn varying one of r, θ and ϕ while keeping the other two coordinates fixed. This gives $dV = r^2 \sin\theta dr d\theta d\phi$, which can be written

$$dV = \sqrt{\det \mathbf{G}}\, dr d\theta d\phi \tag{2.10}$$

a formula that is more generally valid for any curvilinear coordinate system.

Returning to the cartesian case, suppose that we rotate the coordinate system

$$\mathbf{r}' = \mathbf{R}\mathbf{r} \tag{2.11}$$

with $\mathbf{R}$ a 3×3 matrix, which keeps the axes orthogonal and normalized. As is known from basic linear algebra, this means that

$$\mathbf{R}^{-1} = \mathbf{R}^T \tag{2.12}$$

and we know that just rotating the coordinate system should not change distances between points. Thus,

$$ds'^2 = (dx'^1)^2 + (dx'^2)^2 + (dx'^3)^2 = (d\mathbf{r}')^T \mathbf{G} d\mathbf{r}'$$
$$= (\mathbf{R}d\mathbf{r})^T \mathbf{G} (\mathbf{R}d\mathbf{r}) = d\mathbf{r}^T \mathbf{R}^T \mathbf{G}\mathbf{R}d\mathbf{r} = d\mathbf{r}^T \mathbf{G}d\mathbf{r} = d\mathbf{r}^2 = ds^2 \tag{2.13}$$

The distance between neighbouring points is *invariant* under rotations, and this is assured since the rotation matrices fulfil the condition

$$\mathbf{R}^T\mathbf{G}\mathbf{R} = \mathbf{G} \tag{2.14}$$

or in component notation (remember the implicit summation over repeated indices)

$$R_{ki}G_{kl}R_{lj} = G_{ij} \tag{2.15}$$

As an example, the orthogonal matrix that rotates an angle θ around the x^3-axis can be written

$$\mathbf{R}(z;\theta) = \begin{pmatrix} \cos\theta & \sin\theta & 0 \\ -\sin\theta & \cos\theta & 0 \\ 0 & 0 & 1 \end{pmatrix} \tag{2.16}$$

The fact that $\mathbf{R}^{-1} = \mathbf{R}^T$ is obvious in this case, since we see that taking the transpose is the same as letting $\theta \to -\theta$, and rotating by a negative angle around a given axis is obviously the inverse operation to rotating by the same, but positive, angle.

2.3 Minkowski Space

What are the transformations in space-time that generalize rotations in three-dimensional Euclidean space? Of course, even if we consider one and the same inertial frame moving with a fixed velocity, we may still rotate the spatial part of the coordinate system. The four-dimensional (4×4) rotation matrix then factorizes as

$$\Lambda^{rot} = \begin{pmatrix} 1 & 0 \\ 0 & \mathbf{R} \end{pmatrix} \tag{2.17}$$

where $\mathbf{R}$ is a 3×3 rotation matrix.

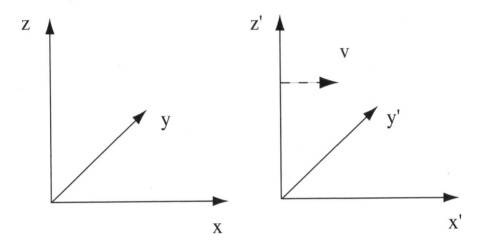

Fig. 2.1. Two inertial frames moving with relative velocity v.

Suppose that instead we keep the directions of the axes fixed, but let a new inertial system move with constant velocity v in the direction of the x^1-axis (see Fig. 2.1). This system is said to be *Lorentz-boosted* along the x^1-axis with respect to the first. Then the four-dimensional 'rotation' (or Lorentz boost) is given by

$$\Lambda(x^1; v/c) = \begin{pmatrix} \gamma & -\beta\gamma & 0 & 0 \\ -\beta\gamma & \gamma & 0 & 0 \\ 0 & 0 & 1 & 0 \\ 0 & 0 & 0 & 1 \end{pmatrix}, \tag{2.18}$$

where

$$\beta = \frac{v}{c} \tag{2.19}$$

$$\gamma = \frac{1}{\sqrt{1 - \beta^2}}; \tag{2.20}$$

We can rewrite it in a convenient form by introducing

$$\beta = \tanh \zeta \tag{2.21}$$

Then

$$\mathbf{\Lambda}(x^1; \zeta) = \begin{pmatrix} \cosh\zeta & -\sinh\zeta & 0 & 0 \\ -\sinh\zeta & \cosh\zeta & 0 & 0 \\ 0 & 0 & 1 & 0 \\ 0 & 0 & 0 & 1 \end{pmatrix} \tag{2.22}$$

We can see (Problem 2.2) that this transformation leaves the form $ds^2 = (dx^0)^2 - (dx^1)^2 - (dx^2)^2 - (dx^3)^2$ invariant; specifically it will also then leave the velocity of a light-ray invariant, as required by the second postulate of special relativity. The four-dimensional metric for an inertial frame that is the analog of the 3×3 matrix G_{ij} in (2.5) in three-dimensional space is thus

$$\eta_{\mu\nu} = \begin{pmatrix} 1 & 0 & 0 & 0 \\ 0 & -1 & 0 & 0 \\ 0 & 0 & -1 & 0 \\ 0 & 0 & 0 & -1 \end{pmatrix} \tag{2.23}$$

where μ and ν run from 0 to 3. This metric is called the Minkowski metric, and space-time with this metric is called Minkowski space. The reader should be warned that there is unfortunately no universal definition of the Minkowski metric. Quite often the definition $ds^2 = -(dx^0)^2 + (dx^1)^2 + (dx^2)^2 + (dx^3)^2$ is used, and then all the components of (2.23) change sign. Of course, no physical result will depend on this convention, however, some intermediate results may look different.

If we define $x_\mu \equiv \eta_{\mu\nu}x^\nu$ (we see that for the Minkowski metric this just amounts to changing the sign of the three space-components), then we can conveniently write $ds^2 = dx^\mu dx_\mu = dx_\mu dx^\mu$.

The condition for any Lorentz transformation (rotation or boost)

$$x^\mu \to x'^\mu = \Lambda^\mu{}_\nu x^\nu \tag{2.24}$$

to preserve the 'distance' ds^2 can be derived in a similar way as for (2.14):

$$\mathbf{\Lambda}^T \eta \mathbf{\Lambda} = \eta \tag{2.25}$$

or

$$\Lambda^\rho{}_\mu \eta_{\rho\sigma} \Lambda^\sigma{}_\nu = \eta_{\mu\nu} \tag{2.26}$$

Taking the determinant of both sides of this equation, one sees that $\det(\Lambda) = \pm 1$. Usually, we will only consider so-called *pure* Lorentz transformations which have $\det(\Lambda) = +1$. They can be continuously connected to the trivial unit transformation through a formal sequence of infinitesimal transformations.

An example of a transformation which has $\det(\Lambda) = -1$ and which therefore cannot be continuously connected to the unit transformation, is the reflection operator Λ_r, which takes $x^\mu = (x^0, \mathbf{r})$ to $x^\mu = (x^0, -\mathbf{r})$. Unlike the pure Lorentz transformations, which depend on a continuous set of parameters (the velocities $\mathbf{v}$ and the rotation angles), the reflection operator is an example of a *discrete* operator. Acting twice with that operator gives back the original coordinates, that is $\Lambda_r^2 = I$, with I the unit operator (that is the unit matrix). From a mathematical point of view, Lorentz transformations form a *group*, and the reflection transformation together with the unit operator corresponds to a discrete subgroup of the full Lorentz group. The term discrete, as opposed to continuous, means that the subgroup contains a finite set of elements – in this case only the two elements I and Λ_r. (This group is sometimes called Z_2.)

When studying properties of the pure Lorentz transformations it is usually enough to look at infinitesimal transformations. We then write

$$\Lambda^\mu{}_\nu = \delta^\mu_\nu + \lambda^\mu{}_\nu \tag{2.27}$$

Here δ^μ_ν is the four-dimensional Kronecker delta function, defined in the same way as the three-dimensional one, meaning that it is unity only for the two indices being the same; otherwise it is zero. The quantites $\lambda^\mu{}_\nu$ are all assumed to be infinitesimally small. If we now demand $a^2 = a^\mu a_\mu = a'^2 = a'^\mu a'_\mu$ for an arbitrary four-vector a^μ, we find by inserting (2.27) in the transformation rule $a'^\mu = \Lambda^\mu{}_\nu a^\nu$

$$a^2 = a'^2 = a^2 + \lambda^\mu{}_\nu a^\nu a_\mu + \lambda_\mu{}^\nu a^\mu a_\nu + \mathcal{O}(\lambda^2) \tag{2.28}$$

Thus,

$$\lambda^\mu{}_\nu a^\nu a_\mu + \lambda_\mu{}^\nu a^\mu a_\nu =$$
$$(\lambda^\mu{}_\nu + \lambda_\nu{}^\mu) a^\nu a_\mu = (\lambda^{\mu\nu} + \lambda^{\nu\mu}) a_\mu a_\nu = 0$$

which means

$$\lambda^{\mu\nu} = -\lambda^{\nu\mu} \tag{2.29}$$

That is, $\lambda^{\mu\nu}$ is an antisymmetric tensor which implies that it has 6 independent components. In particular, the diagonal elements are zero. There are thus 6 independent parameters (for example 3 rotation angles and 3 boost parameters) characterizing the part of the Lorentz group which is continuously connected to the unit transformation. If we also had considered translations ('inhomogeneous transformations') of the coordinates, $x^\mu \rightarrow \Lambda^\mu{}_\nu x^\nu + b^\mu$ with b^μ a constant four-vector, we would have four additional parameters. The full inhomogeneous Lorentz group (or the Poincaré group) is thus described by 10 parameters.

Since the Lorentz transformations are linear transformations of the coordinates

$$x^\mu \rightarrow x'^\mu = \frac{\partial x'^\mu}{\partial x^\nu} x^\nu \tag{2.30}$$

we see that the matrix $\Lambda^\mu{}_\nu$ can be expressed as

$$\Lambda^\mu{}_\nu = \frac{\partial x'^\mu}{\partial x^\nu} \tag{2.31}$$

and the inverse transformation (see Problem 2.6)

$$\Lambda_\nu{}^\mu = \frac{\partial x^\mu}{\partial x'^\nu} \tag{2.32}$$

2.3.1 Causal Structure of Space-Time

The *light-cone* with respect to an event P_1, say, at the origin of an inertial reference frame at time $t = 0$, is defined by $\Delta s^2 = 0$, where Δs is the space-time distance between P_1 and another event P_2 (see Fig. 2.2).

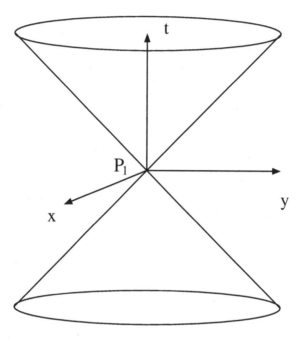

Fig. 2.2. The light-cone through a space-time event P_1. The points inside the cone for $t > 0$ (the future light-cone) can be reached by an observer travelling with a velocity smaller than that of light. Likewise, the points inside the past light-cone ($t < 0$) are causally connected to P_1.

Points inside the light-cone have $\Delta s^2 > 0$, and are said to be causally connected to the observer. This is because $c\Delta t > \Delta r$ so that the event P_2 can be reached from P_1 by signals travelling slower than with the velocity of light. If $\Delta s^2 < 0$ (on the outside of the lightcone) P_1 and P_2 cannot be connected by any means – the events are causally disconnected.

2.3.2 Vectors, Scalars and Tensors

A set of four quantities that transform in the same way as the differential dx^μ under Lorentz transformations is called a *contravariant four-vector*. A set that transforms like dx_μ is called a *covariant four-vector*. A quantity that, like ds^2, is invariant under Lorentz transformations is called a *four-scalar* or

Lorentz scalar. An important example of a covariant four-vector (see Problem 2.5) is provided by the four-gradient of a four-scalar:

$$\phi_{,\mu} \equiv \frac{\partial \phi}{\partial x^\mu} \tag{2.33}$$

In general, we can for any contravariant four-vector A^μ use the metric tensor $\eta_{\mu\nu}$ to 'lower the index' of A^μ to obtain the covariant four-vector $A_\mu = \eta_{\mu\nu} A^\nu$. The usefulness of four-vectors and scalars in special relativity is due to the fact that if we can write the laws of nature in terms of such objects, they will have the same form in all inertial systems (they will be so-called *manifestly covariant*), in agreement of Einstein's first postulate. It is easy to generalize to higher tensors. One example is the direct product of two four-vectors, which is called a second-rank tensor, $T^{\mu\nu} = A^\mu B^\nu$. There are second-rank tensors that cannot be written as simply a direct product of two four-vectors, but in each index the transformation property is as if it was. If $C^{\mu\nu}$ is a second-rank tensor, it thus changes under the Lorentz transformation $x^\mu \to x'^\mu = \Lambda^\mu{}_\nu x^\nu$ as

$$C'^{\mu\nu}(x') = \Lambda^\mu{}_\alpha \Lambda^\nu{}_\beta C^{\alpha\beta}(x) \tag{2.34}$$

Out of two four-vectors one can also form a four-scalar, the scalar product of the two vectors, $S = A.B \equiv \eta_{\mu\nu} A^\mu B^\nu = A^\mu B_\mu = A_\mu B^\mu$. This can also be seen as the *contraction of indices* of the tensor $T^{\mu\nu} = A^\mu B^\nu$, $T^\mu{}_\mu = A^\mu B_\mu$. As a special example, the scalar obtained by contracting a four-vector with itself, $A^2 \equiv A.A = A^\mu A_\mu$ is an invariant under Lorentz transformations: that is, its numerical value is unchanged when we change inertial system.

In special relativity, it may seem rather luxurious to introduce both covariant and contravariant versions of the same four-vector, since they only differ in the sign of the space components, $A_\mu = \eta_{\mu\nu} A^\nu = (A^0, -A^1, -A^2, -A^3)$. However, often it provides a useful check of the relativistic covariance of a set of equations. In general relativity the relation between the two will be less trivial, as we shall see in Chapter 3 and Appendix A.

2.4 Relativistic Kinematics

For a point particle of mass m, an important four-vector is the four-momentum p^μ,

$$p^0 = E/c, \; p^1 = p_x, \; p^2 = p_y, \; p^3 = p_z \tag{2.35}$$

where E is the total energy of the particle

$$E = \gamma m c^2 \tag{2.36}$$

and $\mathbf{p}$ the relativistic linear momentum

$$\mathbf{p} = \gamma m \mathbf{v} \tag{2.37}$$

This means that

$$\mathbf{v} = \frac{c^2 \mathbf{p}}{E} \tag{2.38}$$

We see that the relativistic γ factor for a particle of given energy can be written

$$\gamma(v/c) = \frac{E}{mc^2} \tag{2.39}$$

We saw in (2.20) that γ can also be written as $\gamma = 1/\sqrt{1 - \beta^2}$, with $\beta = v/c$. As $\beta \to 1$ this diverges, which shows that a massive particle can never attain the speed of light, since it would cost infinite energy to accelerate it to that speed. The variation of γ with β, or rather $1 - \beta$, is shown in Fig. 2.3.

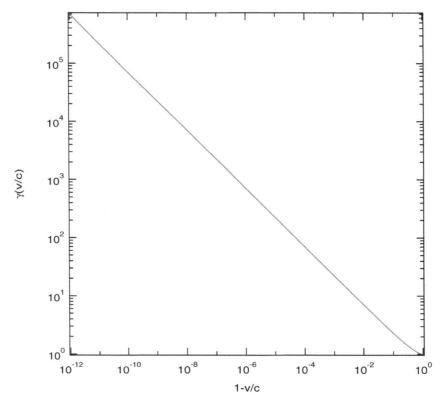

Fig. 2.3. The dependence of the relativistic gamma factor $\gamma(\beta)$, where $\beta = v/c$, on the velocity v. The quantity on the horizontal axis is $1 - \beta = 1 - v/c$. As can be seen, in a logarithmic scale the dependence is close to being a straight line except for β close to zero.

We can also solve for β to find

$$\beta = \frac{v}{c} = \sqrt{1 - \frac{1}{\gamma^2}} \qquad (2.40)$$

For v/c close to one, that is for large gamma factors $\gamma \gg 1$, this can be expanded to give

$$(1 - \beta)_{\gamma \gg 1} \sim \frac{1}{2\gamma^2} \qquad (2.41)$$

Example 2.4.1 At the electron positron collider LEP at CERN outside Geneva, electrons were accelerated to more than 100 GeV. What is the velocity of an electron of 100 GeV total energy? The rest energy of an electron is 511 keV.

Answer: The γ factor is

$$\gamma(v/c) = \frac{E}{m_e c^2} = \frac{100\,\text{GeV}}{0.511 \cdot 10^{-3}\,\text{GeV}} = 1.96 \cdot 10^5$$

From Fig. 2.3 (or (2.41)) we find $1 - v/c \sim 1.3 \cdot 10^{-11}$. These electrons thus travel with 99.999999999 per cent of the speed of light!

Example 2.4.2 Calculate the value of the four-scalar p^2.

Answer: Let us determine the four-momentum in a particularly simple inertial system, namely that in which the particle is at rest. Then $p^\mu = (mc, \mathbf{0})$, which means that $p^2 = m^2 c^2$. Since this is a scalar, it has to have this value in any inertial system. Let us check with the general expression in (2.35): $p^2 = \gamma^2 m^2 c^2 - \gamma^2 m^2 v^2 = m^2 c^2 \gamma^2 (1 - (v/c)^2) = m^2 c^2$.

From this example we see that another four-vector is given by $p^\mu/m = \gamma(c, \mathbf{v})$. This is called the four-velocity v,

$$v^\mu = (\gamma c, \gamma \mathbf{v}) \qquad (2.42)$$

which has the invariant length squared $v^2 = c^2$, and the relation between momentum and velocity is the familiar $p = mv$, with now both p and v being four-vectors (we shall often suppress the Lorentz index; the equation is really $p^\mu = mv^\mu$, which of course means that the equality holds separately for all values of the index μ).

The four-distance between two events on the trajectory of a massive particle moving with constant velocity v is time-like and is most easily calculated in the rest frame of the particle $\Delta s^2 = (\Delta x^0)^2 = (c\Delta\tau)^2$, where τ is the time measured by a clock following the particle, called the *proper time*. Thus, for a clock that moves with the particle $dx^i = 0$, we find in such a frame trivially that

$$ds^2 = c^2 d\tau^2 = \eta_{\mu\nu} dx^\mu dx^\nu \qquad (2.43)$$

and since this is written in an explicitly Lorentz-invariant (or covariant) form, it has to be valid in all inertial frames.

Since the differential of the proper time

$$d\tau = \frac{1}{c}\sqrt{\eta_{\mu\nu}dx^\mu dx^\nu} \tag{2.44}$$

is Lorentz invariant, we can use it to form other four-vectors. As an example, the four-velocity v can also be defined as $v = dx/d\tau$, which is the obvious four-dimensional generalization of the non-relativistic three-velocity $\mathbf{v} = d\mathbf{r}/dt$. Using $p^2 = m^2c^2$, it is easy to see that

$$E = \sqrt{\mathbf{p}^2c^2 + m^2c^4} \tag{2.45}$$

In particular, for a massless particle (such as the photon), $E = c|\mathbf{p}|$.

Just as in elementary mechanics, there are conservation laws in relativistic mechanics, the most important one being the conservation of four-momentum which is valid in the absence of external forces. In any elementary collision process, the sum of the four-momenta of the initial state particles is equal to the sum of the four-momenta of the final state particles. In addition, all particles travelling freely before and after the collision process fulfil the so-called mass-shell condition $p^2 = m^2c^2$.

With $p = mv$, we can write Newton's force law in the four-dimensional form

$$f^\mu = \frac{dp^\mu}{d\tau} \tag{2.46}$$

which defines the four-vector force f. Using $p^\mu = mv^\mu = mdx^\mu/d\tau$, we find

$$f^\mu = m\frac{d^2x^\mu}{d\tau^2} \tag{2.47}$$

In particular, if we have a free particle so that $f^\mu = 0$, it will follow a path given by

$$\frac{d^2x^\mu}{d\tau^2} = 0 \tag{2.48}$$

We can easily solve this equation:

$$x^\mu(\tau) = v_0^\mu\tau + x_0^\mu \tag{2.49}$$

which is the equation for a straight line in Minkowski space.

2.4.1 Kinematics for $2 \to 2$ Processes

A common processes of interest in particle physics, astrophysics and in the physics of the early Universe, is the collision between a pair of particles of masses m_1 and m_2 and four-momenta p_1 and p_2. In relativistic physics, neither the number nor the type of particles need to be preserved. However, let us treat the simplest $2 \to 2$ process, with particles 1 and 2 in the initial state and 3 and 4 in the final state. If we neglect possible spin degrees of freedom of the particles, the kinematic state of the particles is described by their respective four-momenta. According to special relativity, the basic scattering process should be determined by Lorentz-invariant quantities. Starting with four four-momenta, we can form six Lorentz-invariant products $p_i.p_j$ with $i \neq j$. However, overall four-momentum conservation means $p_1 + p_2 = p_3 + p_4$, which imposes four relations (one for each component). Thus, there are only two independent kinematical four-scalars which for historical reasons are usually taken to be

$$s = (p_1 + p_2)^2 \qquad t = (p_1 - p_3)^2 \qquad\qquad (2.50)$$

Example 2.4.3 Show that the invariant $u = (p_1 - p_4)^2$ is not independent of s and t.

Answer: $u = (p_1 - p_4)^2 = p_1^2 + p_4^2 - 2p_1.p_4 = m_1^2 c^2 + m_4^2 c^2 - 2p_1.p_4$. Now, $p_1 = p_3 + p_4 - p_2$, and multiplying this equation by p_4, and using $s = (p_3 + p_4)^2$, $t = (p_1 - p_3)^2$, one finds (fill in the missing steps!)

$$s + t + u = \left(m_1^2 + m_2^2 + m_3^2 + m_4^2\right) c^2$$

which means that u is linearly dependent on s and t. The set $\{s, t, u\}$ is usually referred to as the Mandelstam variables.

Example 2.4.4 Show that $c\sqrt{s}$ is the total energy in the centre of momentum frame of the two incoming particles.

Answer: In the centre of momentum frame, the three-momenta of the particles are of equal magnitude but of opposite direction, thus $p_1^{cm} = (E_1^{cm}/c, \mathbf{p}^{cm})$, $p_2^{cm} = (E_2^{cm}/c, -\mathbf{p}^{cm})$. Thus, $\sqrt{s} = (E_1^{cm} + E_2^{cm})/c$, or $c\sqrt{s} = E_1^{cm} + E_2^{cm}$, which indeed is the total energy in the centre of momentum frame.

Example 2.4.5 Suppose that two protons collide head-on, each with the same energy, 100 GeV. How much energy is needed for a proton that collides with a proton at rest to give the same total energy in the centre of momentum frame. The proton mass is 0.94 GeV/c^2.

Answer: In the first case, $c\sqrt{s} = 200$ GeV. In the second case, $p_1 = (E/c, \mathbf{p})$, $p_2 = (m_p c, 0)$. Thus $s = (E/c + m_p c)^2 - \mathbf{p}^2 = 2m_p^2 c^2 + 2m_p E$, where (2.45) was used. Thus, 200 GeV $= c\sqrt{s} = \sqrt{2 \cdot 0.94^2 + 2 \cdot 0.94 E}$, which solved for E gives $E = 21300$ GeV. The high energy needed explains the popularity in particle physics to use colliders rather than fixed-target accelerators. The physical interpretation of this result is that, due to momentum conservation, in the latter case a lot of energy is 'wasted' on the overall motion of the particles in the laboratory system.

2.4.2 System of Units

Sometimes it is convenient not to have to keep track of all the factors of c, the velocity of light, in relativistic formulae. One way to achieve this is to choose units such that $c = 1$. (Alternatively, this simply means that one measures all velocities in fractions of the light velocity.) In the particle physics literature, this is very common. In addition, $\hbar$, Planck's constant divided by 2π, is conveniently put equal to unity in quantum mechanical problems. Since all physical units can be expressed using combinations of length, time and mass, and we have made two units dimensionless, it means that we can choose just one to fix all physical dimensions. Usually, mass is used for this purpose.

Example 2.4.6 What is the dimension of length and time in a system of units where $\hbar = c = 1$?

Answer: In SI units, $[\hbar]$=kgm^{2}s^{-1} and $[c] = $ ms^{-1}. Since we have chosen c to be dimensionless, length and time must have the same dimensions. Since $\hbar$ is also chosen to be dimensionless, and thus also the ratio $\hbar/c$, with $[\hbar/c]$ =kg·m, we see that mass and length must have inverse dimensions. Thus [time]=[length]=[mass]$^{-1}$.

2.4.3 Some Relativistic Kinematics for $2 \to 2$ Processes

Putting now $c = 1$, we derive some useful kinematical relations for $2 \to 2$ processes. In Example (2.4.5) we saw that we could choose to calculate, for example, s in two different inertial frames. Since we know that s is Lorentz-invariant, this usually gives a convenient way of relating kinematical variables in the two frames. In the *laboratory frame* where particle 2 is at rest, we had

$$s = m_1^2 + m_2^2 + 2E_1^{\text{lab}}m_2 \qquad (2.51)$$

while in the *centre of momentum frame* $\mathbf{p}_1^{\text{cm}} = -\mathbf{p}_2^{\text{cm}}$, so that

$$s = (E_1^{\text{cm}} + E_2^{\text{cm}})^2 \qquad (2.52)$$

Equating the two expressions (2.51) and (2.52) gives, after some algebra (Problem 2.7)

$$|\mathbf{p}^{\text{cm}}| = \frac{m_2|\mathbf{p}^{\text{lab}}|}{\sqrt{s}} \qquad (2.53)$$

From (2.52) one can also derive

$$|\mathbf{p}^{\text{cm}}| = \frac{\sqrt{\lambda(s, m_1^2, m_2^2)}}{2\sqrt{s}} \qquad (2.54)$$

where λ is the 'triangle function'

$$\lambda(x, y, z) = x^2 + y^2 + z^2 - 2xy - 2xz - 2yz \tag{2.55}$$

It is also easy to derive (or guess from symmetry) the corresponding expression for the cm momentum of the final state particles:

$$|\mathbf{p}_3^{\text{cm}}| = \frac{\sqrt{\lambda(s, m_3^2, m_4^2)}}{2\sqrt{s}} \tag{2.56}$$

and the expression for E_3^{lab}

$$E_3^{\text{lab}} = \frac{m_2^2 + m_3^2 - u}{2m_2} \tag{2.57}$$

which is useful for deriving a relation between the invariant t and the scattering angle θ in the lab frame,

$$\cos\theta_{13}^{\text{lab}} = \frac{(s - m_1^2 - m_2^2)(m_2^2 + m_3^2 - u) + 2m_2^2(t - m_1^2 - m_3^2)}{\sqrt{\lambda(s, m_1^2, m_2^2)}\sqrt{\lambda(u, m_2^2, m_3^2)}} \tag{2.58}$$

For a $2 \rightarrow 2$ process, the kinematically allowed region in s is

$$s > (m_3 + m_4)^2 \tag{2.59}$$

which can be understood from energy conservation: In the cm frame, where we have seen that $\sqrt{s}$ corresponds to the total energy, at least the rest mass energy $m_3 + m_4$ has to be provided.

The kinematical limits for t are more complicated and are most easily obtained from the condition $|\cos\theta_{13}^{\text{cms}}| \leq 1$, with

$$\cos\theta_{13}^{\text{cm}} = \frac{s(t - u) + (m_1^2 - m_2^2)(m_3^2 - m_4^2)}{\sqrt{\lambda(s, m_1^2, m_2^2)}\sqrt{\lambda(s, m_3^2, m_4^2)}} \tag{2.60}$$

2.5 Relativistic Optics

When making observations in the Universe, one often deals with moving sources which emit light. In some cases the motion of the observer with respect to some frame also has to be taken into account. Due to the form of the Lorentz transformations this usually means that there will be a change in appearance of a beam of light emitted from a source. There are two immediate effects that appear: aberration and Doppler shift.

2.5.1 Aberration

Consider a light-ray which according to one observer arrives at the origin at an angle θ at $t = 0$. Due to the properties of the Lorentz transformations, this angle of incidence will have another value θ' in a frame that moves with velocity v with respect to the first observer. This change of apparent direction is called aberration (see Fig. 2.4).

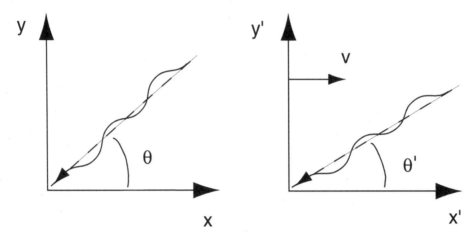

Fig. 2.4. A light-ray arriving at the common origin at time $t = t' = 0$ in two inertial frames. The primed frame is moving with velocity v with respect to the unprimed frame, and the measured angles of incidence are θ' and θ, respectively.

We first derive the transformation properties of velocities (that is, the addition law for relativistic velocities). Looking at a massive particle with velocity $\mathbf{u} = d\mathbf{r}/dt$ in one reference frame, we can compute $\mathbf{u}' = d\mathbf{r}'/dt'$ by taking the differentials of the Lorentz transformation (2.18). This gives (remember that we have put $c = 1$, that is all velocities are measured in units of the velocity of light)

$$u_1' = \frac{u_1 - v}{1 - u_1 v}$$

$$u_2' = \frac{u_2}{\gamma(v)\,(1 - u_1 v)}$$

$$u_3' = \frac{u_3}{\gamma(v)\,(1 - u_1 v)} \tag{2.61}$$

If we now insert $u_1' = -c\cos\theta' = -\cos\theta'$, $u_2' = -\sin\theta'$, $u_1 = -\cos\theta$, $u_2 = -\sin\theta$ in (2.61), we obtain

$$\cos\theta' = \frac{\cos\theta + v}{1 + v\cos\theta}$$

$$\sin\theta' = \frac{\sin\theta}{\gamma(v)\,(1 + v\cos\theta)} \tag{2.62}$$

It is interesting to note the factor $1/\gamma(v)$ which enters the expression for $\sin\theta'$. Also, we see that if we change $v \to -v$ and $u \to -u$, the denominator in (2.61) is unchanged. It means that, for instance, light-rays which are either incident or emitted isotropically in one system are driven to small angles in a moving frame. This 'relativistic beaming' of light is seen, for example, in some of the jets emanating from active galactic nuclei.

2.5.2 Doppler Effect

To discuss the Doppler effect, it is convenient to use four-velocities of observers and the wave four-vector of a photon. The latter is defined as $k^\mu = (\omega, \mathbf{k})$, where $\omega = 2\pi\nu$ is the angular frequency, and

$$\mathbf{k} = \frac{2\pi}{\lambda}\mathbf{n} \tag{2.63}$$

with λ the wave-length and $\mathbf{n}$ a unit vector in the direction of propagation. According to quantum mechanics, $E = \hbar\omega$ and $\mathbf{p} = \hbar\mathbf{k}$, so we see that in our units where $\hbar = 1$, k^μ is simply the momentum four-vector. (Note that $k^\mu k_\mu = 0$ since photons propagate on light-like trajectories; the photon is massless.) Suppose now in the general case that we have a light source with four-velocity

$$u^\mu = \gamma(u)\,(1, \mathbf{u})$$

and an observer with four-velocity

$$v^\mu = \gamma(v)\,(1, \mathbf{v})$$

In the rest frame of the light source, we easily compute $k^\mu u_\mu = \omega_0$, since in that frame $u^\mu = (1, \mathbf{0})$. Since this is a four-scalar and thus an invariant, we find by computing it in the given frame

$$\omega_0 = \gamma(u)\omega\,(1 - \mathbf{n} \cdot \mathbf{u}) \tag{2.64}$$

Similarly for the frequency ω' measured in the rest frame of the observer:

$$\omega' = \gamma(v)\omega\,(1 - \mathbf{n} \cdot \mathbf{v}) \tag{2.65}$$

Taking the ratio between (2.65) and (2.64) we find

$$\frac{\omega'}{\omega_0} = \frac{\gamma(v)\,(1 - \mathbf{n} \cdot \mathbf{v})}{\gamma(u)\,(1 - \mathbf{n} \cdot \mathbf{u})} \tag{2.66}$$

Here ω can be exchanged for ν, or if we instead consider wavelengths,

$$\frac{\lambda'}{\lambda_0} = \frac{\gamma(u)\,(1 - \mathbf{n} \cdot \mathbf{u})}{\gamma(v)\,(1 - \mathbf{n} \cdot \mathbf{v})} \tag{2.67}$$

As an example, suppose that we observe light emitted from a source moving away from us. Then $v = 0$, $\mathbf{n} \cdot \mathbf{u} = -u_r$ where u_r is the radial velocity (that is the projection of the recession velocity on the radial direction) and we measure a wavelength λ which is longer (that is redshifted) by a factor D (the Doppler factor),

$$D = \gamma(u)\,(1 + u_r) \tag{2.68}$$

Note that even for $u_r = 0$, there is a redshift, the so-called transverse Doppler effect, which is of relativistic origin and can be traced to the time dilation (the frequency of radiation can be viewed as the ticking of a 'clock', and we know that time goes more slowly in a moving frame). The factor $\gamma(u)$ gives an effect of second order in v/c, and is therefore often less important than the first-order Doppler factor D_1, which from (2.67) is, with c reinserted,

$$D_1 \sim 1 + u_r/c \tag{2.69}$$

2.6 Electromagnetic Vectors and Tensors

According to the postulates of special relativity, physical laws should have the same form in all inertial frames. We have seen that if we can formulate these laws in terms of four-scalars, vectors and tensors, this relativistic covariance is automatic. Such laws are sometimes said to be manifestly Lorentz invariant.

To build a theory which fulfills the requirements of special relativity, we should thus try to identify the relevant scalars, vectors and tensors. In electromagnetism, an important four-vector is the four-potential

$$A^\mu = (\phi, \mathbf{A}) \tag{2.70}$$

where ϕ is the electrostatic potential and $\mathbf{A}$ the three-vector potential. The electric field vector $\mathbf{E}$ and magnetic field vector $\mathbf{B}$ are, one the other hand, parts of an antisymmetric second-rank tensor $F^{\mu\nu}$,

$$F^{\mu\nu} = \begin{pmatrix} 0 & E_x & E_y & E_z \\ -E_x & 0 & B_z & -B_y \\ -E_y & -B_z & 0 & B_x \\ -E_z & B_y & -B_x & 0 \end{pmatrix} \tag{2.71}$$

$F^{\mu\nu}$ can be expressed in terms of A^μ through

$$F^{\mu\nu} = \partial^\mu A^\nu - \partial^\nu A^\mu \tag{2.72}$$

The current four-vector is given by

$$j^\mu = (\rho, \mathbf{j}) \tag{2.73}$$

where ρ is the electrostatic charge density and $\mathbf{j}$ is the three-current density. Maxwell's equations can be summarized by

$$\partial_\mu F^{\mu\nu} = j^\nu \tag{2.74}$$
$$\partial^\lambda F^{\mu\nu} + \partial^\mu F^{\nu\lambda} + \partial^\nu F^{\lambda\mu} = 0 \tag{2.75}$$

We see that the antisymmetry of $F^{\mu\nu}$ gives from (2.74) $\partial_\nu j^\nu = 0$. This is the continuity equation for the electric current, which integrated over all space expresses the conservation of global electric charge. Thus, consistency requires that we only couple the electromagnetic field to conserved currents.

A special tensor is of course $\eta_{\mu\nu}$, which has the same constant value in all inertial frames. Another important tensor is the four-dimensional generalization of the Levi-Civita tensor ϵ_{ijk}, namely $\varepsilon_{\alpha\beta\gamma\delta}$, which is antisymmetric in all indices and has $\varepsilon_{0123} = 1$. With its help we can, for instance, define the so-called dual tensor $^*F$ to F:

$$* F^{\mu\nu} = \frac{1}{2} \varepsilon^{\mu\nu\alpha\beta} F_{\alpha\beta} \qquad (2.76)$$

Then the second Maxwell equation (2.75) can simply be written as

$$\partial_\mu {}^* F^{\mu\nu} = 0 \qquad (2.77)$$

In terms of A^μ the first set of Maxwell equations becomes

$$\Box A^\mu - \partial^\mu (\partial_\nu A^\nu) = j^\mu, \qquad (2.78)$$

where the *d'Alembertian operator* is defined as

$$\Box \equiv \partial^\mu \partial_\mu = \left(\frac{1}{c^2} \frac{\partial^2}{\partial t^2} - \nabla^2 \right). \qquad (2.79)$$

Since (2.72) is unchanged if we make the *gauge transformation*

$$A^\mu \rightarrow A^\mu + \partial^\mu f(\mathbf{r}, t) \qquad (2.80)$$

we can use this freedom to impose a *gauge condition* on A^μ, for example, $\partial_\mu A^\mu = 0$ (Lorentz gauge), $\nabla \cdot \mathbf{A} = 0$ (Coulomb gauge) or $A^0 = \phi = 0$ (axial gauge). In the absence of charges, both the axial and Coulomb gauge conditions can be chosen simultaneously (radiation gauge). Then the Maxwell equations for the freely propagating electromagnetic vector field A^μ in vacuum simply become

$$\Box A^\mu = 0 \qquad (2.81)$$

which is a relativistic wave equation describing propagation with the speed of light. Solutions can easily be found of the form

$$A^\mu (\mathbf{r}, t) = \epsilon^\mu e^{\pm i(\omega t - \mathbf{k} \cdot \mathbf{r})} = \epsilon^\mu e^{\pm i k \cdot x} \qquad (2.82)$$

Here $\mathbf{k}$ is the wave-vector which describes the direction of propagation and where the four-vector index is carried by the constant *polarization vector* ϵ^μ. Inserting this into (2.81) gives $k^\mu k_\mu = 0$, which we have seen is the mass-shell condition for a massless particle, the photon. It can thus be seen that the reason that the photon is massless is the property of gauge invariance, or gauge symmetry, of the Maxwell theory.

The axial and Coulomb gauge conditions mean that of the four components of the vector potential A^μ, only two remain as physical degrees of freedom. We see that $A^0 = 0$ and $\nabla \cdot \mathbf{A} = 0$ translate into $\epsilon^0 = 0$ and $\mathbf{k} \cdot \epsilon = 0$, showing that only the two physical degrees of freedom are transverse to the direction of propagation. Choosing the latter as the z-direction, we can for example use $\epsilon_1^\mu = (0, 1, 0, 0)$ and $\epsilon_2^\mu = (0, 0, 1, 0)$ as the basis states in which we can express an arbitrary polarization. Sometimes it is, however, more convenient to use the so-called circular polarization four vectors $\epsilon_\pm^\mu = (0, 1, \mp i, 0)/\sqrt{2}$ as basis states.

The fact that A^μ is a four-vector means that it couples to vector-type sources: for example, a varying electric dipole field. In Chapter 15 we shall use a similar reasoning to show that there exist gravitational waves as well.

In that case, it is a tensor field which is propagating which means that a quadrupole moment instead of a dipole moment will be acting as a source.

2.7 Summary

- Special relativity is based on two postulates, formulated by Einstein. The first states that the laws of nature take the same form in all inertial frames. The second states that the velocity of a light-ray in vacuum has the universal value of around $3 \cdot 10^8$ m/s in all inertial frames, irrespective of the velocity of the source or the observer.
- Space and time can be treated in a unified way by introducing space-time which is parameterized by the position four-vector $x^\mu = (x^0, x^1, x^2, x^3)$, where $x^0 = ct$. If other physical quantities are also grouped into similar four-vectors, scalars or tensors, the laws of physics can be written in a form which is the same in all inertial frames in agreement with Einstein's postulate.
- The line element in special relativity is

$$ds^2 = \left(dx^0\right)^2 - \left(dx^1\right)^2 - \left(dx^2\right)^2 - \left(dx^3\right)^2$$

It has the same value (is invariant) in all inertial reference frames related by Lorentz transformations.

- In collision processes, it is convenient to choose Lorentz invariant kinematical variables. One of the most important is

$$s = \left(p_1 + p_2\right)^2$$

which is related to the available energy in the centre of momentum system of particles 1 and 2 with four-momenta p_1 and p_2, respectively.

- The relativistic Doppler factor for a source moving with four-velocity u is given by

$$D = \gamma(u)\left(1 + u_r\right)$$

where u_r is the projection of the recession velocity on the radial direction.

- In electromagnetism, an important four-vector is the four-potential

$$A^\mu = (\phi, \mathbf{A}) \tag{2.83}$$

where ϕ is the electrostatic potential and $\mathbf{A}$ the three-vector potential. The electric field vector $\mathbf{E}$ and magnetic field vector $\mathbf{B}$ are parts of an antisymmetric second-rank tensor,

$$F^{\mu\nu} = \partial^\mu A^\nu - \partial^\nu A^\mu$$

The current four-vector is given by

$$j^\mu = (\rho, \mathbf{j})$$

where ρ is the electrostatic charge density and $\mathbf{j}$ is the three-current density.

2.8 Problems

2.1 Show that $\mathbf{R}(z;\theta)$ of (2.16) fulfills (2.12).

2.2 Show that the Lorentz transformation (2.18) leaves ds^2 invariant.

2.3 Let A^μ be a light-like four-vector. Show that A^0 has the same sign in all inertial frames.

2.4 Show that the sum of two orthogonal space-like four-vectors is space-like.

2.5 Show that the four-gradient of a four-scalar transforms as a covariant four-vector.

2.6 Show from the expressions (2.31) and (2.32) that $\Lambda_\nu{}^\mu$ is the inverse of $\Lambda^\mu{}_\nu$, meaning that $\Lambda^\mu{}_\sigma \Lambda_\nu{}^\sigma = \delta^\mu_\nu$, where δ^μ_ν is the Kronecker delta.

2.7 Derive (2.53) from (2.51) and (2.52).

2.8 When a charged π^- meson with very low velocity reacts with a proton, a neutron and a neutral pi meson are produced, $\pi^- + p \to \pi^0 + n$. Suppose that the proton, neutron and π^- masses are known: $m_p = 938.3$ MeV, $m_n = 939.6$ MeV, $m_{\pi^+} = 139.6$ MeV. Determine the mass of the π^0, if the neutron kinetic energy is measured to be $T_n \equiv E_n - m_n = 0.4$ MeV.

2.9 According to quantum electrodynamics, there is a small but finite probability that two photons may scatter against each other. Suppose two photons, with wavelengths λ_1 and λ_2, collide head-on. Compute the wavelength of one of the photons after the collision, if its scattering angle is θ. (The energy of a photon with wavelength λ is $\hbar\omega$, where $\omega = 2\pi c/\lambda$.)

2.10 What is the value of H_0 (~ 70 km s^{-1}Mpc^{-1}) expressed as a mass, in a system of units where $\hbar = c = 1$?

2.11 Show that a free electron and a free positron, both with mass m_e, cannot annihilate into a single freely propagating photon.

2.12 The cosmic microwave background radiation (CMBR) consists of photons of typical energy $3 \cdot 10^{-4}$ eV. How high an energy must a cosmic gamma photon γ_c have if pair production on this background $\gamma_c + \gamma_{CMB} \rightarrow e^+ e^-$ is to be kinematically allowed?

2.13 In special relativity, one can define the kinetic energy T of a particle of mass m and velocity v as $T = E - m = \gamma(v)m - m$ (we have put $c = 1$). Use four-momentum conservation to determine the final kinetic energy of a particle of mass m, velocity v, which collides elastically with an equally heavy particle at rest, if the scattering angle is θ.

2.14 Consider a particle of four-momentum $p_a = (m_a, 0)$ which decays at rest into three other particles of masses m_b, m_c and m_d with four-momenta p_b, p_c and p_d, respectively. Introduce the variable $s_{bc} = (p_b + p_c)^2$. Use four-momentum conservation and:
(a) Determine the maximum and minimum values that s_{bc} can have.
(b) For a given value of s_{bc}, calculate the energy of particle d in the rest frame of the decaying particle.

3 General Relativity

3.1 Introduction

We have seen in Chapter 2 that inertial systems seem to play an important role in physics. The postulates of special relativity tell us that the laws of physics should look the same in all inertial frames. However, there are many cases when we would like to treat non-inertial frames. For instance, a compact neutron star may spin with a rotation velocity which is high enough that matter on the equator moves with a speed close to the speed of light. The 'inertial forces' in a frame fixed to the surface of the neutron star can certainly not be neglected. Likewise, near such a star there are strong gravitational forces, and it is not obvious how to treat such a situation using only special relativity.

It was Einstein who, in a magnificent work, succeeded in formulating a theory, the general theory of relativity, which can deal also with accelerated frames and with gravitation. In fact, Einstein even realised how the laws of gravitation had to be modified from Newton's form, and at the same time gave a new geometrical interpretation of gravitational forces.

3.2 The Equivalence Principle

Einstein used a number of thought experiments to formulate and illustrate the basic principles of general relativity. Suppose first that we are on board a window-less rocket far from the solar system and other sources of gravitation. Assume that it accelerates with a constant acceleration which is numerically exactly equal to the gravitational acceleration g at the Earth's surface. Can we then, by performing experiments on board this spacecraft, determine if the rocket is moving or just standing still on the surface of the Earth? Certainly the two situations should be very similar. If, for example, we duplicated Galileo's experiments with falling bodies (say one iron ball and one wooden ball) inside the rocket, we would obtain his results. Since after releasing the test bodies the floor of the rocket accelerates towards the balls with the value g, they will simultaneously hit the floor, just as Galileo's experiment on Earth showed (neglecting air resistance, of course).

This brings us to a strange fact about the Newtonian law of gravity. According to the force law, a test body with 'inertial' mass m_i in the Earth's gravitational field will be accelerated by

$$a = \frac{F}{m_i} \qquad (3.1)$$

where the gravitational force F according to Newton is

$$F = m_g g \qquad (3.2)$$

with m_g the 'gravitational' mass on which the gravitational force acts. It is easy to solve for the acceleration:

$$a = \frac{m_g}{m_i} g \qquad (3.3)$$

A priori, there is nothing that would require that the ratio $\frac{m_g}{m_i}$ is the same for all bodies irrespective of the composition. However, numerous experiments verify this equality, and therefore it is natural to choose units such that $m_g = m_i$. This equality of the gravitational and inertial mass is called the *weak equivalence principle*.

The equality of the inertial and gravitational mass has been tested to a level of one part in 10^{12} in modern free-fall experiments.[1]

Returning to the example of the rocket, we can also ask what would happen if we let it hover at a fixed height above the Earth and then suddenly stop the engines and let the spacecraft fall freely back towards the Earth. Since according to the weak equivalence principle everything inside the rocket will be accelerated towards the Earth with identical value of the acceleration the situation will be very similar to one without any gravitational field at all (this is in fact what gives the 'weightlessness' of astronauts in orbit – such an orbit is a free-fall situation). However, there is a difference. Since the gravitational field has the Earth as its source, the trajectories of objects inside the spacecraft will converge slightly (see Fig 3.1), a phenomenon known as tidal attraction. The larger the separation of the bodies, the bigger the tidal attraction.

If one restricts oneself to nearby objects, meaning to perform *local* mechanical experiments, the gravitational field can be regarded as homogeneous and there will be no difference between free fall and the situation of no gravitational field. Einstein elevated this principle to the postulates of the *strong equivalence principle*:

- The results of all local experiments in a freely falling frame are independent of the state of motion, and the results are the same for all such frames. The results in freely falling frames are consistent with the special theory of relativity.

---

[1] For a review of tests of the weak equivalence principle, see [12].

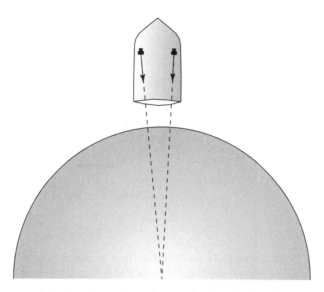

Fig. 3.1. Tidal attraction of two bodies in free fall inside a spacecraft.

3.3 Gravitational Redshift and Bending of Light

With the postulates of the strong equivalence principle, we can without much mathematics derive two of the most astonishing consequences of general relativity: the fact that times goes slower in the presence of a gravitational field and that a straight light-ray is bent by a gravitating body near its path. (Our numerical estimates will then indicate that on the surface of the Earth, both of these effects are very tiny but non-zero.)

The concept of time is difficult to define rigorously. However, some processes in nature, such as the oscillation of certain molecules and the pulsations of distant magnetized neutron stars, give us a practical way to define time as, for example, what is measured by an atomic clock. Given one clock, we can synchronize all other processes with it. In particular, biological processes will occur at a pace which has a fixed relation to the atomic clock (since biological activity is governed by atomic and molecular processes). Thus, if we can show that an atomic clock goes slower in a given situation it means that all other clocks synchronized with it go slower; in particular, aging of humans will be slower.

Consider a rocket in free fall with acceleration g, starting with zero velocity in the Earth's gravitational field. An atomic clock at the floor of the rocket controls a pulsed laser. Each pulse defines a 'tick' of a clock and thus a time as measured at the location of the laser. Suppose then that the pulses are received at a height Δh above the source (Fig. 3.2). What is the time interval between the arrival of the pulses at height Δh; that is, what is the relation be-

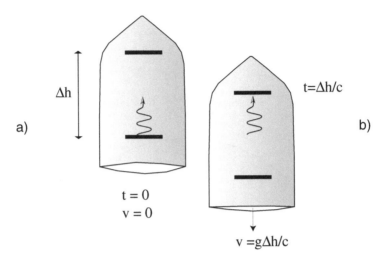

Fig. 3.2. (a) Light transmitter and receiver in a rocket which starts to fall freely in the Earth's gravitational field at the time $t = 0$, when the emitter sends a light pulse (a 'tick' of a clock). (b) The same rocket at the time $\Delta t = \Delta h/c$ when the light pulse reaches the receiver at a height Δh above the transmitter.

tween the pace of time at height Δh and at the floor of the rocket? According to the strong equivalence principle, light travels with the constant velocity c in all directions in the freely falling rocket, so after a time $\Delta t = \Delta h/c$ as measured in the rocket, light reaches the receiver. During this time, the rocket has accelerated to a velocity

$$\Delta v = g\Delta t = \frac{g\Delta h}{c} \tag{3.4}$$

so that an outside observer at the same position as the receiver but *at rest with respect to the Earth's surface* will see the pulses Doppler shifted to lower frequency by the relative amount

$$\frac{\Delta \nu}{\nu} = \frac{\Delta v}{c} = \frac{g\Delta h}{c^2} \tag{3.5}$$

We assume that $\Delta v/c \ll 1$, so that we need only include the first-order Doppler shift (2.69). Also, special relativistic effects such as the Lorentz contraction of the rocket are of second order in v/c and can thus be neglected.

Comparing this slower rate with the rate of a stationary atomic clock at the position of the receiver, the observer will find that the stationary clock makes more ticks in a given time period than the number picked up from the emitter. This must mean that the atomic clock, and therefore all clocks synchronized with it, runs slower than a stationary clock at the height Δh above it by the fractional amount $g\Delta h/c^2$. Time runs slower in the basement!

To further clarify this result, suppose that the gravitational field is such that $g\Delta h/c^2$ is 1 per cent. Let an observer synchronize his clock with a clock in the basement and then climb the gravitational field. During movement, there may of course be special relativistic time dilation effects, but these can be made very small by moving slowly. Staying at the higher level for, say, 24 hours and then descending again to the basement, the observer finds his clock to show a time which is about 14 minutes more advanced. Had he stayed in the basement, he would have been 14 minutes younger!

Example 3.3.1 Estimate how much slower a clock runs at the bottom of the Harvard University tower (height 22.6 m) than a clock at the top

Answer:

$$\frac{\Delta\nu}{\nu} = \frac{g\Delta h}{c^2}$$

with $g = 9.8$ ms^{-2}, $h \sim 22.6$ m, $c = 3 \cdot 10^8$ ms^{-1} gives $\Delta\nu/\nu = 2.46 \cdot 10^{-15}$, that is, less than 1 second in 100 million years.

The tiny gravitational timeshift was measured by Pound and Rebka in 1960 in a classic experiment at Harvard University. Using the Mössbauer effect, they were able to measure a gravitational redshift over 22.6 m of $\Delta\nu/\nu = (2.57\pm0.26)\cdot10^{-15}$, which compares well with the prediction of Example 3.3.1 of $2.46 \cdot 10^{-15}$.

Let us return to the freely falling rocket. Suppose that a passenger on board shines a laser beam horizontally across the rocket (Fig. 3.3 (a)). An observer at rest with respect to the Earth will see the rocket move downwards a little before the light-ray hits the rocket's wall (Fig. 3.3 (b)). To the observer in the Earth's gravitational field it thus seems that light follows a bent path. According to the laws of optics, light follows the path that takes the shortest time (Fermat's principle), and we have seen that in a gravitational field this path is not straight. This is a first indication that a massive body like the Earth makes space-time curved instead of flat.

We have a very simple analogy of this situation when determining the shortest flight route between two cities on Earth. For instance, flying between Stockholm and New York by the shortest route takes the plane well into the polar region. This trajectory, the shortest path between two points on a sphere, is certainly not a straight line. It is a part of a great circle with the same curvature as the curvature of the Earth. The fact that the shortest path between two points on Earth is not a straight line is of course a direct consequence of the fact that the Earth is not flat (although it may look quite flat locally).

To analyse curvature caused by gravity more quantitatively we need, however, to find the equation that relates the curvature of space-time to the energy and momentum of matter that is present in a given situation. Before doing this, we should make ourselves familiar with curved spaces, starting with the two-dimensional case such as the surface of a sphere or a cylinder, where our everyday intuition can be used as a guide.

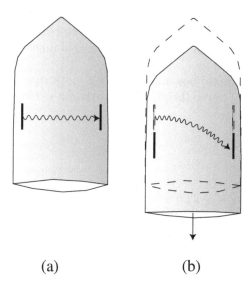

(a)   (b)

Fig. 3.3. (a) Path of a light beam in a freely falling rocket in the Earth's gravitational field, as seen by an observer in the rocket. (b) The same light beam seen by an external observer at rest with respect to the surface of the Earth.

3.4 Curved Spaces

We are all familiar with the difficulty of representing the Earth on a planar map. The surface of a two-dimensional sphere cannot, in contrast to that of a cylinder, be cut and laid flat on a plane. The curvature of the sphere can be determined in several ways. One very important way is shown schematically in Fig. 3.4 (a). Starting, for example, at the North Pole N of the sphere, we choose an arbitrary direction of a vector V and move down the path P_1 following the meridian along which V points down to the equator. This line of longitude is part of a great circle on the sphere, thus it is the shortest possible way between N and the equator. Such a shortest possible line on a curved surface is called a *geodesic*. We keep the direction, that is, we parallel-transport the vector to a point on the equator (which is another great circle). We then parallel-transport the vector along the path P_2 on the equator. Finally, we parallel-transport the vector back to the North Pole N along the meridian P_3. When we compare the direction of the resulting vector V' after the round-trip with the original direction we find that its direction has been rotated by an amount θ which is, in this case, equal to the enclosed area divided by the radius squared of the sphere.

On the other hand, making a round-trip on the surface of a cylinder (Fig. 3.4 (b)), we see that no displacement of the direction occurs. As Gauss showed, this construction of round-trips with parallel-transported vectors can be generalized to determine the curvature of arbitrarily curved two-dimensional surfaces. If the curvature is not constant, we will of course have to make small closed curves and take suitable limits of the infinitesimal displacement divided by the infinitesimal enclosed area.

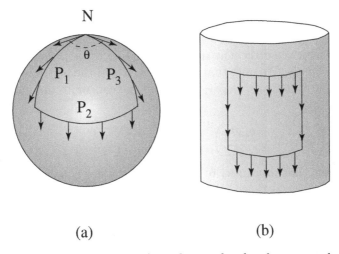

(a)   (b)

Fig. 3.4. (a) A vector, parallel-transported around a closed curve on the surface of a sphere, starting at the North Pole N and going along the paths $P_1 P_2 P_3$, will be rotated by an angle θ which depends on the enclosed area. (b) A vector, parallel-transported around a closed curve on a cylinder will point in the original direction after a full loop.

3.5 Coordinates and Metric

We are used to introducing coordinates to label points in a manifold like the Euclidean plane or the surface of a sphere. However, it is not only the labels of points we are interested in, but also *distances* between points in the manifold. Just like we did for Minkowski space in Section 2.3, a *metric* $g(P_1, P_2)$ is introduced to give a real number, the square of the distance, between two arbitrary points P_1 and P_2. For instance, on the sphere with radius a we may introduce spherical coordinates θ and ϕ, and the square of the distance between points (θ, ϕ) and $(\theta + d\theta, \phi + d\phi)$ is given by the line element or *metric equation*

$$ds^2 = a^2 d\theta^2 + a^2 \sin^2 \theta d\phi^2 \tag{3.6}$$

which is quadratic in $d\theta$ and $d\phi$. Since ϕ and θ are orthogonal coordinates, there are no cross terms of the type $d\theta d\phi$.

The surface of a sphere is an example of a *Riemann space* since it has a metric equation between neighbouring points which is quadratic in the coordinate separations. In the two-dimensional case, with general coordinates u and v, we can write the most general metric equation as

$$ds^2 = g_{11} du^2 + 2g_{12} dudv + g_{22} dv^2 \tag{3.7}$$

or in matrix form

$$ds^2 = (du, dv) \begin{pmatrix} g_{11} & g_{12} \\ g_{21} & g_{22} \end{pmatrix} \begin{pmatrix} du \\ dv \end{pmatrix} \tag{3.8}$$

where the metric matrix may always be chosen to be symmetric, $g_{12} = g_{21}$, since $dudv=dvdu$. We note that, if we have scaled the metric such that g_{11} is positive, and if $\det(g) = g_{11}g_{22} - g_{12}^2 \neq 0$, we can write the metric equation (3.7)

$$ds^2 = \left(\sqrt{g_{11}} du + \frac{g_{12} dv}{\sqrt{g_{11}}} \right)^2 + \left(g_{22} - \frac{g_{12}^2}{g_{11}} \right) dv^2 \tag{3.9}$$

which means that we can write it locally as the flat equation

$$ds^2 = d\xi_1^2 + d\xi_2^2 \tag{3.10}$$

with

$$d\xi_1 = \sqrt{g_{11}} du + \frac{g_{12} dv}{\sqrt{g_{11}}}; \quad d\xi_2 = \sqrt{g_{22} - \frac{g_{12}^2}{g_{11}}} dv \tag{3.11}$$

However, this only works if $\det(g) > 0$; if $\det(g) < 0$ we have to extract a minus sign and write

$$ds^2 = d\xi_1^2 - d\xi_2^2 \tag{3.12}$$

In both cases, we get locally flat spaces. This means that we can locally match the curved space with a flat 'tangent space'. In the first case, the metric is positive definite and the tangent space is just like the Euclidean plane. In the second case, the metric is not positive definite (similar to the metric of space-time in special relativity), and the tangent space is called pseudo-Euclidean (and the curved space pseudo-Riemannian).

To repeat, we have found the important result that given a point P on our two-dimensional Riemannian manifold, we can find a local coordinate system (ξ_1, ξ_2) with corresponding metric $g_{ik}(P) = \eta_{ik}$ (we will generically denote a flat Euclidean or pseudo-Euclidean metric by the symbol η), where $g_{ik}(P)$ does not depend on (ξ_1, ξ_2), that is,

$$\frac{\partial g_{ik}}{\partial \xi_1}(P) = \frac{\partial g_{ik}}{\partial \xi_2}(P) = 0 \tag{3.13}$$

Example 3.5.1 Perform the construction leading to (3.10) for the unit sphere with spherical coordinates θ and ϕ.

Answer: Since the metric corresponding to the line element (3.6) is diagonal, with $g_{\theta\theta} = a^2$ and $g_{\phi\phi} = a^2 \sin^2 \theta$, we just choose $d\xi_1 = a d\theta$ and $d\xi_2 = a \sin\theta_0 d\phi$ as locally flat coordinates near a point $P : (\theta_0, \phi_0)$. Since the determinant $\det g = g_{\theta\theta} g_{\phi\phi} \geq 0$, the line element then becomes $ds^2 = d\xi_1^2 + d\xi_2^2$. The only problem appears for $\theta = 0$, where ϕ is undefined. This is a so-called coordinate singularity, since it can be avoided by choosing another point as the North Pole on the sphere.

Of course it is important to notice that although the metric of a sphere can be made locally Euclidean (which is why our ancestors thought for such a long time that Earth is flat), it is impossible to cover an extended region of the sphere using Euclidean (cartesian) coordinates. We need a continuum of locally Euclidean coordinate systems which have to be patched together consistently to fully describe the curved space of the sphere.

We have spent a long time on this seemingly trivial example because as we shall see, the curved space-time of general relativity can be constructed similarly by patching together locally pseudo-Euclidean free-fall frames.

3.5.1 Measures of Curvature

Staying with two-dimensional examples, we shall investigate another method of measuring the curvature besides the one using parallel transport of a vector which we discussed in Section 3.4.

Look first at a circle drawn out on a sphere (Fig. 3.5 (a)). Will an observer confined to live on the surface of the sphere be able to deduce from local measurements that the surface is curved? Yes, we saw that parallel-transporting a vector around a closed curve and checking whether the direction has changed after a full loop gives such an intrinsic measurement of curvature. Alternatively, one may measure the length of the circumference of a circle, and its radius. Since the latter is a part of a great circle on the sphere, it is clear from the figure that the ratio of the circumference to the radius will be less than the Euclidean value 2π. It may be seen from Fig. 3.5 (b) that if we make the corresponding construction on the surface of a saddle, the value of the ratio will be greater than 2π. In fact, given the measured radius s of the circle on the surface and the length of the circumference c one may define the local curvature of a two-dimensional surface as the limit

$$K = \frac{3}{\pi} \lim_{s \to 0} \left(\frac{2\pi s - c}{s^3} \right) \tag{3.14}$$

Example 3.5.2 Suppose a circle is drawn around the North Pole on a sphere with radius a (see Fig. 3.5 (a)). Show that the curvature of the sphere defined as in (3.14) is equal to $1/a^2$.

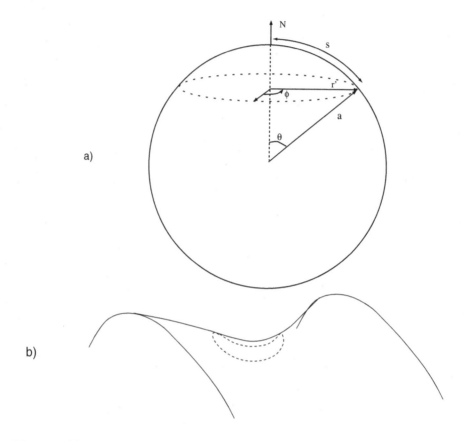

a)

b)

Fig. 3.5. (a) The circumference $2\pi r'$ and radius s of a circle drawn out on a sphere have a ratio less than 2π. (b) The circumference and radius of a circle drawn out on a saddle surface have a ratio greater than 2π.

Answer: Let the circle subtend a half-angle of $\theta = s/a$ (see Fig. 3.5 (a)). Then the circumference of the circle is

$$c = 2\pi r' = 2\pi a \sin\theta \approx 2\pi a \left(\frac{s}{a} - \frac{1}{6}\frac{s^3}{a^3} + \ldots\right) = 2\pi s \left(1 - \frac{s^2}{6a^2} + \ldots\right)$$

Thus, the curvature K is

$$K = \frac{3}{\pi} \lim_{s \to 0} \left(\frac{2\pi s - 2\pi s \left(1 - \frac{s^2}{6a^2}\right)}{s^3}\right) = \frac{1}{a^2}$$

We may introduce coordinates (r', ϕ) on the sphere with the property (see Fig. 3.5 (a)) that the circumference of a circle around the North Pole has the value exactly equal to $2\pi r'$. With the auxiliary angle $\theta = s/a$, we see that $r' = a \sin\theta$, so

$$s = a \arcsin(r'/a) \tag{3.15}$$

The fact that the sphere is curved then appears in the metric for the coordinates (r', ϕ). Keeping r' fixed (that is $dr' = 0$) and varying ϕ we see that $ds = r'd\phi$. Keeping instead ϕ fixed and varying r' we find by differentiating (3.15) $ds = dr'/\sqrt{1 - (r'/a)^2}$, so that

$$ds^2 = g_{r'r'}dr'^2 + r'^2 d\phi^2 \tag{3.16}$$

with, introducing the curvature $K = 1/a^2$,

$$g_{r'r'} = \frac{1}{1 - Kr'^2} \tag{3.17}$$

For flat space, the same formula applies, with the curvature $K = 0$. A similar construction will show that a two-dimensional space with constant negative curvature $K = -1/a^2$ also obeys this metric equation. Since in the curved case the dimensioned parameter a is used, it may be convenient to measure all lengths in units of a, that is: we introduce the dimensionless variable

$$r = \frac{r'}{a}$$

which gives the metric equation

$$ds^2 = a^2 \left(\frac{dr^2}{1 - kr^2} + r^2 d\phi^2 \right) \tag{3.18}$$

with $k = Ka^2 = -1, +1, 0$ depending on whether the space is negatively curved, positively curved or flat.

A couple of features are worth noticing for this two-dimensional example. First of all, we see that a appears as an overall scale factor in the metric equation. We derived the equation for $k = +1$ using a third coordinate since we could embed the sphere in Euclidean three-dimensional space. However, there is no need to do so – we could use (3.18) as the definition of the metric for the two coordinates (r, ϕ) without reference to a third coordinate. In fact, it turns out that the space with constant negative curvature cannot be embedded in Euclidean three-space, yet the metric is described by (3.18) with $k = -1$.

Let us compute the path length s when going from $r = 0$ to a finite value of r along a meridian $d\phi = 0$:

$$s = a \int_0^r \frac{dr}{\sqrt{1 - kr^2}} = a\mathcal{S}(r) \tag{3.19}$$

Here

$$\mathcal{S}(r) = \begin{cases} \arcsin(r) & k = +1 \\ r & k = 0 \\ \text{arcsinh}(r) & k = -1 \end{cases} \tag{3.20}$$

This result of course agrees with (3.15) for the closed case $k = +1$.

The circumference $2\pi r' = 2\pi a \sin(s/a)$ of the circle on the sphere is first increasing with s until $s = \pi a/2$, and then it is decreasing until reaching zero at $s = \pi a$. The reason for this behaviour is obvious when we look at a sequence of circles on a sphere, starting at the North Pole, then growing in size towards the equator, and then shrinking again towards the South Pole. This phenomenon is typical for a positively curved, closed space like the sphere. For a space of negative curvature, the situation is quite different (see Problem 3.2). There the circumference grows indefinitely with increasing s, showing that such a space has infinite extent (i.e., it is an open space).

As a final note, we see that we could have parameterized the sphere also by using coordinates (x, y, z) and using the constraint $x^2 + y^2 + z^2 = a^2$ to eliminate z from the expressions for the path length (see Problem 3.3). Also, since we have been dealing with homogeneous spaces, we could have taken any point (and not only the North Pole on the sphere, for example) as the starting point with identical results.

3.5.2 Three-Dimensional Space

In three dimensions, the situation is similar in principle but rather more complicated in practice since there are more numbers needed to describe the curvature than just the one number K in the two-dimensional case. We know how to parametrize the usual flat Euclidean three-space using cartesian coordinates (x, y, z) or spherical coordinates (r, θ, ϕ), where we have again made the radial coordinate dimensionless by dividing out a scale factor a. In the latter case, the square of the line element distance (the metric equation) is given by

$$ds^2 = a^2(dr^2 + r^2 d\Omega^2) \tag{3.21}$$

with

$$d\Omega^2 \equiv d\theta^2 + \sin^2\theta d\phi^2 \tag{3.22}$$

This three-dimensional flat space is, as we know, homogeneous and isotropic. That is, it looks the same at every point and in every direction, and the local curvature is the same (in this case zero) at all points. In cosmology, this feature is usually assumed for the Universe as a whole: as a kind of extended Copernican principle our position in the Universe should not have a particular significance. This is called the *cosmological principle*. Apart from local inhomogeneities like stars, planets, galaxies, clusters etc., it seems that on average, matter is indeed quite smoothly distributed everywhere. This is particularly true on the very large scales probed by the microwave background observations where inhomogeneities are of the order of only a few times 10^{-5}.

In a curved isotropic homogeneous space we can make the same type of analysis as in the previous section. Thus, we choose angular variables θ and ϕ, and define a dimensionless radial coordinate r such that a surface passing through the coordinate r will have area $4\pi(ar)^2$. Analogously with the two-dimensional case, the metric equation will then contain an r-dependent g_{rr}:

$$ds^2 = a^2(g_{rr}dr^2 + r^2d\Omega^2) \tag{3.23}$$

For a positively curved three-space such as the so-called three-sphere parameterized by the equation $x^2 + y^2 + z^2 + w^2 = a^2$ we can essentially repeat everything we did for the two-sphere identically to obtain the metric equation

$$ds^2 = a^2\left(\frac{dr^2}{1-r^2} + r^2d\Omega^2\right) \tag{3.24}$$

In fact, one can show that every isotropic, homogeneous three-space can be parameterized (perhaps after performing a coordinate transformation) with coordinates of this form giving the metric equation

$$ds^2 = a^2\left(\frac{dr^2}{1-kr^2} + r^2d\Omega^2\right) \tag{3.25}$$

with $k = -1, 0, +1$. For a given value of k, we see that this defines a one-parameter family of similar spaces, where the scale factor a is the parameter.

We have now, in fact, come a long way towards building a cosmological model that has a chance of describing our Universe. At a given value of time t, the space-like part of the four-dimensional space-time metric should be given by (3.25), with a possibly depending on t, so that we have to write $a = a(t)$. However, to specify the model completely we need both the value of k and at each time t the value of the scale factor $a(t)$. To do this we will need to develop more powerful methods to formulate the theory of general relativity.

3.6 Curved Space-Time

In the previous section, we saw how three-dimensional space curvature could be described in the isotropic and homogeneous case through the introduction of the metric equation (3.25), which we generalize to four-dimensional space-time by allowing the scale factor a depend on time, $a = a(t)$:

$$ds^2 = dt^2 - a^2(t)\left(\frac{dr^2}{1-kr^2} + r^2d\theta^2 + r^2\sin^2\theta d\phi^2\right) \tag{3.26}$$

where from now on we set $c = 1$.

We recall that r is dimensionless and k can be chosen to be $+1$, 0, or -1 depending on whether the constant curvature is positive, zero or negative, respectively. For $k = 0$ we recognize flat Minkowski space. The coordinates (3.26) are such that the circumference of a circle corresponding to t, r and θ all being constant is $2\pi a(t)r$, the area of a sphere corresponding to t, r constant is $4\pi a^2(t)r^2$, but the physical radius of this circle and sphere is given by (see (3.19))

$$r_{\text{phys}} = a(t) \int_0^r \frac{dr}{\sqrt{1 - kr^2}} \tag{3.27}$$

In particular, for $k = +1$, the Universe will be closed (but without boundary, just like the two-sphere), and $a(t)$ can be interpreted (see (3.19)) as the radius of the Universe at time t. If $k = 0$ or $k = -1$ the Universe would be open and plausibly of infinite extent.

We will return to this metric, the *Robertson Walker* line element, as a model for the space-time of our Universe in Section 4.2. However, we need to develop some machinery for treating differentiation of space-time dependent quantities in curved space-time. This is done in Appendix A. As this is fairly technical, we summarize the main results here for students who do not want to enter too deeply into the subject of general relativity.

According to the equivalence principle, we can always *locally* at a space-time point P find a reference frame with coordinates $\xi^\mu = (\xi^0, \xi^1, \xi^2, \xi^3)$ such that (see (3.10), (3.13))

$$ds^2 = g_{\mu\nu} d\xi^\mu d\xi^\nu \tag{3.28}$$

with

$$g_{\mu\nu}(P) = \eta_{\mu\nu} \tag{3.29}$$

$$\frac{\partial g_{\mu\nu}}{\partial \xi^\rho}(P) = 0 \tag{3.30}$$

This is the free-fall frame at P. To transfer to such a frame from any arbitrary space-time coordinate system with coordinates $x^\mu = (t, x^1, x^2, x^3) = (x^0, x^1, x^2, x^3)$ we need to perform a coordinate transformation of a more general type than the Lorentz transformations of special relativity. This is derived in Appendix A. Free-fall motion in the ξ^μ coordinates is simply (see (2.48))

$$\frac{d^2 \xi^\mu}{d\tau^2} = 0 \tag{3.31}$$

where, as in (2.43)

$$d\tau^2 = \eta_{\mu\nu} d\xi^\mu d\xi^\nu \tag{3.32}$$

In the general coordinate system x^μ, (3.31) is replaced by the *geodesic equation*

$$\frac{d^2 x^\sigma}{d\tau^2} + \Gamma^\sigma_{\mu\nu} \frac{dx^\mu}{d\tau} \frac{dx^\nu}{d\tau} = 0 \tag{3.33}$$

where the *metric connections* (sometimes called affine connections or Christoffel symbols) $\Gamma^\sigma_{\mu\nu}$ are given by derivatives of the metric $g_{\mu\nu}(x)$:

$$\Gamma^\sigma_{\mu\nu} = \frac{g^{\rho\sigma}}{2} \left(\frac{\partial g_{\nu\rho}}{\partial x^\mu} + \frac{\partial g_{\mu\rho}}{\partial x^\nu} - \frac{\partial g_{\nu\mu}}{\partial x^\rho} \right) \tag{3.34}$$

where $g^{\mu\nu}$ is the inverse of $g_{\mu\nu}$:

$$g_{\rho\mu}g^{\mu\nu} = \delta^\nu_\rho \tag{3.35}$$

The metric connections may be used for performing *covariant derivatives* usually written in the so-called semicolon convention:

$$V^\mu_{;\nu} = \frac{\partial V^\mu}{\partial x^\nu} + \Gamma^\mu_{\nu\rho}V^\rho \tag{3.36}$$

For instance, the covariant divergence of V^μ is given by

$$V^\mu_{;\mu} = \frac{\partial V^\mu}{\partial x^\mu} + \Gamma^\mu_{\mu\rho}V^\rho \tag{3.37}$$

This can in fact be rewritten in a form that is usually easier to use in practice:

$$V^\mu_{;\mu} = \frac{1}{\sqrt{-g}}\frac{\partial}{\partial x^\mu}\left(\sqrt{-g}V^\mu\right) \tag{3.38}$$

where $g = \det(g_{\mu\nu})$.

Unlike the first term on the right-hand side of (3.36), $V^\mu_{;\nu}$ transforms as a tensor quantity in both the indices μ and ν. This is the basic key to setting up formulae that are general relativistic; that is, in accordance with the strong equivalence principle. According to this principle, the laws of nature in a free-fall frame are the usual tensor formulae of special relativity. To make them applicable to any frame, we just have to substitute the Minkowski metric $\eta_{\mu\nu}$ by $g_{\mu\nu}$, and the ordinary derivatives like (2.33) by covariant derivatives like (3.36).

By comparing the geodesic equation of motion (3.33) with Newton's equations of motion in a gravitational field (see A.2), one finds in the weak-field limit (and for velocities much smaller than the speed of light)

$$g_{00} = 1 + \frac{2\phi}{c^2} \tag{3.39}$$

where ϕ is the ordinary Newtonian gravitational potential, $\phi = -\frac{GM}{r}$.

Example 3.6.1 Show that in the weak-field limit, the gravitational time dilation is given by

$$dt = d\tau\left(1 - \frac{GM}{rc^2}\right)^{-1} \tag{3.40}$$

Answer: Using the definition of proper time (2.44) with the metric of the weak-field limit (3.39) we find that two events with coordinate separations $(cdt, 0, 0, 0)$ give

$$d\tau = \sqrt{g_{00}}\, dt = \sqrt{1 + \frac{2\phi}{c^2}}\, dt \tag{3.41}$$

Thus, for very weak fields, $\phi \ll c^2$, we find we arrive at (3.40).

From the affine connections (3.34), a measure of the curvature of space-time can be obtained, by a combination of partial derivatives and index contractions (see Appendix A). The object obtained, the *Riemann curvature tensor* $R^{\mu}{}_{\sigma\beta\alpha}$, has the geometrical interpretation of telling how much the direction a vector is changed when it is parallel-transported around a closed curve (so that it is zero for flat space-time).

The Riemann tensor can be a complicated object, and in fact for a given metric $g_{\mu\nu}(x)$ it is wise to use any of the symbolic algebra computer programs available for its calculation. However, due to a large number of symmetries, the number of independent components which naively looks like $4^4 = 256$ is reduced to 20. These symmetries are most easily summarized for the associated tensor $R_{\alpha\beta\gamma\delta} = g_{\alpha\rho}R^{\rho}{}_{\beta\gamma\delta}$ formed by lowering the first index:

• Symmetry in the exchange of the first and second pairs of indices:

$$R_{\alpha\beta\gamma\delta} = R_{\gamma\delta\alpha\beta} \tag{3.42}$$

• Antisymmetry:

$$R_{\alpha\beta\gamma\delta} = -R_{\beta\alpha\gamma\delta} = -R_{\alpha\beta\delta\gamma} \tag{3.43}$$

• Cyclic property:

$$R_{\alpha\beta\gamma\delta} + R_{\alpha\delta\beta\gamma} + R_{\alpha\gamma\delta\beta} = 0 \tag{3.44}$$

Through contraction of the first and third index of the Riemann tensor one obtains the *Ricci tensor*

$$R_{\mu\nu} = g^{\alpha\gamma}R_{\alpha\mu\gamma\nu} \tag{3.45}$$

which is symmetric in its indices. By contracting the two indices of the Ricci tensor one gets the *Ricci scalar*

$$R = g^{\mu\nu}R_{\mu\nu} \tag{3.46}$$

From the Ricci tensor and the Ricci scalar one can form another symmetric tensor, which by construction has vanishing covariant divergence, the *Einstein tensor* $G_{\mu\nu}$:

$$G_{\mu\nu} = R_{\mu\nu} - \frac{1}{2}g_{\mu\nu}R \tag{3.47}$$

3.6.1 The Energy-Momentum Tensor

The basic idea behind general relativity is that the presence of matter makes space-time curved. This curvature can be mathematically summarized in the Riemann tensor $R^{\mu}{}_{\sigma\beta\alpha}$ (see (A.48) for its exact form) and its contractions. Before setting up the Einstein equations we must now find a way to describe the energy and momentum content of matter (and radiation) in tensor form. It is instructive to first consider special relativity only. For a collection of N

point particles, with four-momenta p_i^μ, which follow trajectories $\mathbf{r} = \mathbf{r}_i(t)$, we can define a momentum density by

$$T^{\mu 0} \equiv \sum_{i=1}^{N} p_i^\mu(t) \delta^{(3)}(\mathbf{r} - \mathbf{r}_i(t)) \tag{3.48}$$

and the corresponding momentum 'current' by

$$T^{\mu k} \equiv \sum_{i=1}^{N} p_i^\mu(t) \frac{dx_i^k(t)}{dt} \delta^{(3)}(\mathbf{r} - \mathbf{r}_i(t)) \tag{3.49}$$

or, written as one tensor equation (from now on we set $c = 1$, and since $dx_i^k/dt = p_i^k/E_i$; see (2.38)):

$$T^{\mu\nu} = \sum_{i=1}^{N} \frac{p_i^\mu p_i^\nu}{E_i} \delta^{(3)}(\mathbf{r} - \mathbf{r}_i(t)) \tag{3.50}$$

We see from (3.50) that the *energy-momentum tensor* $T^{\mu\nu}$ is symmetric,

$$T^{\mu\nu} = T^{\nu\mu} \tag{3.51}$$

For free (non-interacting) particles, one can show (see Problem 3.4) that the energy-momentum tensor fulfils a conservation equation

$$\frac{\partial}{\partial x^0} T^{\mu 0} + \nabla_i T^{\mu i} = \frac{\partial}{\partial x^\nu} T^{\mu\nu} \equiv T^{\mu\nu}{}_{,\nu} = 0 \tag{3.52}$$

For particles that interact, for example, through electromagnetic forces one has to include contributions from the electromagnetic fields to $T^{\mu\nu}$ (see, for example [51]).

If there are many particles present, a hydrodynamical treatment should be adequate for the 'fluid' of particles. A so-called perfect fluid has the property that viscous forces (that would correspond to non-vanishing T^{ij} with $i \neq j$) are absent, and in the rest frame of a fluid element the energy-momentum tensor takes the form

$$T^{ij} = p\delta_{ij}$$
$$T^{i0} = 0$$
$$T^{00} = \rho \tag{3.53}$$

To get the expression for $T^{\mu\nu}$ in a frame where the four-velocity of the fluid element is u^μ we just have to write (3.53) in terms of tensor quantities. This could be done by making a Lorentz transformation on (3.53), or just by guessing: from the available tensors $\eta^{\mu\nu}$ and $u^\mu u^\nu$ we see that

$$T^{\mu\nu} = (p + \rho)u^\mu u^\nu - p\eta^{\mu\nu} \tag{3.54}$$

reduces to (3.53) in the rest frame of the fluid (where $u^\mu = (1, 0, 0, 0)$). Since it is a tensor, it has this form in all inertial systems. The continuity equation of hydrodynamics is embedded in the conservation equation

$$T^{\mu\nu}_{\ ,\nu} = 0 \qquad (3.55)$$

In the presence of gravity, we know from the principle of covariance how to construct the corresponding energy-momentum tensor. We just replace $\eta^{\mu\nu}$ by $g^{\mu\nu}$:

$$T^{\mu\nu} = (p + \rho)u^{\mu}u^{\nu} - pg^{\mu\nu} \qquad (3.56)$$

and replace the four-divergence in (3.55) by the covariant divergence

$$T^{\mu\nu}_{\ ;\nu} = 0 \qquad (3.57)$$

3.7 Einstein's Equations of Gravitation

If we, as Einstein did, want to view matter and energy as the source of the curvature of space-time, we now have a suitable tensor that could serve as the source of the curvature and therefore of the gravitational field – the energy-momentum tensor. Since it is symmetric and divergenceless, the equations we require should be of the form

$$G^{\mu\nu} = \text{const} \cdot T^{\mu\nu} \qquad (3.58)$$

where we know that

- $G^{\mu\nu}$ should be symmetric since $T^{\mu\nu}$ is symmetric.
- $G^{\mu\nu}_{\ ;\nu} = 0$ since $T^{\mu\nu}_{\ ;\nu} = 0$.
- The constant on the right-hand side should be such that Newton's law of gravity is recovered in the Newtonian limit.

We have previously noticed that the Einstein tensor (3.47) is symmetric and divergence-free. This is why Einstein conjectured that

$$G^{\mu\nu} = R^{\mu\nu} - \frac{1}{2}g^{\mu\nu}R \qquad (3.59)$$

is what should be proportional to $T^{\mu\nu}$. By demanding that the Newtonian limit is correctly obtained, the value of the constant in (3.58) is found to be $8\pi G_N$ (or $8\pi G_N/c^4$ if we had not put $c = 1$), where G_N is Newton's constant. Thus, the celebrated *Einstein equations* of general relativity can be written

$$R_{\mu\nu} - \frac{1}{2}g_{\mu\nu}R = 8\pi G_N T_{\mu\nu} \qquad (3.60)$$

This can be written in a slightly different form. First, we raise one index and contract with the other:

$$R - \frac{1}{2}Rg^{\mu}_{\ \mu} = 8\pi G_N T^{\mu}_{\ \mu} \qquad (3.61)$$

which, since $g^{\mu}_{\ \mu} = \eta^{\mu}_{\ \mu} = 4$ (it is a scalar and can therefore be evaluated in a free-fall frame with the Minkowski metric $\eta_{\mu\nu}$) gives

$$R = -8\pi G_N T^{\mu}_{\ \mu}. \qquad (3.62)$$

This inserted into (3.60) gives

$$R_{\mu\nu} = 8\pi G_N (T_{\mu\nu} - \frac{1}{2} T^{\rho}{}_{\rho} g_{\mu\nu}) \tag{3.63}$$

Since there will be no more risk of confusion between the Einstein tensor G and Newton's constant, we shall return to calling the latter G from now on.

3.7.1 The Schwarzschild Solution

In (3.39) we have given the Newtonian limit of general relativity in a static gravitational potential ϕ. In fact, the Einstein equations for the space-time metric outside a static massive body, mass M, were solved exactly by Schwarzschild in 1916. The solution is (see Problem 3.5)

$$ds^2 = (1 - r_S/r)dt^2 - \frac{1}{1 - r_S/r}dr^2 - r^2(d\theta^2 + \sin^2\theta d\phi^2), \tag{3.64}$$

with the Schwarzschild radius given by

$$r_S = 2GM/c^2. \tag{3.65}$$

A number of interesting features can be derived from the Schwarzschild solution (see, for example [51]):

- A test particle which orbits the central mass on an elliptical or-bit will undergo 'perihelion motion', meaning that the long axis of the ellipse rotates slowly with respect to distant stars. The calculated perihelion motion for the innermost planet Mercury agrees with observation and was one of the earliest triumphs of general relativity.
- A passing light-ray which travels at a closest distance b from the central body will be deflected by an angle $\Delta\theta = 4GM/b$ (gravitational bending of light). In 1919 an expedition to observe a total solar eclipse from the island of Principe, led by A.S. Eddington, set out to measure the deflection of starlight near the obscured Sun during the eclipse. The value, 1.74 arcseconds, agreed with the Einstein prediction and brought immediate worldwide fame to Einstein.
- Look at a photon with $ds^2 = 0$ travelling radially in the Schwarzs-child metric. Then $cdt = dr/(1-r_S/r)$, and the time taken to leave from $r = r_S$ to an outside point becomes formally infinite. Thus, if an object is so dense that its radius is inside the Schwarzschild radius, the object does not emit any light – it becomes a *black hole*.

3.8 Summary

- General relativity relies on a postulate of Einstein, *the strong equivalence principle*, namely, that there is no way by local experiments to distinguish different free-fall frames.
- Gravitational redshift and bending of light are important predictions of general relativity which have been verified by experiments.
- A curved space-time needs a rank-four tensor, the Riemann tensor, for its description. It is formed by various partial derivatives of the metric tensor $g_{\mu\nu}$, which defines invariant local distances in curved space-time through the line element

$$ds^2 = g_{\mu\nu}dx^\mu dx^\nu.$$

- By contracting two indices of the Riemann tensor, the Ricci tensor $R^{\mu\nu}$ is obtained. In the presence of matter or other forms of energy and momentum defined by the energy-momentum tensor $T^{\mu\nu}$, *Einstein's equations* of general relativity are given by

$$R^{\mu\nu} = 8\pi G \left(T^{\mu\nu} - \frac{1}{2}T^\rho{}_\rho g^{\mu\nu} \right),$$

  where G is Newton's constant of gravitation.
- The Schwarzschild radius of a black hole is given by

$$r_S = \frac{2GM}{c^2}.$$

3.9 Problems

3.1 Estimate the gravitational redshift of a photon of given frequency ν emitted from the surface of the Sun (mass $2 \cdot 10^{30}$ kg, radius $7 \cdot 10^8$ m)

3.2 (a) Derive the expression for the circumference of a circle on a surface with constant negative curvature, that is $k = -1$ in (3.18). (b) If $K = -5$ m^{-2}, what is the circumference of a circle with physical radius 3 m?

3.3 Derive an expression for the line element ds^2 on the two-sphere by using $x^2 + y^2 + z^2 = a^2$ and eliminating z.

3.4 Use the definition (3.50) of the energy-momentum tensor for free particles to derive (3.52).

3.5 (This problem requires study of Appendix A for its solution.) We want to construct a solution to Einstein's equations for space-time outside a static massive body (for example, a star) of mass M. From symmetry, we can guess that the line element should be of the form

$$ds^2 = A(r)dt^2 - B(r)dr^2 - r^2(d\theta^2 + \sin^2\theta d\phi^2)$$

(a) Which values should A and B approach as $r \to \infty$ if we want to have an asymptotically flat solution?

(b) Calculate the connections $\Gamma^\mu_{\nu\rho}$ (many of them will vanish).

(c) Use the Einstein equations $R_{\mu\nu} = 0$ for the vacuum outside the massive body to obtain a set of equations for $A(r)$ and $B(r)$.

(d) By combining the equations obtained in (c), show that $A'(r)B(r) + B'(r)A(r) = 0$, that is, $A(r)B(r) = const$. Use the result in (a) to fix this constant. Show that $\frac{d^2}{dr^2}(rA(r)) = 0$, and solve this to give $A(r) = (1-r_S/r)$, with r_S a constant. By comparing with the Newtonian limit, show that $r_S = 2G_N M$.

In this way, you have constructed the so-called Schwarzschild solution, which has many important applications in relativistic astrophysics.

4 Cosmological Models

4.1 Space without Matter – the de Sitter Model

The Einstein equations (3.60) when applied to cosmology in a homogeneous and isotropic Universe, do not permit static solutions, only expanding or contracting. This was considered a failure at the time, since it was generally taken for granted that the Universe would be static and eternal. This prompted Einstein to try to modify his equations in such a way that static solutions would be possible. He found that nothing really forbids a term proportional to $g_{\mu\nu}$ in the equations. (The only argument against this would be that the equations become less simple and therefore 'uglier'.) The full Einstein equations, including such a *cosmological constant* term can thus be written

$$R_{\mu\nu} - \frac{1}{2}g_{\mu\nu}R - \Lambda g_{\mu\nu} = 8\pi G T_{\mu\nu} \qquad (4.1)$$

If we wish, we can move the cosmological constant term over to the right-hand side of the equation, and then we see that in a local free-fall frame (where $g_{\mu\nu} = \eta_{\mu\nu}$) it acts exactly like a contribution to the energy-momentum tensor of the form

$$T^{\Lambda}_{\mu\nu} = \begin{pmatrix} \rho_{\Lambda} & 0 & 0 & 0 \\ 0 & -\rho_{\Lambda} & 0 & 0 \\ 0 & 0 & -\rho_{\Lambda} & 0 \\ 0 & 0 & 0 & -\rho_{\Lambda} \end{pmatrix} \qquad (4.2)$$

with $\rho_{\Lambda} = \Lambda/(8\pi G)$. Comparing this to (3.53) we see that this vacuum energy has very unusual properties: if ρ_{Λ} is positive it corresponds to negative pressure! Below we will see that this indeed means that it will cause the Universe to expand at an ever accelerating rate.

Later, when Hubble discovered the expansion of the Universe, Einstein is said to have regretted ever introducing the cosmological constant. However, since no symmetry principle forbids it, it has been considered wise to keep it and to use cosmological observations to try to bound or eventually give a numerical value to Λ. In fact, as we shall see, many observations today must be interpreted as favouring a non-zero value. Also very important is the role such a constant may have played in the early Universe.

Let us try to find solutions to Einstein's equations with no matter or radiation (that is, $T_{\mu\nu} = 0$), as was first done by de Sitter.

The equations to solve are thus

$$R_{\mu\nu} - \frac{1}{2}g_{\mu\nu}R = \Lambda g_{\mu\nu} \tag{4.3}$$

To satisfy the cosmological principle (see Section 3.5.2), we try to find solutions that are isotropic and homogeneous in space, and which look like Minkowski space locally. This means that the line element should have the form (3.26)

$$ds^2 = dt^2 - a^2(t)\left(\frac{dr^2}{1 - kr^2} + r^2 d\theta^2 + r^2 \sin^2\theta d\phi^2\right) \tag{4.4}$$

Since this metric is diagonal, $g^{\mu\mu} = 1/g_{\mu\mu}$ (no summation over μ), and we read from (4.4)

$$g_{\mu\nu} = \begin{pmatrix} 1 & 0 & 0 & 0 \\ 0 & -a^2/(1-kr^2) & 0 & 0 \\ 0 & 0 & -a^2r^2 & 0 \\ 0 & 0 & 0 & -a^2r^2\sin^2\theta \end{pmatrix}. \tag{4.5}$$

The metric connections are easy to calculate according to (3.34), for example,

$$\Gamma^1_{01} = \frac{g^{11}}{2}\left(\frac{\partial g_{11}}{\partial x^0} + \frac{\partial g_{01}}{\partial x^1} - \frac{\partial g_{10}}{\partial x^1}\right)$$

$$= \frac{g^{11}}{2}\frac{\partial g_{11}}{\partial t} = \left(-\frac{1-kr^2}{2a^2}\right)\left(-\frac{2a\dot{a}}{1-kr^2}\right) = \frac{\dot{a}}{a} \tag{4.6}$$

With some more algebra (Problem 4.1) one finds the 00 and 11 components of (4.3) to give

$$3\left(\frac{\dot{a}}{a}\right)^2 + \frac{3k}{a^2} = \Lambda \tag{4.7}$$

and

$$2\frac{\ddot{a}}{a} + \left(\frac{\dot{a}}{a}\right)^2 + \frac{k}{a^2} = \Lambda \tag{4.8}$$

with the 22 and 33 components giving the same equation as (4.8).

We see that for $k = 0$ and $\Lambda > 0$, it is easy to find a solution to the equation

$$H^2(t) \equiv \left(\frac{\dot{a}}{a}\right)^2 = \frac{\Lambda}{3}, \tag{4.9}$$

namely

$$a(t) = e^{Ht} \tag{4.10}$$

with $H = \sqrt{\Lambda/3}$. This means an exponentially increasing scale factor, a behaviour caused by the negative pressure of the cosmological constant term, which is called *inflation*. It is customary to include, as we did, Λ in $T_{\mu\nu}$, and it turns out that in theories of particle physics there appear scalar fields whose potential energy may include just such a term, which could have driven inflation at an early epoch, if that contribution to $T_{\mu\nu}$ dominated over matter and radiation. We return to this later, when we shall also see that even after a short period of inflation the curvature term proportional to k in (4.7) may in fact be neglected, even for $k = 1$ or $k = -1$. Physically, the explanation for this is that the exponential growth of the scale factor makes the Universe look flat (just like a small sphere when expanded by a huge factor will locally look very flat).

4.2 The Standard Model of Cosmology

We now derive the equations that govern the *Standard Model* of present-day cosmology, namely, the isotropic and homogeneous Friedmann-Lemaître-Robertson-Walker model (or FLRW model, for short). Like the de Sitter model it is based on a line element of the type (4.4), but it is more general in that also it allows for other forms of energy than vacuum energy, such as matter and radiation. The properties of the FLRW model will be seen to provide the basis for the *Hot Big Bang model* that has been so successful in explaining many important features of the observable Universe.

    With little further work, we can in fact derive the Einstein equations for the FLRW model (sometimes called the Friedmann equation). The terms that enter the left-hand side of the Einstein equations (4.1) have already been calculated in the previous section. It just remains to add the contribution to the energy-momentum tensor $T_{\mu\nu}$ from matter and radiation. In the early Universe, the energy density was very smooth, as witnessed by the isotropy of the cosmic microwave background radiation. It should therefore be adequate to use the perfect fluid hydrodynamical approximation of the cosmic fluid. In fact, even today when there are large density contrasts in the form of galaxies and other structures, the overall expansion of the Universe is still well described by the equations of (in this case pressureless) hydrodynamics. In today's cosmological research this approximation has been tested by making supercomputer calculations of the evolution of the particles that make up the 'cosmic fluid'.

    In a perfect fluid, there exists a frame, the *comoving frame*, where the fluid looks perfectly isotropic. For the FLRW line element

$$ds^2 = dt^2 - a^2(t) \left(\frac{dr^2}{1 - kr^2} + r^2 d\theta^2 + r^2 \sin^2\theta d\phi^2 \right) \tag{4.11}$$

there is indeed such a preferred frame, which corresponds to constant values of the coordinates r, θ and ϕ. It is easy to show (Problem 4.2) that a particle at rest in the comoving frame satisfies the geodesic equation

$$\frac{d^2x^i}{ds^2} + \Gamma^i_{\mu\nu}\frac{dx^\mu}{ds}\frac{dx^\nu}{ds} = 0 \tag{4.12}$$

with the line element given by $ds^2 = dt^2$. The world line of such a particle therefore corresponds to free fall in the cosmic fluid. It can also be shown that a particle which has a 'peculiar velocity' with respect to the comoving frame will come to rest as the Universe expands. At every point a comoving frame can be found, where the Universe (in particular, the microwave background) looks maximally isotropic. This is an attractive feature of a homogeneous model like the FLRW model: all observers in the Universe will see an isotropic Universe (and therefore it will appear that they are all at the 'centre' of the Universe!) from wherever they look, if they are at rest (that is having constant r, θ and ϕ) in the local comoving frame.

We saw in Section 3.6.1 that the general relativistic energy-momentum tensor for a perfect fluid is of the form (see (3.56))

$$T_{\mu\nu} = (p+\rho)u_\mu u_\nu - pg_{\mu\nu} \tag{4.13}$$

which fulfils the vanishing covariant divergence condition

$$T^{\mu\nu}_{\;;\mu} = 0 \tag{4.14}$$

The 00 component of the Einstein equations will then only differ from (4.7) by the addition of the term $8\pi\rho$ on the right-hand side, where ρ is the total energy density in matter and radiation ρ_m and ρ_{rad}. Or, by defining the vacuum energy density

$$\rho_{vac} = \rho_\Lambda = \frac{\Lambda}{8\pi G} \tag{4.15}$$

we can write the *Friedmann equation*

$$\left(\frac{\dot{a}}{a}\right)^2 + \frac{k}{a^2} = \frac{8\pi G}{3}\rho_{tot}, \tag{4.16}$$

with

$$\rho_{tot} = \rho_m + \rho_{rad} + \rho_{vac}. \tag{4.17}$$

The second equation (see (4.8)) becomes

$$\frac{2\ddot{a}}{a} + \left(\frac{\dot{a}}{a}\right)^2 + \frac{k}{a^2} = -8\pi Gp \tag{4.18}$$

It is linearly related to (4.16) and the relation that comes from the vanishing covariant divergence (4.14). As the two independent equations, (4.16) is usually chosen together with $T^{\mu\nu}_{\;;\mu} = 0$, which gives (see Problem 4.3):

$$\dot{p}a^3 = \frac{d}{dt}\left(a^3[\rho + p]\right) \tag{4.19}$$

where ρ and p are the total energy density and the total pressure of the fluid. By writing the left-hand side of (4.19) as

$$\dot{p}a^3 = \frac{d}{dt}(pa^3) - p\frac{d}{dt}a^3 \tag{4.20}$$

we obtain

$$\frac{d}{dt}(\rho a^3) = -p\frac{d}{dt}a^3 \tag{4.21}$$

which has a very simple physical interpretation: the rate of change of total energy in a volume element of size $V = a^3$ is equal to minus the pressure times the change of volume, $-pdV$.

Usually, there is a simple relation, the so-called *equation of state*, between ρ and p, generally

$$p = \alpha \cdot \rho \tag{4.22}$$

where α is a constant. For non-relativistic matter, ρ is dominated by the rest mass energy mc^2 ($=m$, since we put $c = 1$) which is huge compared to the pressure which is proportional to the velocity $v \ll c$. Thus, to a good approximation non-relativistic matter is pressureless and $\alpha = 0$. (This is sometimes called a *dust* Universe.) For radiation, $v = c = 1$, and $p = \rho/3$, since the pressure is averaged over the three spatial directions, i.e. $\alpha = 1/3$. For vacuum energy, $p = -\rho$ so $\alpha = -1$. For $p = \alpha\rho$, (4.21) gives (see Example 4.2.1)

$$\rho = \text{const} \cdot a^{-3(1+\alpha)}, \tag{4.23}$$

with the corresponding relation for energy conservation given by

$$\dot{\rho} + 3(1 + \alpha)H\rho = 0 \tag{4.24}$$

Thus, for a radiation dominated Universe (which we shall see was the case the first several thousand years after the Big Bang)

$$\rho \sim \frac{1}{a^4} \quad \text{(radiation domination)} \tag{4.25}$$

and for a matter dominated Universe

$$\rho \sim \frac{1}{a^3} \quad \text{(matter domination)} \tag{4.26}$$

In fact, this behaviour of ρ is not difficult to understand. Since stable matter (like baryons) is not spontaneously created or destroyed, a given density at a particular time will be diluted proportionally to the volume factor $a^3(t)$ as the scale factor $a(t)$ increases. For photons, there is an additional factor of $a(t)$ since the energy of each photon gets redshifted due to the expansion. (See (4.57) and (4.58) below).

The cosmological constant gives a scale factor-independent contribution to the energy density, so that

$$\rho \sim \text{const} \quad \text{(vacuum energy domination)} \tag{4.27}$$

Example 4.2.1 Show that (4.23) and (4.24) follow from (4.21) and (4.22).

Answer: Introducing (4.22) into (4.21) and performing the time derivatives one obtains, collecting terms,

$$\dot{\rho}a^3 = -3(1+\alpha)\rho a^2 \dot{a}$$

or

$$\frac{\dot{\rho}}{\rho} = -3(1+\alpha)\frac{\dot{a}}{a} = -3(1+\alpha)H,$$

which integrated gives

$$\log \rho = -3(1+\alpha)\log a + \text{const}$$

Exponentiating both sides of this equation gives

$$\rho = \text{const} \cdot a^{-3(1+\alpha)}$$

which is (4.23).

Taking the difference between (4.16) and (4.18) gives

$$\frac{\ddot{a}}{a} = \frac{-4\pi G}{3}(\rho + 3p) \tag{4.28}$$

This can be used together with the equation of state (4.22) to determine the time evolution of the scale factor. By making the ansatz $a \sim t^\beta$ (which should be valid for 'small' t but not necessarily asymptotically) and inserting into (4.28) one obtains (Problem 4.4)

$$a(t) \sim t^{\frac{2}{3(1+\alpha)}} \tag{4.29}$$

that is,

$$a(t) \sim \sqrt{t} \quad \text{(radiation domination)} \tag{4.30}$$

and for a matter dominated Universe

$$a(t) \sim t^{2/3} \quad \text{(matter domination)} \tag{4.31}$$

As we saw in (4.10), vacuum energy domination drives an exponentially increasing scale factor,

$$a(t) \sim e^{Ht} \quad \text{(vacuum energy domination)}. \tag{4.32}$$

The more general solution for an arbitrary mixture of matter, radiation and vacuum energy cannot be given in closed form. In Section 4.3.2 we will study some simple cases.

Example 4.2.2 What is the time dependence of the density components ρ_m and ρ_{rad} for radiation and matter domination?

Answer:

- During *radiation domination* the scale factor changes as $a(t) \sim t^{\frac{1}{2}}$. We had also seen that $\rho_{rad} \propto a^{-4} \propto t^{-2}$ and $\rho_m \propto a^{-3} \propto t^{-\frac{3}{2}}$.
- For *matter domination*, $a(t) \sim t^{\frac{2}{3}}$, i.e., $\rho_{rad} \propto a^{-4} \propto t^{-\frac{8}{3}}$ and $\rho_m \propto a^{-3} \propto t^{-2}$.

4.3 The Expanding Universe

From Hubble's observations of redshifts we know that galaxies move away from each other. Thus, if $a(t_0)$ is, say, the distance between the Milky Way and another galaxy far away, we know that this distance is increasing: that is, $\dot{a}(t_0) > 0$. If we want to describe the Universe by an isotropic, homogeneous model like the FLRW model of the last section we thus have to use this fact as a condition on the metric

$$ds^2 = dt^2 - a^2(t) \left(\frac{dr^2}{1 - kr^2} + r^2 d\theta^2 + r^2 \sin^2 \theta d\phi^2 \right) \tag{4.33}$$

However, to completely specify the model, we need to know the curvature k and the behaviour of $a(t)$ for all times, both in the past and in the future. Since the evolution of a is governed by (4.16) and (4.18) which depend on the matter content ρ (and p which is, however, in general calculable from ρ through the equation of state) we also need to know $\rho(t)$. In addition, as we shall see later, structure formation in the Universe depends not only on the total value of ρ but on the various contributions to it (ordinary matter, dark matter, radiation and vacuum energy). It is an important task of modern cosmology to try to determine these parameters.

It may seem that we treat time very differently from space in (4.33). However, we can make a transformation of the time coordinate to *conformal time* η defined by $d\eta = dt/a(t)$ to obtain

$$ds^2 = a^2(\eta) \left(d\eta^2 - \frac{dr^2}{1 - kr^2} - r^2 d\Omega^2 \right) \tag{4.34}$$

This is sometimes useful when one analyses the causal properties of space-time. For instance, in the flat case $k = 0$ space-time looks just like ordinary Minkowski space times the conformal factor $a(\eta)$.

Let us point out some important basic facts about the FLRW model. It was constructed assuming spatial isotropy and homogeneity, which means that it looks the same to all observers at a specific cosmic time t. That is, an observer in a galaxy billions of light years away from us would see the same pattern as Hubble observed: all other galaxies move away from that galaxy with the same slope of the velocity *versus* distance relation as that found by Hubble. Of course, the observations would have to be made at the same time, so the question arises of how to synchronize clocks that are so far away. Fortunately, there are phenomena in the Universe that can be used as cosmic clocks. For instance, the microwave background radiation has a definite temperature which has been monotonically decreasing since it was emitted a few hundred thousand years after the Big Bang. Since there is a definite relation between this temperature and time (we will return to this later), we can in principle specify the cosmic time by specifying the temperature.

Similarly, there is a preferred frame (comoving frame or cosmic rest frame) at each point, which has the property that the microwave background radiation looks maximally isotropic in that frame. Indeed, from the Milky Way this background is isotropic to within a few parts in 10^{-3}, so we are almost at rest in the cosmic frame. It is interesting that the few parts in 10^{-3} anisotropy is of the dipole type (that is, the radiation has higher frequency in one specific direction and lower frequency in the diametrically opposite direction). The natural interpretation of this is that we are not exactly at rest in the cosmic frame. We know that the Earth moves around the Sun, the Sun moves around the Milky Way, and the whole Milky Way is gravitationally attracted by the Virgo cluster of galaxies, which in turn is attracted to an even larger assembly of galaxies (the so-called 'Great Attractor'). Our velocity with respect to the cosmic rest frame, the peculiar velocity, is most likely caused by this gravitational pull from 'nearby' concentrations of matter.

The form of the FLRW metric (4.33) means that for a fixed-time 'slice' $t = t_0$ of the Universe, the Universe looks like one of the fundamental models with $k = -1, 0, 1$. In particular, we could associate, say, to each galaxy i a set of coordinates (r_i, θ_i, ϕ_i). As the Universe evolves, the galaxies stay at the same coordinates (r_i, θ_i, ϕ_i) (apart from the small peculiar motions). The only thing that happens is that all cosmological distances (large enough that local peculiar motions are insignificant) are stretched by the factor $a(t)$. A useful two-dimensional analogy is that of a balloon where the location of 'galaxies' at a certain cosmic time t_1 is indicated. (see Fig. 4.1). At a later time t_2, the pattern of galaxies remains the same, but all distances have been stretched by the ratio of scale factors $a(t_2)/a(t_1)$. It is easy to see that in a uniformly expanding Universe Hubble's law is valid,

$$v = Hd \tag{4.35}$$

with the Hubble parameter

$$H(t) = \frac{\dot{a}(t)}{a(t)}. \tag{4.36}$$

(See Example 4.3.1 for the two-dimensional case.) Sometimes the Hubble parameter $H_0 \equiv H(t_0)$ at the present cosmic time t_0 is called the *Hubble constant*. As we explained in Chapter 1.1, present observations give

$$H_0 = h \cdot 100 \ \mathrm{km\,s^{-1}\,Mpc^{-1}}, \qquad\qquad (4.37)$$

where $h = 0.70 \pm 0.10$.

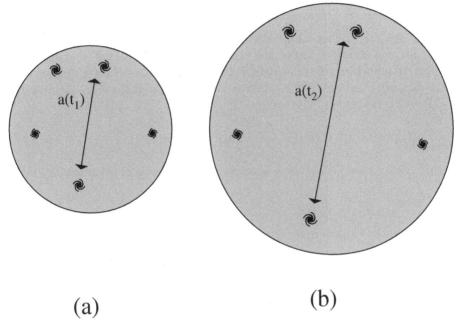

(a)   (b)

Fig. 4.1. (a) A two-dimensional (closed) 'Universe', where the the locations of 'galaxies' at a given cosmic time t_1 are indicated. (b) The same 'Universe' at a later time t_2. The size and pattern of location of individual galaxies are the same, but all distances have been stretched by the ratio of scale factors $a(t_2)/a(t_1)$.

Example 4.3.1 Show that Hubble's law is fulfilled for the balloon model.

Answer: Choose one arbitrary point on the balloon as the location of our Galaxy. Let another galaxy at time t_1 be distant by an amount $d_1 = a(t_1)s$, where s is the arclength distance between us and the other galaxy in normalized coordinates (that is, rescaling to the unit sphere). At the time t_2 the distance to the other galaxy is $d_2 = a(t_2)s$, so the recession velocity is

$$v = \frac{d_2 - d_1}{t_2 - t_1} = \frac{a(t_2) - a(t_1)}{t_2 - t_1} s.$$

Taking the limit $\Delta t = t_2 - t_1 \to 0$ gives

$$v = \frac{\dot{a}}{a}(as) = Hd,$$

which is Hubble's law with $H = \dot{a}/a$ (and the physical distance $d = as$).

4.3.1 The Deviation from the Linear Hubble Law

Equation (4.28)

$$\frac{\ddot{a}}{a} = -\frac{4\pi G}{3}(\rho + 3p) \tag{4.38}$$

shows that if matter alone drives the expansion, the expansion rate is presently decelerating (since $p \sim 0$ and $\rho > 0$). However, if vacuum energy plays an important role (as we shall see that some observations indicate) then the expansion may in fact accelerate. We may expand the scale factor $a(t)$ in a Taylor series around the present time t_0:

$$a(t) = a(t_0) \left(1 + H_0(t - t_0) - \frac{1}{2} q_0 H_0^2 (t - t_0)^2 + \ldots \right) \tag{4.39}$$

which defines the *deceleration parameter* (the name is of historical origin) q_0:

$$q_0 = -\frac{\ddot{a}}{a H_0^2} \tag{4.40}$$

Define the *critical density* ρ_{crit} of the Universe by

$$\rho_{crit} \equiv \frac{3H^2}{8\pi G} \tag{4.41}$$

(As we saw in (1.3), its present value is $\rho_{crit}^0 = 1.9 \cdot 10^{-32} h^2 \text{ kg cm}^{-3}$.) Then we can write (4.16) in the form

$$\frac{k}{H^2 a^2} + 1 = \frac{\rho}{\left(\frac{3H^2}{8\pi G}\right)} \equiv \Omega_t, \tag{4.42}$$

where

$$\Omega_t \equiv \frac{\rho}{\rho_{crit}} \tag{4.43}$$

or

$$\frac{k}{H^2 a^2} = \Omega_t - 1 \tag{4.44}$$

This means that the overall geometry of the Universe is determined by the total energy density parameter Ω_t. We see that for $\Omega_t > 1$, k is positive, which means that the Universe is closed (and $a(t)$ can be chosen as the physical 'radius' of the Universe which is still without boundary; see the two-dimensional spherical analogy). If $\Omega_t < 1$, the Universe has constant negative curvature and is open and could be of infinite extent (although one can construct models where the *global* topology is non-trivial and even an open Universe could be of finite extent). The limiting case $\Omega_t \equiv 1$ would mean that the Universe is flat on large scales (of course local aggregations of matter like galaxies, stars etc., still cause local curvature). It is important to note that Ω_t in general depends on time. The present value is sometimes denoted by Ω_0 or Ω_T. We shall see that in standard cosmology the left-hand side of (4.44) goes to zero rapidly as $t \to 0$. This means that the curvature effects can be neglected in many of the early-Universe calculations.

From (4.38), using the equation of state $p_i = \alpha_i \rho_i$, where the subscript i represent a type of *fluid*, e.g. non-relativistic matter, radiation or vacuum energy density, and all energy densities refer to the present time values one finds:

$$q_0 = \sum_i \frac{\Omega_{0i}}{2} \left(1 + \frac{3p_i}{\rho_i} \right) = \frac{1}{2} \sum_i \Omega_{0i} \left(1 + 3\alpha_i \right) \tag{4.45}$$

Since a cosmological constant corresponds to $\alpha = -1$, q_0 may be negative (the expansion accelerates) if the energy density of the Universe is dominated by ρ_Λ. In general, acceleration of the Universe is possible for $\alpha_i < -\frac{1}{3}$.

4.3.2 The Fate of the Universe

The full solutions to the Friedmann equation (4.16) for the scale factor $a(t)$ at all times t can be quite complicated due to the different a-dependence of the individual terms contributing to ρ_{tot} (and due to the partly unknown physics in the earliest Universe). However, it is instructive to study the behaviour of a in a few generic cases. In particular, it may be of interest to determine the long-term behaviour of $a(t)$ for $t \to \infty$, that is the fate of the Universe. One immediate simplification is then that we know that radiation plays no major role today (and due to the continuing redshift of the cosmic microwave background will be even less important in the future). We thus only need to consider the matter contribution ρ_m and vacuum energy ρ_{vac}.

We start by considering the case where the cosmological constant $\Lambda = 0$ (that is, $\rho_{vac} = 0$). Since $\rho_m = \rho_0 a_0^3 / a(t)^3$ with ρ_0 the present matter density and a_0 the scale factor now, that is at $t = t_0$, the Friedmann equation becomes

$$\dot{a}^2(t) = \frac{8\pi G \rho_0 a_0^3}{3a(t)} - k \tag{4.46}$$

Suppose (unrealistically, of course!) that $\rho_0 = 0$. Then real-valued solutions for $a(t)$ demand $k = -1$. This is the so-called Milne model. The solutions are simply

$$a_{Milne}(t) = t, \tag{4.47}$$

which means a linearly expanding Universe. (Since the Friedmann equation is symmetric under $t \to -t$, $a \sim -t$ is also a solution, which can be interpreted as a contraction since $\dot{a} < 0$. In the following we will not explicitly display such time-symmetric solutions.)

For $\rho_0 > 0$ and small t, where the ansatz $a \sim t^\beta$ should be valid, we have seen in (4.31) that the solution behaves as $\beta = 2/3$, for all k. The solutions for larger t will depend on k. For $k = 0$, we easily solve (4.46) and find that

$$a(t) = a_0 \left(\frac{t}{t_0} \right)^{\frac{2}{3}} \tag{4.48}$$

is the solution also for large values of t. This cosmological model is usually called the Einstein de Sitter model.

For $\rho_0 > 0$ and negative curvature, $k = -1$, (4.46) shows that $\dot{a}^2$ is always positive, which means an ever-increasing $a(t)$. This in turn means that eventually the matter term on the right-hand side will be negligible compared to the k (curvature) term. During this latter stage of curvature domination, the Universe will expand like the Milne model with $a(t) \sim t$.

If $k = +1$, that is positive space curvature, and $\rho_0 > 0$, we see that $\dot{a}^2 > 0$ only up to a critical (largest) value of $a(t)$ given by

$$a_{crit} = \frac{8\pi G \rho_0 a_0^3}{3}. \tag{4.49}$$

Since (4.18) shows that $\ddot{a} \leq 0$ for all a it means that the Universe will start to contract at that point and enter a period of contraction to a 'Big Crunch'.[1]

For $\Lambda \neq 0$, there is a richer 'zoo' of cosmological models. We now have to solve

$$\dot{a}^2(t) = \frac{8\pi G \rho_0 a_0^3}{3a(t)} - k + \frac{\Lambda a(t)^2}{3} \tag{4.50}$$

It is often useful to resolve the contributions to Ω_t today (Ω_T or Ω_0) from non-relativistic matter and from the cosmological constant:

$$\Omega_M = \frac{8\pi G \rho_0}{3H_0^2}, \tag{4.51}$$

$$\Omega_\Lambda = \frac{\Lambda}{3H_0^2}, \tag{4.52}$$

where ρ_0 is the matter density at t_0. Note that Ω_M can also be written

$$\Omega_M = \frac{\rho_0}{\rho_{crit}^0}, \tag{4.53}$$

where ρ_{crit}^0 is the present value of the critical density (see (1.3)). As we shall see later the contributions to Ω_T from relativistic matter,

$$\Omega_R = \frac{8\pi G \rho_{rad}}{3H_0^2} \tag{4.54}$$

is negligible at the present epoch.

An amazing property of the cosmological constant Λ is seen from (4.50) to be that it was completely negligible for small $a(t)$, that is, in the early Universe, but it will eventually dominate over all other forms of matter (and curvature) for large $a(t)$. This is in fact one of the main puzzles today, when

---

[1] There could in principle be a new period of expansion after that: that is, an oscillating Universe. There are, however, thermodynamical arguments against the hypothesis that our Universe is of this type - at each period of expansion and contraction more and more heat would be generated, making nucleosynthesis and the CMBR quite different from what is observed.

observations may favour a non-zero value of Λ. Why is it that we live in such a special epoch when ρ_{vac} just happens to be of similar magnitude as ρ_0?

If $\Lambda < 0$, (4.50) shows that $a(t)$ cannot become arbitrarily large, since $\dot{a}(t)$ has to be real. For the largest possible value a_{crit} of a (when the right-hand side of (4.50) is zero), we again find from (4.18) that $\ddot{a} < 0$, so that we have an oscillating Universe. (This conclusion does not depend on k, although the value of a_{crit} does.)

For $\Lambda > 0$, and $k = 0$ or $k = -1$ we see directly that for large $a(t)$ the Universe will enter a period of exponential expansion: that is, the cosmological model will be similar to the de Sitter model of Section 4.1. For $\Lambda > 0$, and $k = +1$, the interplay between the three terms in (4.50) is more subtle. It is possible to fine-tune Λ such that $\dot{a} = 0$ and $\ddot{a} = 0$ simultaneously (exercise: do this!). This means a static solution, in fact the one that motivated Einstein to introduce the cosmological constant in the first place. Of course, this model does not agree with observations (in particular, it predicts no redshifts). For Λ larger than this fine-tuned value, the repulsive force provided by the cosmological constant will dominate over the gravitational attraction due to the matter density, and the Universe will expand forever. For smaller Λ, there is a range of a where the right-hand side of (4.50) is negative, and which is thus forbidden. Depending on the initial conditions, the Universe will then either oscillate with a scale factor between zero and the lower limit of this range, or always expand (or perhaps contract with a 'bounce') with a scale factor larger than the upper limit of this range. (In the latter case, there would be no Big Bang.) In Fig. 4.2 the solutions for $\Lambda = 0$ with $\Omega_M = 0.2$, 1 and 2 are displayed, corresponding to $k = -1$, 0, and 1, respectively. Also shown is the exponentially growing solution, 'the lonely Universe', for $k = 0$, with $\Omega_M = 0.2$ and $\Omega_\Lambda = 0.8$ (rather close to what is presently favoured by observations). The solutions have been arbitrarily normalized to unity at $t_H = H_0^{-1}$.

4.3.3 Particle Horizons

Light travels, as we have seen, on light-like geodesics $ds^2 = 0$. Let us choose our local cosmic coordinates such that we are located at $r = 0$. Consider a light-ray that moves radially towards us, that is, $\theta, \phi = const$. If this light-ray was emitted from $r = r_E$ at time $t = t_E$ it will reach us at a time t_0 given by (since $ds^2 = 0$ gives $dt = a(t)dr/\sqrt{1 - kr^2}$)

$$\int_{t_E}^{t_0} \frac{dt}{a(t)} = \int_0^{r_E} \frac{dr}{\sqrt{1 - kr^2}} \tag{4.55}$$

Choosing the time coordinate such that $t = 0$ when $a = 0$ (the moment of the 'Big Bang'), we see that the farthest physical distance d_H we can observe today (called the *horizon distance*) is given by multiplying (4.55) by today's physical scale factor $a(t_0)$, and taking the limit $t_E \to 0$ (see (3.27)):

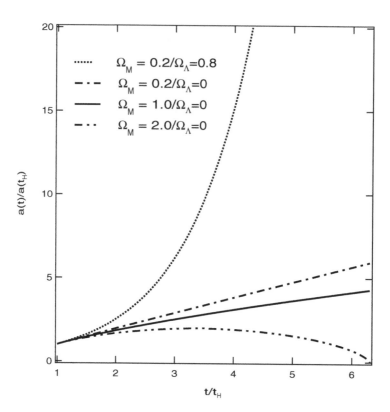

Fig. 4.2. The schematic behaviour of $a(t)$ for vanishing cosmological constant, $\Lambda = 0$, and $\Omega_M = 1$, 2 and 0.2, corresponding to $k = 0, 1$ and -1. Also shown is the case $\Omega_M = 0.2$, $\Omega_\Lambda = 0.8$ (that is k=0), which gives an exponentially growing solution.

$$d_h(t_0) = a(t_0) \int_0^{r_E} \frac{dr}{\sqrt{1 - kr^2}} = a(t_0) \int_0^{t_0} \frac{dt}{a(t)} \tag{4.56}$$

Since in standard cosmology $a(t) \to 0$ slower than t as $t \to 0$ (see (4.29)), d_H is finite ($d_H \sim t$), and our past light-cone is limited by a particle horizon. For example, in a radiation dominated epoch $a(t) \sim t^{1/2}$, which inserted in (4.56) gives $d_H = 2t$. In a matter dominated Universe (with $\Omega = 1$), $d_H = 3t$.

The geodesic equation of motion for a particle with peculiar four-momentum p can be shown to imply that $|\mathbf{p}|$ decreases as $1/a$ (the peculiar motion is said to 'redshift away'). In particular, since $|\mathbf{p}| = \hbar \omega$ for a photon, the frequency (and therefore the energy) will decrease as the Universe expands.

Example 4.3.2 Derive the redshift formula $\lambda_{obs}/\lambda_{emit} = a(t_{obs})/a(t_{emit})$ directly by considering (4.55) for two emission times t_{emit} and $t_{emit} + \delta t_{emit}$ corresponding to two successive wave-crests of radiation.

Answer: The right-hand side of (4.55) does not change for a source at given comoving coordinates. However, on the left-hand side the integration limits are t_{obs} and t_{emit} for the first wave crest, and $t_{obs} + \delta t_{obs}$ and $t_{emit} + \delta t_{emit}$ for the second. Thus

$$\int_{t_{obs}}^{t_{emit}} = \int_{t_{obs} + \delta t_{obs}}^{t_{emit} + \delta t_{emit}} .$$

Comparing here the right-hand side with the left-hand side, we see that we have lost the integration interval from t_{obs} to $t_{obs} + \delta t_{obs}$ but gained the interval from t_{emit} to $t_{emit} + \delta t_{emit}$. Since the full integral is unchanged, it means that

$$\int_{t_{obs}}^{t_{obs} + \delta t_{obs}} = \int_{t_{emit}}^{t_{emit} + \delta t_{emit}} ,$$

or

$$\frac{\delta t_{obs}}{a(t_{obs})} = \frac{\delta t_{emit}}{a(t_{emit})}$$

Since $\delta t_{obs} / \delta t_{emit} = \lambda_{obs} / \lambda_{emit}$ the redshift law follows.

Usually, the redshift parameter z is introduced, defined by

$$1 + z \equiv \frac{\lambda_{obs}}{\lambda_{emit}} \tag{4.57}$$

where we have seen that in the FLRW model the redshift law is simply

$$1 + z = \frac{a(t_{obs})}{a(t_{emit})} \tag{4.58}$$

For small distances, the redshift in the FLRW model can be interpreted as a Doppler effect. The radial recession velocity is $v_r = \dot{a}(t_0)r$, and the first order Doppler shift is given just by this value (or v_r/c if we reinstate the velocity of light), see (2.69). From (4.58) we find for small $t_{obs} - t_{emit}$

$$z \approx \frac{\dot{a}(t_{obs})(t_{obs} - t_{emit})}{a(t_{obs})} \approx r\dot{a}(t_{obs}) \approx v_r \tag{4.59}$$

4.4 Cosmological Distances: Low Redshift

There are various ways of determining the present values of the cosmological parameters. Most of them rely on various observations of distant light (or other electromagnetic radiation). It is therefore important to discuss how a light source at large distance will appear here on Earth, and in particular how its properties depend on the cosmological model.

4.4.1 Luminosity Distance

We have seen that in the FLRW class of models a redshift appears caused by
the expansion of the Universe with the simple rule (4.58). However, distance
is a derived concept (we have no direct means of obtaining distances to cosmological objects) while redshift, flux and angular diameter of sources are
directly measurable. To obtain information on the cosmological parameters
we can use measurements such as the intensity of a source of known intrinsic
strength, a *standard candle*, as a function of the redshift. Then we can use
our underlying general relativistic framework to compare these measurements
with what would be expected for a class of models with different values of
the cosmological parameters.

    The total power of light received by a telescope on Earth from a standard
candle of the total emitting power L at the source (called luminosity) can
be calculated as follows. Suppose first that a 'flash' with a given number of
photons N_γ was emitted isotropically at one particular time t_{emit} from the
source with radial (dimensionless) coordinate r in our cosmic frame. If there
were no expansion of the Universe, the telescope with area A would intercept
a fraction $A/4\pi(a(t_{emit})r)^2$ of the photons. However, due to the expansion
of space, at the detection time $t_0 = t_{obs}$ the area of the spherical shell where
the photons are has increased to $4\pi(a(t_{obs})r)^2$, so

$$\frac{N_{detected}}{N_\gamma} = \frac{A}{4\pi(a(t_{obs})r)^2} \tag{4.60}$$

    To compute the power we must, however, take two additional effects into
account. First, all photons are redshifted by the factor $1+z = a(t_{obs})/a(t_{emit})$,
which means that their energy has decreased by the corresponding factor.
Second, if the time between 'flashes' was $\delta\tau$ at the source, this time interval
will also be increased by the redshift factor, so that the flux $\mathcal{F}$ measured at
the telescope, i.e. the total power per unit area at the telescope will be

$$\mathcal{F} = \frac{L}{4\pi a^2(t_0)r^2(1+z)^2} \equiv \frac{L}{4\pi d_L^2} \tag{4.61}$$

which defines the *luminosity distance*

$$d_L = \sqrt{\frac{L}{4\pi\mathcal{F}}} = a(t_0)r(1+z) \tag{4.62}$$

to the source.

    In (4.61), the coordinate r is unknown, but we can eliminate it in terms
of the redshift z by using (4.55), (4.58) and (4.62). The general solution will
be derived in (4.85), but let us first use an approximation for small z based
on (4.39). Inserting (4.58) in (4.39) gives

$$z = H_0(t_0 - t_1) + \left(1 + \frac{q_0}{2}\right)H_0^2(t_0 - t_1)^2 + \dots \tag{4.63}$$

which can be inverted to give

$$t_0 - t_1 = H_0^{-1} \left[z - \left(1 + \frac{q_0}{2} \right) z^2 + \ldots \right] \qquad (4.64)$$

From (4.56) one finds, by writing $1/a(t) = 1/a(t_0)(a(t_0)/a(t))$ and using (4.39) (see Problem 4.5)

$$r = \frac{1}{a(t_0)} (t_0 - t_1 + \frac{1}{2} H_0 (t_0 - t_1)^2 + \ldots) \qquad (4.65)$$

which, inserting (4.64), finally gives

$$r = \frac{1}{H_0 a(t_0)} \left(z - \frac{1}{2} (1 + q_0) z^2 + \ldots \right) \qquad (4.66)$$

From this expression and (4.62) we obtain

$$d_L = \frac{1}{H_0} \left(z + \frac{1}{2} (1 - q_0) z^2 + \ldots \right) \qquad (4.67)$$

We see that (4.67) can be interpreted as a difference from the linear Hubble law, and since q_0 depends on the cosmological model (see (4.45)), this gives an observational way of determining the fate of the Universe.

4.4.2 Angular Distance

Another frequently used measurement of distance is the *angular distance, d_A*:

$$d_A = \frac{D}{\delta \theta} \qquad (4.68)$$

where $\delta \theta$ is the angular size of an object of known proper size D. According to the FLRW metric (4.4), this is given by

$$d_A = a(t_1) r = \frac{a(t_0)}{(1 + z)} r \qquad (4.69)$$

which means that there is only a difference of a factor $(1 + z)^2$ between the angular and luminosity distances:

$$d_A = \frac{d_L}{(1 + z)^2} \qquad (4.70)$$

Another way to see this is that the latter 'loses' a factor $(1 + z)$ since energy of the photons become redshifted in the expansion, and yet another factor $(1+z)$ because the rate of photon emission is slowed when viewed from an observer at the Earth (see the way that (4.61) was derived).

Unfortunately, we can thus not learn anything about the cosmological parameters by just comparing the distance measures d_L and d_A to a single object. However, if we have a number of sources at different redshifts, we can use the dependence on z of either d_L, d_A or both, to determine the cosmological parameters.

4.5 Cosmological Distances: High Redshift

Let us return to the Friedmann equation (4.16), written in a slightly different form. At the present epoch, the energy density in the form of radiation ρ_{rad} can be neglected in comparison with the matter density ρ_m, so we can write

$$H^2 \equiv \left(\frac{\dot{a}}{a}\right)^2 = \frac{8\pi G}{3}(\rho_m + \rho_{vac}) - \frac{k}{a^2} \tag{4.71}$$

Thus the expansion rate of the Universe depends on the matter density, the cosmological constant and the geometry of the Universe. It is also customary to rewrite (4.71) so that it instead contains the fractional energy density contributions at the present epoch – see (4.43,4.51 and 4.52) – where we now also introduce the contribution,

$$\Omega_K = \frac{-k}{a_0^2 H_0^2} \tag{4.72}$$

from the curvature term.

There are only two independent contributions to the energy density. Equation (4.71) implies that (Problem 4.6):

$$\Omega_M + \Omega_\Lambda + \Omega_K = 1 \tag{4.73}$$

As the matter density scales with the inverse third power of the scale factor $a(t)$, by using the definition of redshift and (4.71) we find that the expansion rate of the Universe at any epoch at redshift less than about 1000 can be related to the one at the present epoch by (Problem 4.7):

$$H^2 = H_0^2[\Omega_M(1+z)^3 + \Omega_K(1+z)^2 + \Omega_\Lambda] \tag{4.74}$$

4.5.1 The Lookback Time and the Age of the Universe

We can now derive the expression for the lookback time: that is, the time difference between the present epoch t_0 and the time of an event that happened at a redshift z. From the definitions of the Hubble parameter and redshift it follows that

$$H = \frac{d}{dt}\log\left(\frac{a(t)}{a_0}\right) = \frac{d}{dt}\ln\left(\frac{1}{1+z}\right) = \frac{-1}{1+z}\frac{dz}{dt} \tag{4.75}$$

Combining (4.74) and (4.75) we thus find:

$$\frac{dt}{dz} = \frac{-(1+z)^{-1}}{H_0[\Omega_M(1+z)^3 + \Omega_K(1+z)^2 + \Omega_\Lambda]^{\frac{1}{2}}} \tag{4.76}$$

Thus, we arrive at the formula for the lookback time from the present:

$$t_0 - t_1 = H_0^{-1} \int_0^{z_1} \frac{dz}{(1+z)[\Omega_M(1+z)^3 + \Omega_K(1+z)^2 + \Omega_\Lambda]^{\frac{1}{2}}} \tag{4.77}$$

By choosing $t_1 = 0$ (that is, $z \to \infty$) in this equation, we obtain the present age of the Universe as a function of H_0, Ω_M, and Ω_Λ (neglecting the 'short' early period of a few hundred thousand years when the Universe was radiation dominated and the even shorter possible period of inflation). Note that the scale of this age is set by H_0^{-1} (sometimes called the Hubble time), with the integral giving a number of order unity.

Example 4.5.1 Calculate the age of the Universe in a $\Omega_M = 1, \Omega_\Lambda = 0$ (Einstein de Sitter) Universe.

Answer: We simply integrate (4.77) with from the present epoch $z = 0$ to the beginning of time at $z = \infty$:

$$H_0\tau = \int_0^\infty \frac{1}{(1+z)^{5/2}} dz = \frac{2}{3} \tag{4.78}$$

that is, the age of the Universe is

$$t_0 = \frac{2}{3H_0}$$

For $H_0 = 65 \text{ km s}^{-1}\text{Mpc}^{-1}$ this gives an age of around 10 billion years.

In Fig. 4.3 we show the lookback time (or the expansion age) of the Universe for the two cases $\Omega_\Lambda = 0$ and $\Omega_M + \Omega_\Lambda = 1$ (a flat Universe), respectively, as a function of Ω_M. As can be seen, the age of the Universe for a given Ω_M is always larger in the flat case (which means in the presence of a cosmological constant). Also, the age is always larger for smaller values of Ω_M. If dating of the Universe by other means (for example through the analysis of the age of stellar systems like globular clusters, or isotope analysis of long-lived radioactive nuclei) would give a value substantially larger than $2H_0^{-1}/3$, one could rule out a matter-dominated Universe with $\Omega_M \sim 1$. At present, these alternative dating methods are not reliable enough to enforce this conclusion.

We finally note from (4.77) that to determine the age at a certain epoch with $z \gg \Omega_M, \Omega_\Lambda$, only the third power of z in the square bracket in that equation becomes important. This gives the estimate of the age of the Universe at the epoch with redshift z

$$t_U(z) \sim \frac{2}{3H_0\sqrt{\Omega_M}} \frac{1}{(1+z)^{3/2}} \tag{4.79}$$

The fact that Ω_Λ does not enter into the equation is related to the dominance according to (4.74) of the matter contribution to the expansion rate.

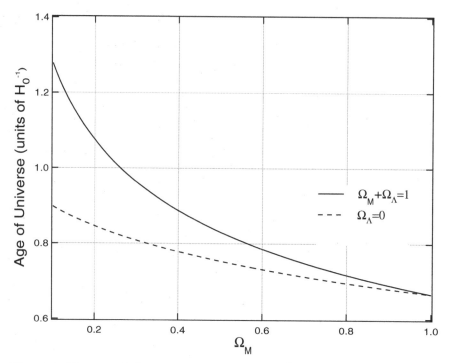

Fig. 4.3. The age of the Universe in units of H_0^{-1} as as function of the matter density Ω_M for the two cases $\Omega_\Lambda = 0$ and $\Omega_M + \Omega_\Lambda = 1$. The value $2/3$ at $\Omega_M = 1$ corresponds to an age of around 10 billion years, if the Hubble parameter h is 0.65.

4.5.2 Measuring Cosmological Parameters

We have seen in Section 4.4, (4.62) that the luminosity distance to an event at $r = r_1$, $t = t_1$ can be written as

$$d_L = a(t_0)(1+z)r_1 \tag{4.80}$$

where a, r and t are related by the equation of the radial, null (light-like) geodesic of the FLRW metric (that is, $d\theta = d\phi = 0$):

$$\frac{dr}{dt} = \frac{\sqrt{1 - kr^2}}{a(t)} \tag{4.81}$$

Multiplying (4.81) by $a_0 \equiv a(t_0)$, using (4.58) and rearranging terms we write:

$$a_0 \frac{dr}{\sqrt{1 - kr^2}} = (1+z)dt \tag{4.82}$$

Using (4.76) to change variables from t into z we thus find:

$$a_0 \int_0^{r_1} \frac{dr}{\sqrt{1-kr^2}} = \int_0^{z_1} \frac{dz}{H_0[\Omega_M(1+z)^3 + \Omega_K(1+z)^2 + \Omega_\Lambda]^{\frac{1}{2}}} \qquad (4.83)$$

where the left-hand-side integral has the solution:

$$a_0 \times \begin{cases} \frac{\arcsin(r_1\sqrt{k})}{\sqrt{k}} & \text{(for } k > 0) \\ r_1 & \text{(for } k = 0) \\ \frac{\operatorname{arcsinh}(r_1\sqrt{-k})}{\sqrt{-k}} & \text{(for } k < 0) \end{cases} \qquad (4.84)$$

Thus, we can extract an expression for r_1 as required by the definition of luminosity distance in (4.80).

Finally, by inserting the definition $\Omega_K = -k/a_0^2 H_0^2$ and writing explicitly the speed of light c we arrive at the expression of the luminosity distance *versus* redshift (Problem 4.8)

$$d_L(z; H_0, \Omega_M, \Omega_\Lambda) = \frac{c(1+z)}{H_0\sqrt{|\Omega_K|}} \mathcal{S}\left(\sqrt{|\Omega_K|} \int_0^z \frac{dz'}{H'(z')}\right) \qquad (4.85)$$

where $\mathcal{S}(x)$ is defined as $\sin(x)$ for $\Omega_K = 1 - \Omega_M - \Omega_\Lambda < 0$ (*closed Universe*), $\sinh(x)$ for $\Omega_K > 0$ (*open Universe*) and $\mathcal{S}(x) = x$, and the factor $\sqrt{|\Omega_K|}$ is removed for the *flat Universe*; and

$$H'(z) = \frac{H(z)}{H_0} = \sqrt{\Omega_M(1+z)^3 + \Omega_K(1+z)^2 + \Omega_\Lambda} \qquad (4.86)$$

In the general case, replacing Ω_Λ by a fluid Ω_X with an non-constant equation of state parameter $\alpha_X(z)$, (4.86) becomes

$$H'(z) = \sqrt{\Omega_M(1+z)^3 + \Omega_K(1+z)^2 + \Omega_X f(z)}, \qquad (4.87)$$

where the redshift dependence $f(z)$ is given by

$$f(z) = \exp\left[3\int_0^z dz' \frac{1+\alpha_X(z')}{1+z'}\right]. \qquad (4.88)$$

4.5.3 Redshift Dependence of the Particle Horizon

We can now return to the discussion about particle horizons. We generalize (4.56) to any time t in a Friedmann-Lemaître-Robertson-Walker Universe:

$$d_H(t) = a(t) \int_0^t \frac{dt'}{a(t')} = a(t) \int_0^{a(t)} \frac{da(t')}{\dot{a}(t')a(t')} \qquad (4.89)$$

Adding the energy density contribution from radiation in the expression in (4.87) and inserting it into (4.89) together with the definition of redshift yields:

$$d_H(z) = \frac{1}{H_0(1+z)} \int_z^\infty f(z')dz' \qquad (4.90)$$

with

$$f(z) = \frac{1}{\sqrt{g(z, \Omega_R, \Omega_M, \Omega_K, \Omega_X)}} \tag{4.91}$$

where

$$g(z, \Omega_R, \Omega_M, \Omega_K, \Omega_X) =$$
$$\Omega_R(1+z)^4 + \Omega_M(1+z)^3 + \Omega_K(1+z)^2 + \Omega_X(1+z)^{3(1+\alpha)} \tag{4.92}$$

We have here been very general and also included the (presently small) contribution from radiation Ω_R and a hypothetical 'X-matter' which we just parametrize by its equation of state $p = \alpha \cdot \rho$ (see (4.23)).

Example 4.5.2 Evaluate the deceleration parameter $q(z)$ (neglect radiation).

Answer: Generalizing the definition in (4.40) to apply at all times we find, $q(z) = -\frac{\ddot{a}}{aH^2}$, which together with (4.23) and (4.45) leads to

$$q(z) = \frac{1}{2} \left[\Omega_M(1+z)^3 + (1+3\alpha)\Omega_X(1+z)^{3(1+\alpha)} \right] \tag{4.93}$$

We note that with the concordance model, $\Omega_M \sim 0.3$ and $\Omega_\Lambda \sim 0.7$, the expansion did slow down until $z \sim 0.7$ at which point it began to accelerate (i.e., $q < 0$).

4.6 Observations of Standard Candles

The evidence of structures beyond our own Galaxy, the so-called *spiral nebulae*, originated in observations by Charles Messier in the 18th century. It took until 1925 before Edwin Hubble demonstrated that these were galaxies, just like the Milky Way. Hubble also showed that all galaxies did not have spiral shape but form three morphological classes: spirals, ellipticals and irregulars. Establishing the distances to other galaxies has been one of the main efforts of observational cosmology since then. Hubble studied Cepheid stars in neighbouring galaxies to deduce their distance from Earth. These pulsating stars have a known correlation between the period with which they brighten and fade and their intrinsic brightness, which may be combined with their apparent brightness to deduce their distance. Still today Cepheids play a critical role in establishing the extragalactic distance scale. The Hubble Space Telescope (HST) has been used to obtain distances to galaxies up to 25 Mpc away. Secondary methods are used to extend the distance ladder further. This is necessary to reach the *Hubble flow*, i.e. the distance range where the velocities of the galaxies are dominated by the cosmological expansion. For nearby galaxies, the peculiar velocities due to the gravitational interactions with neighbours, typically a few hundred km·s^{-1}, introduce significant uncertainties.

One the most common secondary methods exploits the homogeneity of the peak brightness of Type Ia supernovae (SNe Ia). These powerful explosions are empirically found to behave as *standard candles* with approximately a 20% brightness intrinsic dispersion. Type Ia supernovae are believed to result from accreting white dwarfs in a binary system. A thermonuclear explosion is triggered when the white dwarf mass exceeds the so-called Chandrasekhar mass, $M \sim 1.4 M_\odot$. Unlike Cepheids, the absolute brightness of SNe Ia cannot be deduced directly with sufficient precision. However, they have been found in galaxies also hosting Cepheid stars and therefore with known distances. Thus an indirect estimate of the intrinsic intensity of SNe Ia is possible. The extragalactic distance scale is then established by the relative distances provided by the measured fluxes from supernovae in more distant galaxies. This makes it possible to build a Hubble diagram, as shown in Fig. 4.4 for a sample of 80 Type Ia supernovae with recession velocities $v_r > 2500$ km/s. The fitted slope yields $H_0 = 66 \pm 7$ km s^{-1}Mpc^{-1}.

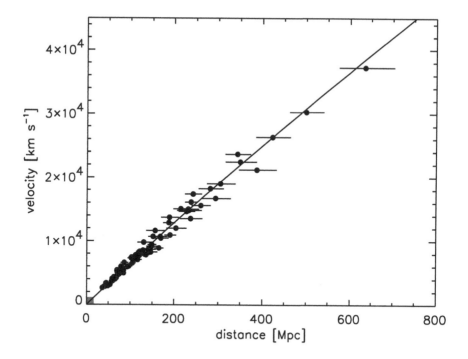

Fig. 4.4. Hubble diagram for 80 Type Ia supernovae with recession velocities exceeding 2500 km/s. The fitted slope (solid line) yields $H_0 = 66 \pm 7$ km s^{-1}Mpc^{-1}. The small shaded box in the left bottom corner shows the extent of Hubble's original diagram. Courtesy of Saurabh Jha.

Optical astronomers measure luminosities in logarithmic units, *magnitudes*.[2] Thus, the Hubble relation expressed in bolometric magnitudes becomes (with d_L measured in Mpc):

$$m(z) = M + 5 \log d_L + 25 \tag{4.94}$$

where M is the *absolute magnitude* (apparent magnitude of the object if it would be at 10 pc distance from the observer) and the Hubble parameter is measured in units of km s^{-1}Mpc^{-1}. By measuring the apparent magnitudes of standard candles as a function of redshift one can solve for the unknown cosmological parameters.

The quantity $m - M = 5 \log d_L + 25$ is called the *distance modulus*.

The *magnitude versus redshift relation* for Type Ia supernovae (believed to be standard candles) measured with broad optical filters out to $z \sim 1$ is shown in Fig. 4.5. The data favours a Universe with a positive cosmological constant, $\Omega_\Lambda = 0.75 \pm 0.08$ in a flat Universe [25]. The current data set of high-redshift Type Ia supernovae is insufficient to break the degeneracy of the density terms, as shown in Fig. 4.6. The results can be approximated by the linear combination $0.8\Omega_M - 0.6\Omega_\Lambda \approx -0.2 \pm 0.1$ [35]. The estimates of the energy density terms, depend solely on *relative* distances and are therefore *independent* of the value of the Hubble parameter.

4.7 Meaning of the Cosmological Constant

The cosmological constant, whatever its origin might be, could play a major role in the dynamics of the Universe. Next we will re-examine the Friedmann equations (4.16) and (4.18) separating the contributions from mass density and the vacuum energy density to the time evolution of the scale factor of the Universe. Thus, we write explicitly:

$$\left(\frac{\dot{a}}{a}\right)^2 + \frac{k}{a^2} = \frac{8\pi G}{3}(\rho_m + \rho_{vac}), \tag{4.95}$$

and (see (4.18))

$$\frac{2\ddot{a}}{a} + \left(\frac{\dot{a}}{a}\right)^2 + \frac{k}{a^2} = 8\pi G \rho_{vac}, \tag{4.96}$$

where we have inserted the equation of state for vacuum energy, $p_{vac} = -\rho_{vac}$. The pressure term from non-relativistic matter is negligible and was therefore dropped. The difference between the two equations becomes (see (4.28)):

---

[2] The brightness of an object through a filter f in the magnitude system is defined as $m_f = -2.5 \log \left(\int_0^\infty T_f(\nu)\mathcal{F}(\nu)d\nu \right) + C_f$, where $T_f(\nu)$ is the transmission function of the filter used and C_f is a constant specific to the choice of filter. Note that brighter objects have *lower* magnitudes.

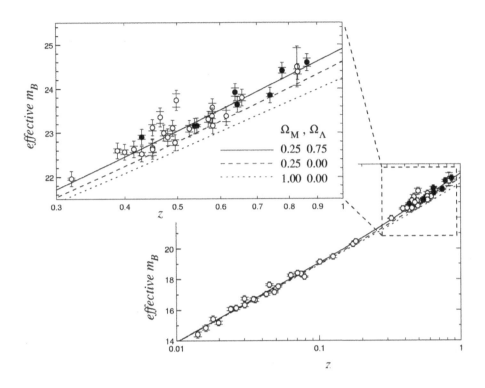

Fig. 4.5. Hubble diagram for high-redshift Type Ia supernovae from the *Supernova Cosmology Project* [25], and low-redshift Type Ia supernovae from the CfA [38] and Calán/Tololo Supernova Survey [19]. The inner error bars show the uncertainty due to measurement errors, while the outer error bars show the total uncertainty when the intrinsic luminosity dispersion (0.17 mag) is added in quadrature. Filled circles represent supernovae measured with the Hubble Space Telescope, i.e. in general with higher accuracy. The curves are the theoretical effective $m_B(z)$ for a range of cosmological models with and without a cosmological constant. It is called 'effective m_B' because the measured intensity corresponds to the restframe B-band (blue). Because of cosmological redshift, the photons are observed at longer wavelengths. The best fit to the data (for a flat Universe) corresponds to the FLRW Universe with $(\Omega_M, \Omega_\Lambda)=(0.25, 0.75)$, as shown in [25].

$$\frac{\ddot{a}}{a} = -\frac{4\pi G}{3}\rho_m + \frac{\Lambda}{3}, \tag{4.97}$$

where $\Lambda = 8\pi G\rho_{vac}$. Let us now consider a sphere of space with comoving radial coordinate r: that is, of physical radius ar (neglecting curvature effects). A test particle with mass m at the boundary of the sphere will accelerate as space evolves as:

$$\frac{d^2(ar)}{dt^2} = -\frac{4\pi G}{3}\rho_m(ar) + \frac{\Lambda}{3}(ar), \tag{4.98}$$

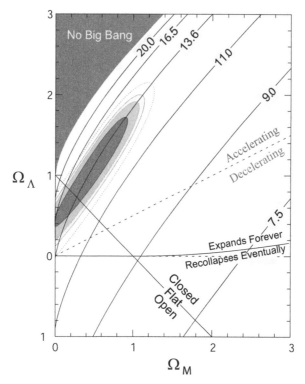

Fig. 4.6. Best-fit coincidence regions in the Ω_M–Ω_Λ plane from the analysis of the *Supernova Cosmology Project* supernova Hubble diagram shown in Fig. 4.5. The dark and light ellipses show the 68 per cent and 90 per cent confidence regions. The outer ellipses show the 95 and 99 per cent confidence levels. A flat Universe ($\Omega_K = 0$) would fall on top of the diagonal solid line passing through the Einstein de Sitter solution. To the right of that line the Universe is closed, and to the left it is open. The dashed line shows the division between acceleration and deceleration for the Universe. Also shown are isochrones of constant age of the Universe in units of billion of years (h=0.71 was assumed). The data suggests that the rate of expansion of the Universe is currently accelerating. Courtesy of Robert Knop.

By recognizing that the mass of the sphere is $M = \frac{4\pi(ar)^3}{3}\rho_m$ we find:

$$\frac{d^2(ar)}{dt^2} = -\frac{GM}{(ar)^2} + \frac{\Lambda}{3}(ar) \tag{4.99}$$

The first term on the right-hand side is the familiar Newtonian force, attractive and inversely proportional to distance squared. The second term shows that the force associated with the cosmological constant is repulsive and proportional to distance. Thus, it is only at the largest scales (comparable to the size of the Universe) that the effects of the vacuum energy density could become sizeable for $\Omega_\Lambda = \mathcal{O}(1)$.

To explain the *smallness* of the cosmological constant is one the most outstanding challenges in modern theoretical physics. In *natural units*, with $\hbar = c = 1$, the vacuum energy density for $\Omega_\Lambda \sim 1$ becomes $\rho_{vac} \approx 10^{-46}$ GeV^4. Physically, this should correspond to m^4, where m is some fundamental mass scale in nature. As will be discussed in Chapter 6, the mass scale for gravity in natural units is the Planck mass, $m_{Pl} = 1.2 \cdot 10^{19}$ GeV. For dimensional reasons, one could therefore expect a vacuum energy density of the order of $m_{Pl}^4 \sim 10^{76}$ GeV^4 to emerge from an eventual quantum theory of gravity. The observed vacuum energy density is thus smaller by 122 orders of magnitude! In physics, this is regarded as a severe fine-tuning problem: How is it that a number, naively expected to be of order 10^{122}, comes out to be of order unity, (but still non-zero)? Usually, it is much easier to explain why a number is exactly zero rather than very small but non-zero. For instance, we saw in Section 2.6 that the photon mass is believed to be exactly zero because of a symmetry, gauge invariance. There have been hopes of finding some symmetry which would give $\Lambda \equiv 0$, but none has so far been found. The present observational indications of $\Omega_\Lambda \neq 0$, if confirmed, have added to the seriousness of the cosmological constant problem. It is conceivable that the solution to this problem is intimately related to the so far unsolved problem of finding a correct theory of quantum gravity.

4.8 Summary

- Standard cosmology is based on Einstein's general relativity, and the Universe is treated as homogeneous and isotropic.
- Events and objects in the Universe are characterized by their redshift. The relation between the emitted and observed wavelength scales with the scale ratio of the Universe:

$$\frac{a_0}{a_e} = \frac{\lambda_o}{\lambda_e} = 1 + z$$

- The time evolution of the scale factor and thus of the entire Universe is a function of the cosmological parameters H_0^2, Ω_Λ, Ω_M and Ω_K, and is given by the Friedmann equation:

$$\begin{cases} H^2 = H_0^2[\Omega_M(1+z)^3 + \Omega_K(1+z)^2 + \Omega_\Lambda] \\ \Omega_M + \Omega_K + \Omega_\Lambda = 1 \end{cases}$$

At redshifts of the order of 1000 or greater, a contribution from radiation should also be included, which scales like $(1+z)^4$.

- The cosmological parameters can be extracted from distance measurements: for example, through the magnitude redshift relation for standard candles.

4.9 Problems

4.1 Derive (4.7) and (4.8) for the de Sitter model from Einstein's equations (4.1).

4.2 Show that a particle at rest with respect to the comoving coordinates r, θ and ϕ satisfies the geodesic equation (4.12). (Hint: Since $dx^i = 0$, only Γ^i_{00} has to be computed.)

4.3 Use the definition of the covariant divergence

$$T^{\mu\nu}{}_{;\mu} = T^{\mu\nu}{}_{,\mu} + \Gamma^\mu_{\mu\sigma}T^{\sigma\nu} + \Gamma^\nu_{\rho\sigma}T^{\rho\sigma}$$

and the condition $T^{\mu\nu}{}_{;\mu} = 0$ to derive (4.19).

4.4 Derive (4.29).

4.5 Fill in the missing steps leading to (4.67).

4.6 Show that the relation $\Omega_M + \Omega_\Lambda + \Omega_K = 1$ follows from (4.71).

4.7 Show that the Hubble parameter at redshift z is related to the present epoch cosmological parameters as shown in (4.74).

4.8 Fill in the details to arrive at (4.85).

4.9 Show that the Friedman-Lemaître-Robertson-Walker line element can also be written

$$ds^2 = dt^2 - a^2(t)\left(d\chi^2 + \Sigma^2(\chi)d\Omega^2\right),$$

with $\Sigma^2(\chi) = \sin^2 \chi$, (for $k = 1$), $\Sigma^2(\chi) = \chi^2$ (for $k = 0$) or $\Sigma^2(\chi) = \sinh^2 \chi$ (for $k = -1$).

4.10 Compute the Ricci scalar for the FLRW metric.

4.11 How large was the presently observable Universe (the volume within the present horizon) one millisecond after the Big Bang? Of course, you only need to give an order of magnitude estimate. Assume standard FLRW cosmology, $\Omega_M + \Omega_R = 1$, $\Omega_\Lambda = 0$, and $H_0 = 65$ km/s per Mpc (that is $h = 0.65$).

4.12 In a flat FLRW Universe, the physical coordinate distance d is related to the redshift z through the formula

$$d = \frac{2}{H_0}\left(1 - \frac{1}{\sqrt{1 + z}}\right)$$

The apparent angular diameter $\delta\theta_A$ is given by the true diameter D divided by the coordinate distance times $(1 + z)$:

$$\delta\theta_A = \frac{D}{d}(1 + z)$$

(a) What is the relation between apparent angular diameter of a fixed-size object and redshift in this cosmological model?

(b) At what redshift does the apparent angular diameter have a minimum? Do you have any idea why objects which are further away than this can actually appear to be bigger?

4.13 A galaxy has an apparent magnitude of 18 and an absolute magnitude of -17. Show that its distance to us is around 100 Mpc.

4.14 Show that given an equation of state $p_X = \alpha\rho_X$, where the energy density ρ does not include the vacuum energy, the evolution equation of the Hubble parameter becomes:

$$H^2(z) = H_0^2[(1 + z)^2 \left(1 + \Omega_X(1 + z)^{1+3\alpha} - 1\right)) - z(2 + z)\Omega_\Lambda]$$

5 Gravitational Lensing

5.1 The Bending of Light

The idea that light may be bent gravitationally when passing near massive bodies is old, stemming at least back to Newton and Laplace. In a corpuscular model of light, such as the one of Newton, it is natural that the gravitational attraction will make an otherwise straight light path bend like the trajectory of any material body (see Fig. 5.1).

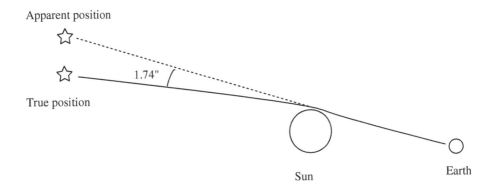

Fig. 5.1. The deflection of light from a star behind the Sun. Since the ray of light is bent towards the Sun, the apparent position of the star is shifted by 1.74 arcseconds according to general relativity. This prediction was first verified by Eddington in 1919.

The gravitational bending of light by the Sun computed in Newtonian theory for a massive photon, with the limit of the mass going to zero, turns out to be 0.87 arcseconds – exactly one half of the value predicted by general relativity. When Eddington measured the true value during a solar eclipse observed from the island of Principe in 1919, he obtained a result which agreed with that of Einstein's theory within error bars, whereas the Newtonian prediction was a factor of 2 too small. It was this success ('Newton was wrong – Einstein was right') that brought world fame to Einstein. Today, the general relativistic value of the deflection angle has been proven correct to the five per cent level.

Much later it was realised that in fact the bending of light by massive objects provides a powerful technique to measure the mass, for example, of galaxies and galaxy clusters. In analogy with the refraction of light by optical lenses, individual galaxies, stars, or planets may act as *gravitational lenses*. Even though the lensing may be too weak to give noticeable multiple images, the increase in light caused by the focusing of a gravitational lens may be observable. A particularly important example is provided by a sub-solar mass dark object in the Milky Way galactic halo (a so-called MACHO: Massive Astrophysical Compact Halo Object) which could be a candidate for the galactic dark matter. If such an object would transit, for example, the line of sight towards one of the stars in the nearby galaxy the Large Magellanic Cloud (LMC), the luminosity of the background star would appear to rise during the passage. Since bending of light according to general relativity is a purely geometrical effect it does not depend on the energy – that is, the wavelength – of the photons. Therefore the signature of such a so-called *microlensing* event is a time-symmetric achromatic luminosity bump. We shall return to this and other applications in Section 5.2.

We now investigate how the gravitational bending of light is calculated in general relativity.

We saw in (3.33) the geodesic equation that a free, massive particle has to follow in a given metric. There is a problem in using this directly for photons, however, since $ds = cd\tau$ vanishes along a light-like path. Therefore, we have to choose another parameter, a so-called affine parameter λ which parametrizes the path of the photon. One such choice is to make use of the local four-momentum vector k^α, and define λ by

$$\frac{dx^\alpha}{d\lambda} = k^\alpha \tag{5.1}$$

The geodesic equation for a light-ray then becomes

$$\frac{d^2x^\sigma}{d\lambda^2} + \Gamma^\sigma_{\mu\nu}\frac{dx^\mu}{d\lambda}\frac{dx^\nu}{d\lambda} = 0 \tag{5.2}$$

If we want to investigate the bending of light caused by a spherically symmetric body, such as the Sun, we may employ the Schwarzschild solution (3.64). Since the effects we are considering are very small, we can write the metric approximately as

$$ds^2 = (1 + 2\varphi(r))dt^2 - (1 - 2\varphi(r))dr^2 - r^2(d\theta^2 + \sin^2\theta d\phi^2) \tag{5.3}$$

since

$$r_S/r = 2GM/r = -2\varphi(r) \tag{5.4}$$

with $\varphi(r) = -GM/r$ the ordinary Newtonian gravitational potential, and where we have used

$$\frac{1}{1 - r_S/r} = 1 + r_S/r + \mathcal{O}\left(\frac{r_s^2}{r^2}\right) \tag{5.5}$$

Since we have spherical symmetry we can let the light-ray move at a constant polar angle $\theta = \pi/2$: that is, in the equatorial plane. Then the angular part of the metric is given by $-r^2 d\phi^2$. The solution of (5.2) is now straightforward, but tedious (see [51]). Introducing $w = 1/r$, the geodesic equation for a photon becomes

$$\frac{d^2 w}{d\phi^2} + w = 3GMw^2 \tag{5.6}$$

The effect of space-time curvature is given by the right-hand side, which is very small for: for example, the mass of the Sun. We can therefore solve this equation by iteration. A zero-order solution is given by putting $M = 0$, which gives

$$w_0(\phi) = \frac{\cos(\phi + \phi_0)}{b} \tag{5.7}$$

where we can rotate the ϕ coordinate to put $\phi_0 = 0$ without loss of generality. The equation $w = 1/r = \cos(\phi)/b$ describes the trajectory of a straight light-ray in the xy-plane, passing at a closest distance (the so-called impact parameter) b, as ϕ varies from $\pi/2$ to $-\pi/2$. To solve (5.6), we insert w_0 on the right-hand side. The resulting equation

$$\frac{d^2 w}{d\phi^2} + w = 3GMw^2 = \frac{3GM\cos^2(\phi)}{b^2} \tag{5.8}$$

has the particular solution

$$w_{part}(\phi) = \frac{2GM(1 - \cos^2(\phi)/2)}{b^2} \tag{5.9}$$

and the full solution is the sum of this particular solution and the solution to the homogeneous equation with $M = 0$:

$$w(\phi) = w_0(\phi) + w_{part}(\phi) = \frac{\cos(\phi)}{b} + \frac{2GM(1 - \cos^2(\phi)/2)}{b^2} \tag{5.10}$$

At distant points on the photon trajectory, $w = 1/r \to 0$ and ϕ is very close to $\pm\pi/2$. Therefore, the $\cos^2(\phi)$ terms can be neglected and (5.10) to first order becomes

$$\cos(\phi)_{\phi \simeq \pm \pi/2} = \frac{-2GM}{b} \tag{5.11}$$

This means that the asymptotic values of ϕ are $+\pi/2 + 2GM/b$ and $-\pi/2 - 2GM/b$, and the deflection caused by the mass M is given by

$$\Delta\phi = \frac{4GM}{b} \tag{5.12}$$

For the Sun, if one takes b equal to the solar radius (that is, if we consider rays from distant stars that graze the solar surface), this deflection becomes 1.74 arcseconds.

There is another way to derive the law of deflection which is of general applicability. This starts from a general relativistic version of Fermat's principle, which states that light-rays follow paths which minimize the travel time. From the optics point of view, the gravitational potential φ causes a time delay (called the Shapiro delay) which can be represented by a refractive index

$$n = 1 - 2\varphi \tag{5.13}$$

so that light travels slower than c by a factor n, $v_{eff} = c/n \simeq c(1-2|\varphi|)$. Just as in optics, the deflection is the integral of the gradient of n perpendicular to the direction of propagation, along the light path:

$$\Delta\phi = -\int \nabla_\perp n dl = 2\int \nabla_\perp \varphi dl \tag{5.14}$$

Returning to our example of a point mass, the Newtonian potential for a light-ray travelling along the y axis in the xy-plane is given by

$$\varphi(b,y) = \frac{-GM}{\sqrt{b^2+y^2}} \tag{5.15}$$

which inserted into (5.14) gives $\Delta\phi = 4GM/b$ as before.

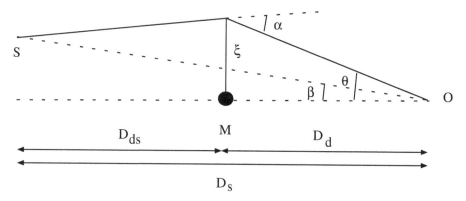

Fig. 5.2. The geometry of gravitational lensing. The direction of the source in the absence of lensing is specified by the angle β, the deflection angle is α and the observed direction is θ. The distance to the source is D_s, the distance to the lens is D_d, and the distance between the lens and the source is D_{ds}. The ray passes the lens at a transverse distance ξ, which for small deflections is equal to the nearest distance between the ray and the lens, the impact parameter b. It is seen that $\beta D_s = \xi D_s/D_d - \alpha D_{ds}$, from which the lens equation (5.16) follows by using $\xi \sim \theta D_d$ and $\alpha \sim 4GM/\xi$.

Now that we know the deflection caused by a mass M, simple geometry (see Fig. 5.2) produces the lens equation

$$\beta(\theta) = \theta - \frac{D_{ds}}{D_d D_s} \frac{4GM}{\theta} \tag{5.16}$$

where D_s, D_d and D_{ds} are the distances to the source, the lens (deflector) and the distance between the source and the lens respectively, and β is the undeflected angle. In general, this equation may have several solutions: that is, the source object has multiple images. In particular, setting $\beta = 0$, azimuthal symmetry requires that we acquire a ring-like image, the Einstein ring, with angular radius

$$\theta_E = \sqrt{\frac{4GM}{c^2} \frac{D_{ds}}{D_d D_s}} \tag{5.17}$$

as shown in Fig. 5.3. For clarity, we have here explicitly written the speed of light, c, otherwise set to unity throughout the text. If the source is not exactly behind the lens (that is, $\beta \neq 0$), one obtains two solutions for a point-mass lens,

$$\theta_{\pm} = \frac{1}{2} \left(\beta \pm \sqrt{\beta^2 + 4\theta_E^2} \right) \tag{5.18}$$

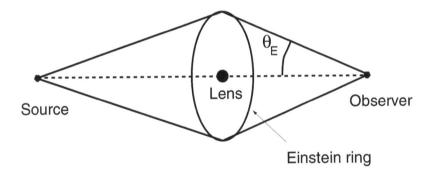

Fig. 5.3. A circularly symmetric lensing object exactly in the line of sight to a distant star causes the source to be imaged as a ring with angular radius θ_E.

In many cases, the angular separation between these two images is too small to be detectable by telescopes. However, the focusing effect of the gravitational lens means that the intensity of the image is magnified. In the case of a double image from a point source, the total magnification is given by

$$\mu = \frac{u^2 + 2}{u\sqrt{u^2 + 4}} \tag{5.19}$$

where $u = \beta/\theta_E$. This magnification effect is important for the microlensing experiments to be discussed in Section 5.2.1. As can be seen, the Einstein angular radius θ_E sets an angular scale which determines whether a noticeable magnification occurs. If the line of sight passes far outside the Einstein radius of an intervening object, lensing effects from that object become very weak.

Compared to the optical lenses found in cameras and binoculars, gravitational lenses are very bad if considered as optical instruments, with essentially all types of distortions and aberrations present. However, they are perfectly achromatic (producing the same deflection for light of any wavelength), which is a useful signature when searching for them. The reason for this property is of course that the geodesic deviation of light-rays is a purely geometrical effect, independent of photon energy.

5.2 Observation of Gravitational Lensing

What new can be learned from gravitational lenses? In 1936 Einstein discussed the creation of multiple images from a single source due to lensing effects caused by stars, but concluded that the effect was of little interest because of its rarity. He also noted that the image splitting was too small to be resolved by a ground-based telescope. One year later, Zwicky pointed out that if the lensing mass was a galaxy rather than a star, the angular deflection would be large enough to be observed for very distant sources. Moreover, Zwicky showed that lensing of such sources should occur in about 1 per cent of the cases. Today, lensing both by stars and galaxies has become a major observational field in astronomy.

There are at least three possible scenarios in which lensed systems can add to our knowledge of astrophysics and cosmology.

- Lensing properties depend on the mass distribution of the lens, and the studies of such systems can thus provide information about *dark matter*.
- The lensing probability for individual high redshift objects to be lensed is a function of the cosmological parameters. Thus it allows limits to be set on the mass density Ω_M and the vacuum energy density (from a cosmological constant), Ω_Λ. Studies of the time delay between images originating at cosmological distances are also used to measure the Hubble parameter, H_0.
- Gravitational lensing also helps to enhance the power of telescopes. Because of the magnification effect, one can observe intrinsically faint or very distant objects that would otherwise not be observable.

5.2.1 Galactic Dark Matter Searches: Lensing of Stars

The possibility of studying the population of non-luminous astrophysical objects in the Milky Way through gravitational lensing was proposed by Paczynski in 1986. By monitoring the light from millions of stars in the Magellanic

Clouds and in the galactic bulge it is possible to detect the magnification in light intensity caused by the passage of a compact lensing star in the galactic halo. This type of gravitational lensing, where a star is lensed by another star, is called *microlensing*. The Einstein radius for galactic microlensing (D ~ 10 kpc and $M \sim M_{\odot}$) by a solar mass lensing star is about 1 milliarcsecond, nearly 1000 times too small to be resolved from ground-based telescopes. The lensed images are thus observed together as one image with increased brightness. Several groups (EROS, MACHO, OGLE and DUO) are currently using this technique to map the galactic density of objects such as brown dwarfs (stars with too little mass for fusion to turn on), faint stars, neutron stars, black holes, and possibly even planets.

There are three measurable quantities: the probability of lensing of stars, the time duration, and magnification of the source image.

The rate of microlensing events is measured in terms of the optical depth, defined as the probability at any instant of time that the line of sight of an individual star is within an angle θ_E from a lens. Thus, the expression for optical depth becomes (Problem 5.1):

$$\tau = \int_0^{D_s} \pi D_d^2 \theta_E^2 n_l(D_d) dD_d \tag{5.20}$$

where $n_l(D_d)$ is the number density of lenses.

Inserting the expression for the Einstein radius from (5.17) and assuming Euclidean space locally: that is, $D_{ds} = D_s - D_d$, the expression for optical depth can be simplified to

$$\tau = \frac{4\pi G}{c^2} D_s^2 \int_0^1 \rho_l(x) x(1-x) dx \tag{5.21}$$

where $x = D_d D_s^{-1}$, and ρ_l is the mass density of lenses. Notice that this result does not depend on the mass of the individual objects which make up the mass density.

If ρ_l is constant along the line of sight, the optical depth is

$$\tau(D_s) = \frac{2\pi}{3} \frac{G\rho_l}{c^2} D_s^2 \tag{5.22}$$

Example 5.2.1 Show that if the mass of our Galaxy is mainly in the form of compact objects homogeneously distributed in the halo, the optical depth for lensing at the edge $(D_s = R_{gal})$ becomes $\tau \approx \left(\frac{v}{c}\right)^2$, where v is the rotational speed of the Galaxy.

Answer: Describing the halo as a homogeneous sphere of mass $M_{gal} = \frac{4\pi R_{gal}^3}{3}\rho$, the optical depth becomes (from (5.22))

$$\tau = \frac{GM_{gal}}{2c^2 R_{gal}} \tag{5.23}$$

We can now make use of Kepler's third law to relate the mass of the Milky Way to its rotational speed:

$$\frac{GM_{gal}}{R_{gal}} = v^2 \tag{5.24}$$

Combining (5.23) and (5.24) we thus find $\tau = \frac{1}{2}\left(\frac{v}{c}\right)^2$.

With the simplifications described in the previous exercise we found that the optical depth at the edge of the Milky Way, with a rotational speed ~ 200 km/s, is $\tau \approx 10^{-7}$. It is thus required to study tens of millions of stars in the Magellanic Clouds at about ~ 50 kpc from the Earth in order to be able to detect microlensing events. Only with the help of computerized automatic search methods has such a task become technically feasible.

The time scale for the star magnification induced by a microlensing event is given by the angular size and speed of the lens:

$$T = \frac{D_d \theta_E}{v} \tag{5.25}$$

Inserting the expression of the Einstein radius from (5.17) the microlensing time scale becomes

$$T = 2.6 \text{ months} \times \left(\frac{M_l}{M_\odot}\right)^{\frac{1}{2}} \cdot \left(\frac{D_d}{10\text{kpc}}\right)^{\frac{1}{2}} \cdot \left(\frac{D_{ds}}{D_s}\right)^{\frac{1}{2}} \cdot \left(\frac{200 \text{ kms}^{-1}}{v}\right) \tag{5.26}$$

For stars in the Magellanic Clouds and halo lenses, $\frac{D_{ds}}{D_s}$ is close to 1. As the experiments can resolve microlensing events with durations from an hour to about a year, MACHOs can be searched for in the mass range between $10^{-6} M_\odot$ and $10^2 M_\odot$. If the lensing object moves with constant transverse speed with respect to the line of sight the light-curve of a microlensing event can be calculated from (5.19):

$$u^2(t) = u_{min}^2 + \left(\frac{t - t_0}{T}\right)^2 \tag{5.27}$$

where u_{min} is the angle of closest approach in units of the Einstein radius θ_E. Fig. 5.4 shows the duration of a microlensing event for $u_{min} = 0.1$, 0.5 and 1.0; the smaller the angle separation the larger the magnification.

Several microlensing events have now been observed. A high amplification event towards the Large Magellanic Cloud is shown in Fig. 5.5. In its first year of data collection, the MACHO collaboration observed three events towards the LMC. The corresponding optical depth measured in the mass range: $10^{-4} \leq M \leq 10^{-1} M_\odot$ was $\tau = 9^{+7}_{-5} \cdot 10^{-8}$ [1]. With the low statistics yet available it is still too early to reach a definitive conclusion about the MACHO density of the galactic halo. Among the uncertainties we can mention the possibility that some lenses reside not in the halo of the Milky Way, but are

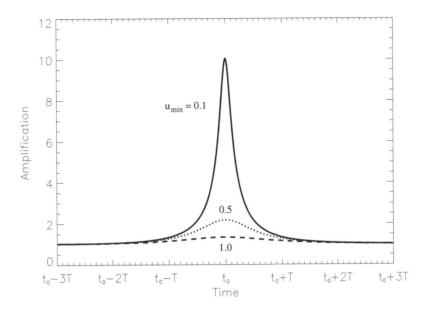

Fig. 5.4. Light-curve shape of microlensed star. The duration of the event is sensitive to the mass of the lens, while the maximum amplification depends on the mass of the lens and the minimum angular distance between the source and the lensing object. The separation $u_{min} = 0.1, 0.5$ and 1.

associated with either the LMC or intervening debris from tidally disrupted stellar clusters.

5.2.2 Lensing of Objects at Cosmological Distances

Quasars (QSOs) can be considered as point sources at cosmological distances. The probability of a QSO being lensed by a point mass in the line of sight can be easily derived following the recipes derived in the previous section. If we assume that the dark matter of the Universe consists of a homogeneous population of point masses with mass density Ω_M, we can compute the optical depth towards a QSO at a redshift z_s. Let us first do this for small redshifts to illustrate the ideas. From (5.22) and the Hubble law, $H_0 D_s = cz_s$ we find that the probability for lensing of a source which by assumption is located at redshift $z_s \ll 1$ is (Problem 5.2)

$$\tau = \frac{1}{4}\Omega_M z_s^2 \qquad (5.28)$$

where, as usual, $\Omega_M = \rho\rho_c^{-1}$. In general, the probability of gravitational lensing by compact objects homogeneously distributed in a matter dominated

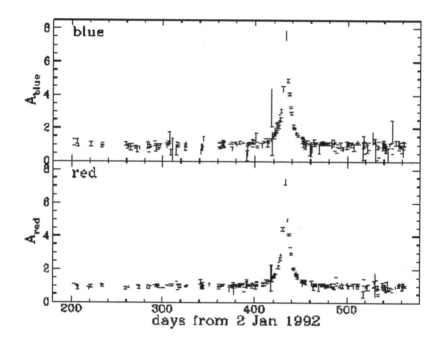

Fig. 5.5. Light curves in two filters, blue and red, of a microlensing event discovered by the MACHO collaboration. The lensed star in the Large Magellanic Cloud was magnified by about a factor 8 at maximum (independently of wavelength) and the time scale was about a month. Credits: The MACHO collaboration. Reprinted by permission from Nature[2] copyright (1993) Macmillan Magazines Ltd.

Universe can be shown to be proportional to the comoving mass density of lenses, Ω_M. For Ω_M close to unity it should happen in a large fraction of the distant sources.

The image splitting is given by (see (5.18))

$$\Delta\theta = \theta_E \sqrt{u^2 + 4} \tag{5.29}$$

where θ_E is the size of the corresponding Einstein radius. For cosmological distances, $D \sim 1$ Gpc, and for typical galaxy masses, $M = 10^{11} M_\odot$, we obtain

$$\theta_E = 0.9'' \times \left(\frac{M}{10^{11} M_\odot}\right)^{\frac{1}{2}} \left(\frac{D_s}{1 \text{ Gpc}}\right)^{-\frac{1}{2}} \left(\frac{D_{ds}}{D_d}\right)^{\frac{1}{2}} \tag{5.30}$$

Gravitationally lensed quasars (by foreground galaxies) are observed. One of the most remarkable images of lensed quasars is the so called *Einstein Cross*, shown in colour Plate 2.[1] The lens is in this case a spiral galaxy

---

[1] The colour plate section is positioned in the middle of the book.

at $z = 0.04$. The four images show uncorrelated time variations which are thought to be caused by microlensing from the passage of individual stars in the foreground galaxy.

When estimating the probability of cosmological objects being lensed by finite size galaxies, a model for the mass distribution within the lens has to be understood. In particular, the lensing properties are sensitive to the galaxy type. However, the general features of the lensing probabilities and how they depend on cosmological parameters can be understood from the point mass approximation.

Next, we generalize the expression of the optical depth for gravitational lensing of a source at any redshift, including the contributions from a non-vanishing vacuum density, Ω_Λ.

The differential optical depth $d\tau$ for a beam encountering a lens in a path from z_l to $z_l - dz_l$ is

$$d\tau = n_l(1 + z_l)^3 \pi \theta_E^2 D_d^2 c \frac{dt}{dz_l} dz_l \tag{5.31}$$

where n_l is the number density of lenses at $z = 0$. Inserting (5.17) and $n_l = \frac{\Omega_l 3 H_0^2}{8\pi G M_l}$ the equation above simplifies to

$$d\tau = \frac{3}{2} \Omega_l (1 + z_l)^3 \frac{H_0}{c} \cdot \frac{D_{ds} D_d}{D_s} \cdot H_0 \frac{dt}{dz_l} dz_l \tag{5.32}$$

where $\frac{dt}{dz_l}$ was derived in (4.76) and the angular distance terms D_d, D_s and D_{ds} can be expressed as a function of cosmological parameters as shown in (4.70) and (4.85):

$$D_{ab}(z_a, z_b) = [(1 + z_b) H_0 \sqrt{|\Omega_K|}]^{-1} S\left(\sqrt{|\Omega_K|} \int_{z_a}^{z_b} \frac{dz'}{H'(z')}\right) \tag{5.33}$$

with S and $H'(z)$ as defined in Section 4.5.2 (if $k = 0$, $|\Omega_K|$ is replaced by 1).

In Fig. 5.6 (a) it is shown how, for any given redshift, the angular distance D_s in (5.33) depends strongly on Ω_Λ and much less on Ω_M. As distance increases with Ω_Λ, the volume available to be filled with lenses with a comoving mass density Ω_l is larger. As a consequence, the optical depth is much larger in a Universe with an energy density dominated by Ω_Λ, as shown in Fig. 5.6 (b). For the highest redshifts where quasars have been observed at $z \sim 4$, the rates of gravitationally lensed sources differ by an order of magnitude for the two extreme values of the cosmological constant. Thus, measurements of the optical depth τ versus redshift can be used to set limits on Ω_Λ.

While the rate of lensed events can be used to establish limits on the energy density terms of the Universe, the time difference between multiply lensed sources is given by the geometric path lengths of the beams and are thus proportional to the distance scale in the Universe, i.e., to the inverse of the Hubble parameter, H_0, as sketched in Fig. 5.7. For simple spherically

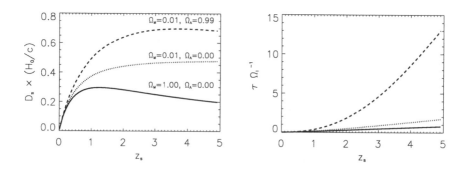

Fig. 5.6. (a) Angular distance to a source at z_s for three combinations of cosmological parameters. (b) Optical depth for the same combination of cosmological parameters. A non-zero cosmological constant has a large effect on the distance estimate for any given redshift.

symmetric lenses, the time difference between two images separated by an angle $\Delta\theta$ and with flux ratio r is given by:

$$\Delta t = \frac{\Delta\theta^2}{2} \frac{D_d D_s}{D_{ds}} (1 + z_l) \mathcal{G}(r), \qquad (5.34)$$

which indicates that $\Delta t \propto H_0^{-1}$. The function $\mathcal{G}(r)$ depends on the mass profile of the lens, which needs to be modeled. Another difficulty with this technique is the measurement of the time difference between lensed images of quasars, as these objects do not normally have regular time structures.

5.2.3 The Mass Density in Galaxy Clusters

Galaxy clusters are the largest well-defined structures in the Universe. With masses around 10^{14} $M_\odot$ galaxy clusters may cause image separations up to one arcminute, as shown schematically in Fig. 5.8 and in a real Hubble Space Telescope image of the galaxy cluster CL0024+1654 at a redshift $z = 0.39$, in colour Plate 3.[2] Clusters are 'weighed' by modeling the distortion of the background galaxies, and it is found that their mass is predominantly 'dark'. Masses may also be determined from cluster images not showing giant luminous arcs as in Fig. 5.8, but rather a statistical behaviour of the background sources. If all galaxies were round, their images would look elliptical due to the tidal components of the gravitational field. This is known as *weak lensing* (discussed further in the next section) . In practice, with galaxies being intrinsically elliptical, the effect is still detectable under the assumption that their orientation is random. Thus, statistical studies of the brightness, shapes and orientation of background galaxies provide an estimate of the tidal gravitational fields and by extention, the mass profile of galaxy clusters.

---

[2] The colour plate section is positioned in the middle of the book.

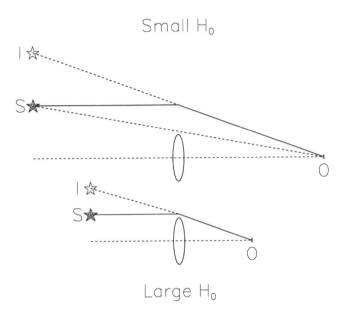

Fig. 5.7. The scale of the Universe and the path length of a gravitationally lensed image is inversely proportional to the Hubble parameter, H_0.

The distribution of mass shows two distinct components. First, around each galaxy there is a halo of dark matter which has a more extended distribution than the stars, gas and dust of the galaxies. There is also a broad dark matter distribution associated with the cluster as a whole. The mass-to-light ratio of these clusters is several hundred times the mass-to-light ratio of the Sun.

In fact, in many of the clusters most of the baryons are not in the form of stars, but rather in gas which is hot and therefore emitting X-rays. The reason for the high temperature is that when the gas molecules move in the gravitational potential of the cluster, they obtain a velocity which grows with cluster mass. Some of the kinetic energy is transferred to radiation, for example through collisions between molecules. By studying the X-ray temperature and intensity one may therefore get an independent estimate of the total cluster mass. This type of estimate typically agrees with that obtained by lensing within a factor of two.

By generalizing the observations of several galaxy clusters to the entire Universe, a mass density $\Omega_M \sim 0.2 - 0.3$ is deduced.

5.2.4 Weak Gravitational Lensing

Since the turn of the millennium, several independent observations of weak lensing by large scale structures have been made. These observations provide a way to probe the mass content in the Universe as a whole, by seeing how ellipticities of galaxies are changed by the lensing potentials. To understand

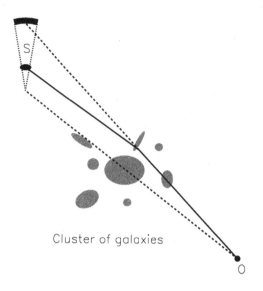

Fig. 5.8. Lensing of distant objects by a foreground galaxy cluster.

this, we must have a closer look at the lensing equation. Defining a length scale ξ_0 in the lens plane and a corresponding length scale $\eta_0 = \xi_0 D_s/D_d$ in the image plane, we can rewrite the lens equation (see the caption of Fig. 5.2) in the coordinates $\eta = D_s\beta$ and $\xi = D_d\theta$ (where we have extended β and θ to two dimensions, i.e., they span the lens and image planes) as

$$\mathbf{y} = \mathbf{x} - \alpha(\mathbf{x}), \tag{5.35}$$

where $\mathbf{y} = \eta/\eta_0$, $\mathbf{x} = \xi/\xi_0$ and

$$\alpha(\mathbf{x}) = \frac{D_d D_{ds}}{D_s} \frac{\widehat{\alpha}(\xi_0 \mathbf{x})}{\xi_0}. \tag{5.36}$$

Here the deflection angle is given by

$$\widehat{\alpha}(\xi) = \int d^2\xi' 4G\Sigma(\xi') \frac{\xi - \xi'}{|\xi - \xi'|^2} \tag{5.37}$$

where $\Sigma(\xi)$ is the projection of the volume density of the deflector onto the lens plane.

In (5.35) we thus have a formula that we can use for any given, continuous mass distribution, such as the predicted large scale structure of galaxies and galaxy clusters (which is currently being probed by large galaxy surveys). Looking at the structure of equation, we can compute its Jacobian matrix:

$$A_{ij}(\mathbf{x}) = \frac{\partial y_i}{\partial x_j} = \begin{pmatrix} 1 - \kappa - \gamma_1 & -\gamma_2 \\ -\gamma_2 & 1 - \kappa + \gamma_1 \end{pmatrix}, \tag{5.38}$$

where the *convergence* κ and the *shear* components γ_1 and γ_2 can be computed from (5.35).

One can convince oneself the the magnification of the mapping is given by

$$\mu(\mathbf{x}) = \frac{1}{\det A(\mathbf{x})} = \frac{1}{(1-\kappa)^2 - \gamma_1^2 - \gamma_2^2}. \tag{5.39}$$

Thus, a circular source will be changed in two ways by weak lensing: it will be magnified with magnification given by μ, and it will become elliptical due to the effects of γ_1 and γ_2. The direction of ellipticity will mostly be tangentially around the lens (as indicated by the extreme case: the Einstein ring). The analysis is complicated by the fact that galaxies may have ellipical shape even before the lensing, but by collecting a large number of images and analyzing their shape statistically, a signal has been found, with magnitude in accordance with the 'Standard Model', $\Omega_M \sim 0.3$.

5.3 Black Holes

The most extreme example of light bending in a gravitational field occurs for *black holes*, objects smaller than their Schwarzschild radius, $r_s = 2.95\left(\frac{M}{M_\odot}\right)$ km (see (3.65)). Within the known laws of physics nothing will prevent an object from becoming a black hole if the gravitational force from its own mass cannot be balanced by an outward pressure. Very heavy stars are thus expected to end their evolution as black holes. As stars collide and coalesce, black holes are also believed to form in the centre of globular star clusters and in the nuclei of galaxies.

For example, the energy release observed from *active galactic nuclei* (AGN) suggests that stars, dust and gas form an accretion disk around black holes with M$\sim 10^8 M_\odot$. The infall of matter towards the black hole would generate the observed electromagnetic radiation and could also be responsible for the highest energy cosmic rays. The detection of high-energy photons, neutrinos and other types of radiation from AGN is of great importance for the understanding of astrophysical acceleration sites. Such black hole candidates have been observed in the centre of the Virgo galaxies M84 and M87, about 50 million light years from the Earth. By looking at the central region of the galaxies with a spectrograph onboard the Hubble Space Telescope (HST), rapid rotation of gas, $v_r \sim 550$ km/s, has been observed 60 light years from the core, as shown in colour Plate 4.[3]

Observational evidence for stellar mass black holes is sought for by searching for collapsed stars in binary systems. It would appear as the visible star orbits a compact, invisible, object with a mass larger than the maximum mass for a neutron star. Candidates for such a system exist: for example, the X-ray source Cygnus X–1.

---

[3] The colour plate section is positioned in the middle of the book.

5.3.1 Primordial Black Holes

It is conceivable that 'mini' black holes ($r_s \approx 10^{-15}$ m) could have formed during the Big Bang. These objects are sometimes called *primordial* black holes. What can be said about the density of such objects today? Hawking [20] showed in a celebrated paper that once quantum effects are taken into account, black holes *do* radiate and therefore have a finite lifetime. According to quantum mechanics, pair production of virtual particles ($\Delta E = 2mc^2$) may take place in the vacuum, provided that they annihilate within the short time allowed by the uncertainty principle: $\Delta t \sim \hbar/2mc^2$. Such *vacuum fluctuations* will take place also in the neighbourhood of black holes. In the presence of the strong gravitational field the particles can be separated fast enough that one of them falls beyond the event horizon. They thus escape annihilation and one of them thereby becomes a 'real' particle. In the process, gravitational energy from the black hole has been converted into rest energy and kinetic energy of the produced particle, so that energy of the order of ΔE has been transferred from the black hole to the outside world. Thus black holes are said to *evaporate*. Hawking also showed that the radiation has a thermal spectrum with temperature inversely proportional to the mass of the black hole:

$$T = \frac{\hbar c^3}{8\pi k G M} = 10^{-7}\left(\frac{M_\odot}{M}\right) \text{ K} \tag{5.40}$$

The lifetime of black holes then becomes (Problem 5.5):

$$\tau \approx 10^{10}\left(\frac{M}{10^{15}\text{g}}\right)^3 \text{ years} \tag{5.41}$$

Thus, black holes with masses much below 10^{15} g have a lifetime shorter than the present age of the Universe and should therefore have already evaporated. For stellar mass black holes and heavier, the evaporation rate is negligible. It is therefore not clear whether this interesting process, which includes elements of special and general relativity, quantum mechanics and thermodynamics, can be observationally confirmed. Black holes and their Hawking evaporation are, however, great 'theoretical laboratories' where current ideas of quantum gravity are tested.

Example 5.3.1 Derive the relation $T \propto M^{-1}$ from the Heisenberg uncertainty principle and the fact that the size of a black hole is given by the Schwarzschild radius.

Answer: The position of quanta inside a black hole can only be known within $\Delta x \approx 2r_s$. Thus, $\Delta t = 2r_s/c$. According to the uncertainty principle, $\Delta E \Delta t \approx \hbar$. Thus,

$$\Delta E \approx \frac{\hbar c^3}{4GM}$$

and the relation for temperature (except for a numerical factor) follows directly.

The energy of the radiated quanta from a 'mini' black hole is $E \sim 100 \left(\frac{10^{15}g}{M} \right)$ MeV (see (5.40)). The observed low flux of gamma-rays with energies around ~ 100 MeV can be translated into an upper limit on the density of primordial black holes to about $\Omega_{BH}(M \leq 10^{15}g) \leq 10^{-8}$.

5.4 Summary

- Light beams are bent, independently of wavelength, as they pass in the neighbourhood of massive objects. For a point-mass lens, the bending angle becomes

$$\Delta\phi = \frac{4GM}{bc^2}$$

  where b is the closest distance from the light-ray to the lens.
- A source can be multiply imaged, in particular if the observer, lens and source are perfectly aligned. The resulting image becomes an *Einstein ring* with radius

$$\theta_E = \sqrt{\frac{4GM}{c^2} \frac{D_{ds}}{D_d D_s}}$$

- Microlensing of stars in the Milky Way and neighbouring galaxies is used to study the population of faint compact objects in the Galaxy.
- Gravitational lensing is a useful tool to measure cosmological parameters.

5.5 Problems

5.1 Calculate the number of lenses within an Einstein radius in a thin layer D_d along the line of sight of observation assuming a number density n_l. Use this to derive (5.20).

5.2 Derive the expression for microlensing of cosmological point sources by point-like lenses (5.28).

5.3 Estimate the Einstein radius and time duration for the lensing of a bright point source at cosmological distance assuming the lens is a faint point object in the Milky Way halo.

5.4 Use dimensional analysis to construct a density out of c, $\hbar$ and G. Compare this 'Planck density' to the mass density of a stellar mass object with r_S.

5.5 Show that the lifetime of black holes $(T \propto M^{-1})$ is proportional to M^3.

6 Particles and Fields

6.1 Introduction

One of the most fascinating aspects of modern cosmology is the interplay between that subject and particle physics. It is perhaps somewhat paradoxical that the study of the smallest things we know, the elementary particles, can have applications in the study of the largest structures in the Universe. This is, however, the case, and the interface between particle physics and cosmology is a growing area of scientific study today. It sometimes goes under the name of astroparticle physics, which usually also includes the study of cosmic rays and relativistic astrophysics. All this is a good motivation for today's astrophysicist to become familiar with some of the basic features of current particle physics models.

One main discovery during the twentieth century was the importance of fields for the understanding of the fundamental interactions in nature. Of course, current theories are formulated such that they are consistent with relativity and quantum mechanics. In this book, we do not present a full-fledged version of relativistic quantum field theory. However, by using the knowledge you have accrued from the previous chapters, and if you at this point also study Appendix B *Relativistic Dynamics* and Appendix C *The Dirac Equation* in this book, you will acquire a rather complete basis for understanding the particle physics relevant to cosmology and relativistic astrophysics. On the other hand, if you are ready to accept some results without derivation, it is enough to have a quick look at the Summary sections of these Appendices. Some topics such as quantized fields in curved space-time are beyond the scope of this book. However, in Appendix E we treat the quantization of scalar field in an expanding (de Sitter) Universe.

6.2 Review of Particle Physics

Let us begin our introduction to particle physics by reviewing some basic facts about the elementary constituents of matter. According to the extremely successful *Standard Model* of particle physics, the basic building blocks of matter are *quarks* and *leptons*, of which six of each are known to exist (see Table 6.1). As can be seen, the quarks and leptons are naturally grouped into

Table 6.1. The quarks and leptons of the Standard Model. Notice that since quarks are confined, the quark mass is not a uniquely defined parameter. The neutrino mass limits are given by laboratory experiments. There are cosmological and astrophysical data which indicate that they are much lower, but non-zero (see Sections 11.6 and 14.4).

Electric charge	$Q = 0$	$Q = -1$	$Q = +2/3$	$Q = -1/3$
First family:	ν_e	e	u	d
Mass:	< 3 eV	511 keV	1.5–4.5 MeV	5–8.5 MeV
Second family:	ν_μ	μ	c	s
Mass:	< 170 keV	106 MeV	1.0–1.4 GeV	80–155 MeV
Third family:	ν_τ	τ	t	b
Mass:	< 18 MeV	1.78 GeV	170–180 GeV	4.0–4.5 GeV

three *families*, each consisting of an electrically neutral lepton (such as the electron neutrino ν_e), one lepton with charge -e (such as the electron), one quark with charge +2/3 e (such as the u-quark), and one quark with charge −1/3 e (such as the d-quark). A curious fact of quarks is that they appear not to exist as free particles. A proton consists of two u-quarks and one d-quark, a neutron of one u-quark and two d-quarks (check that this gives the correct electric charges!). The forces that bind these quarks together are so strong that a single quark cannot be extracted from the bound system. This feature of the strong interaction between quarks is called *confinement*. Nevertheless there are methods to experimentally 'shake' the quarks within a proton and in this way prove that they exist as point-like constituents. The fact that the strongly bound quarks within nucleons (that is, protons or neutrons) can act as free particles during very short time intervals, as experimentally seen, is an intriguing property of the modern theories called *asymptotic freedom*.

The particular pattern of charges and other quantum numbers in a family of two leptons and two quarks means that the theory will be consistent quantum mechanically. If one particle were missing, a so-called anomaly would be generated with catastrophic consequences for the theory. (Essentially, nothing would be calculable, due to infinite quantities appearing which cannot be cancelled, 'renormalized', in a controlled way.) This was the reason for the firm belief among particle physicists that the top quark had to exist even long before it was finally discovered in 1995.

In Table 6.1, we have suppressed some *quantum numbers*: for example, each quark has three internal degrees of freedom called *colours*. The theory of the strong interaction, quantum chromodynamics (QCD), describes how coloured quarks interact. Also, quarks and leptons have *spin* 1/2 (in units of $\hbar$, Planck's constant divided by 2π). They are thus *fermions* which obey the Pauli exclusion principle. Finally, for each known species of particle there exists a corresponding *antiparticle* with the same spin and mass but with opposite sign of the charge. Neutrinos, which have zero electric charge, have still another type of charge, *weak hypercharge*, which means that a neutrino and an antineutrino are different particles. However, all neutrinos seem to have only left-handed spin (or helicity, which is the projection of the spin on the direction of the momentum). Likewise, all antineutrinos are right-handed.

This is the way neutrinos appear in the Standard Model. However, the neutrino sector is very difficult to study experimentally due to the very feeble interactions of neutrinos with ordinary matter. It is possible (see Appendix C) that neutrinos are their own antiparticles – so-called Majorana neutrinos.

6.3 Quantum Numbers

The concept of a quantum number is important in particle physics. Since quantum mechanics tells us that, for example, the internal angular momentum s of a particle is quantized with value being an integer or half-integer times $\hbar$, this internal angular momentum (spin) is a quantum number. Usually the existence of conserved quantum numbers can be traced back to an invariance of the theory under some set of transformations. For example, angular momentum conservation follows from the invariance under rotations of the form discussed in Section 2.3. A given system of particles can have a total angular momentum that is given by the total spin of the constituent particles coupled to the total orbital angular momentum according to the rules of quantum mechanics. One practical use of this conservation is the general rule that a system which has a total angular momentum which is a half-integer, cannot decay to an integer-spin system.

There are other types of transformations, acting on internal degrees of freedom, which may also lead to conserved quantum numbers. One prime example is electric charge, whose conservation follows from the invariance under gauge transformations (2.80). In addition, there are other 'charges' such as baryon number, which seem to be conserved to a very high accuracy (the lifetime of the proton must be at least 10^{33} years to fit with experimental data). There, the invariance causing the conserved baryon number is less well understood (in fact, according to some theories there should not be an exact conservation law for baryon number), but it can be regarded as an empirical law which tells us which processes involving baryons are allowed. The normalization is usually fixed by defining the baryon number of a proton as +1 (and thus −1 for the antiproton). Thus, a quark has baryon number +1/3.

Also for leptons there seem to be quantum numbers that are at least approximately conserved. One may define an 'electron lepton number' which is +1 for an electron and an electron neutrino, and similarly for the other leptons. It seems that each such lepton number is conserved to a very good approximation. Their sum, the total lepton number, seems to be even more well-conserved. However, again there is no strong theoretical motivation for the absolute conservation of lepton number (in contrast to electric charge, where gauge invariance is a very powerful constraint). It is possible that the lepton number is violated at some level, although this has not yet been experimentally verified. The individual lepton numbers are very likely not to be conserved, as we shall see in Chapter 14.

Example 6.3.1 Explain why the following decay processes are not allowed.

    a. $\mu^- \to e^- +$ nothing.
    b. $t \to \bar{s} + \bar{b}$.
    c. $b \to \bar{s} + s + d + e^- + \bar{\nu}_e$. (The bar denotes an antiparticle.)

Answer: (a) Lepton number conservation (and also energy and momentum conservation).
(b) Baryon number conservation.
(c) Charge conservation.

6.4 Degrees of Freedom in the Standard Model

One very useful way to look at the spin degrees of freedom of a particle is to regard a state with any one of the $2s + 1$ possible values of m_s as a separate 'particle'. This is justified because, as we shall see in Chapter 8, each such spin state adds independently to, for example, the energy density. Lorentz transformations act not only on space-time but also on the internal spin states: they mix with each other. Here m_s is the spin projection on a fixed, but arbitrary, axis. Usually it is taken to be the z-axis. However, an even better choice which makes it easy to deal also with massless particles such as (perhaps) neutrinos is to use the helicity, the projection of the spin on the direction of motion.

    Let us count the number g_{fam} of independent helicity states of one family of quarks and leptons. Each quark comes in three colours and two spins. This means $2 \cdot 3 \cdot 2 = 12$ states for the u and d quarks and antiquarks. The charged lepton has two states, and the neutrino one state. Thus, one family consists of $g_{fam} = 15$ helicity states, meaning 45 states. If we also include the antiparticles, we obtain

$$g_{fermions}^{tot} = 90 \tag{6.1}$$

Above a certain temperature, of the order of $100 - 300$ MeV, where the *quark-gluon phase transition* is thought to occur, it is believed that quarks and gluons are not confined anymore but act as free particles.

6.5 Mesons and Baryons

Below the QCD phase transition temperature only colour-neutral systems seem to be allowed. One way to achieve this is for a quark to bind together with an antiquark, forming a strongly bound but colour-neutral system, called a *meson*. The lightest such systems are the π-mesons, or pions. The pion mass (or rest energy) is around 140 MeV. A π^+ particle consists of one u-quark and one d-antiquark. Usually a bar over the particle name is used to denote the antiparticle, so the composition of a π^+ is written $(u\bar{d})$. The π^-, which is the antiparticle of π^+, is of course made of one d-quark and one u-antiquark instead, $(d\bar{u})$. There is also a neutral pion, the π^0, which is a quantum mechanical mixture of $(u\bar{u})$ and $(d\bar{d})$.

Another way to produce a colour-neutral state is to take three quarks, each of a different colour which gives, for instance, the proton or the neutron.

Thus, particles containing quarks and therefore strongly interacting, *hadrons*, are of two types: either *baryons* consisting of three quarks like the nucleons (the proton and neutron), or *mesons* consisting of one quark and an antiquark like the pions. There is speculation over the existence of more 'exotic' particles consisting, for example, of two quarks and two antiquarks, but so far none have conclusively been proven to exist.

When the quark model appeared in the 1960s, it provided a solution to the puzzle of the nature of all the hundreds of particles which had by then been discovered in accelerators. Out of a few quarks (and antiquarks) a large number of meson and baryon states can be constructed, using the rules of quantum mechanics when building up the states. The most important ones are the meson and baryon octets shown in Fig. 6.1, which are the lowest-lying (that is the least massive) meson and baryon states. The classification makes use of a quantum mechanical symmetry $SU(3)$ which is based on the principle that all the three light quarks u, d and s enter on equal footing.[1] A K^+ meson, for instance, consists of a u quark and an s antiquark $(u\bar{s})$, whereas a π^+ meson consists of $(u\bar{d})$. The π^0 and η mesons contain linear combinations of $(d\bar{d})$, $(u\bar{u})$ and $(s\bar{s})$ states.

Additional combinations are produced by excited states of the fundamental ground state mesons and baryons. For example, the proton is the ground state of a system made up by the quark combination (uud), with the quarks having a total angular momentum $\ell = 0$ (S-wave), and total spin $1/2$. Then there should also exist a state with the same quark content but with the total spin equal to $3/2$. Indeed, such a particle exists. It is the Δ^+ baryon, which in fact plays an important role in astrophysics. Since it is so closely related to the proton, it is easy to excite it by colliding, for example, a proton with a photon. We shall see later (Section 12.3) that such interactions between very energetic cosmic ray protons and the cosmic microwave background photons

---

[1] Since the classification in terms of octets, decuplets and singlets assumes an exact symmetry between the light u, d and s quarks, and that symmetry is in practice broken due to somewhat different quark masses (the s quark is heavier), there occurs mixing between states. For instance, the η^0 and η' mesons are mixtures of singlet and octet states.

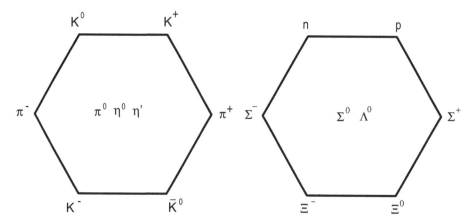

Fig. 6.1. The meson (on the left) and baryon (on the right) octets obtained in the $SU(3)$ classification of three-quark bound states.

determines how far a high-energy proton can travel in intergalactic space without interacting.

For mesons and baryons containing c, b and t quarks, the classification in terms of $SU(N)$ is not very useful, since the rest masses are completely determined by the quark masses. However, the spectroscopy of states can be understood from simple combinatorial mathematics, with baryons always containing three quarks and mesons containing a quark and an antiquark.

As yet there is no deep understanding of why there are precisely three families of quarks and leptons in nature. The solution to this and many other puzzles of particle physics will probably have to await a full theory of all particles and interactions in nature, including quantum gravity. This theory is not yet known, but it is speculated that there may exist a hitherto unknown theory, called M-theory, which under certain circumstances has solutions which appear as strings or higher-dimensional objects, so-called D-branes. Once the correct theory is found, it is hoped that the particle content (such as the number of families), charges, masses and other quantities will be explained by geometric properties in the high-dimensional space in which the theory 'lives'. [2]

6.6 Gauge Fields

A Universe consisting of only quarks and leptons and nothing else would be quite boring. What brings dynamics into the picture is the fact that they in-

---

[2] The reader is directed to the appropriate bulletin boards on the World Wide Web, such as http://arxiv.org/hep-th, for the most recent developments in this rapidly evolving field.

teract with each other: in particular, that they form bound states like hadrons and atoms. An interesting consequence of relativity and quantum mechanics is that interactions must be described by exchange of mediator particles. This is not difficult to understand. Consider two electrically charged particles, let us say a proton and an electron, separated by a finite distance. Since they are both charged, they influence each other through the electromagnetic (Coulomb) interaction. Suppose now that we suddenly move the proton a short distance. Then the electromagnetic field surrounding it changes, and this will influence the electron. However, special relativity tells us that the information about the disturbance cannot reach the electron with a velocity larger than the speed of light. In particular, there can be no instantaneous 'action at a distance'.

The modern description of forces is based on the notion of fields: that is, the disturbances cause 'ripples' in the fields between the proton and electron. According to quantum mechanics, such ripples represent dynamical degrees of freedom and therefore have to be quantized like all other dynamical degrees of freedom. The lowest excitations of the field can be interpreted as particles, and the interaction between a proton and an electron is described by the exchange of these particles. Since the electromagnetic field is known from Maxwell's equations to permit wave solutions, where visible light is one example, these exchange particles are naturally identified with the *photons* that Einstein introduced to explain the photoelectric effect.[3]

The quantum theory describing the interaction between photons and electrons is called Quantum Electro Dynamics, QED. It has been, and still is, a tremendously successful theory which matches experiment to an accuracy better than one part in a million. It turns out that classical electromagnetism, and also QED, can be 'derived' by postulating symmetries, *gauge symmetries*, in the theory of free electrons. QED is therefore the prime example of a *gauge theory*.

Let us see how this works in a simple example, which also introduces the concept of a *field*, that is a function of space-time whose quantization gives rise to the elementary excitations that can be interpreted as particles. We present a description based only on special relativity here (the general relativistic formulation is mathematically rather more involved and is seldom needed except for very extreme conditions, such as near a black hole). A much more complete treatment of relativistic field theory is presented in Appendices B and C

A pion, a π^+ for instance, has spin 0, and should be describable by a scalar field $\phi(x)$ (where x stands for the space-time coordinate x^μ). Under a Lorentz transformation $x \to x'$ the field should then transform as

$$\phi'(x') = \phi(x) \tag{6.2}$$

---

[3] The fact that dynamical degrees of freedom of a system are always quantized is familiar from condensed matter physics, where the quantized vibrational states of a lattice are called *phonons*.

In relativistic field theory, we should also be able to simultaneously describe the antiparticle π^-. This is because in energetic processes, pairs of pions could appear 'from nowhere'. For example, in proton-proton collisions, the process $p + p \rightarrow p + p + \pi^+ + \pi^-$ is possible if the total kinetic energy of the initial proton pair in the centre of momentum frame is greater that $2m_\pi c^2$. In principle, we could introduce two scalar fields $\phi_1(x)$ and $\phi_2(x)$ to describe π^+ and π^- respectively. However, a more elegant way is to let $\phi_1(x)$ and $\phi_2(x)$ be the real and imaginary parts of a complex field $\phi(x) = \rho(x)e^{i\theta(x)} = (\phi_1(x) + i\phi_2(x))/\sqrt{2}$. If the field has no interactions, it should satisfy the relativistic equation of motion

$$\Box\phi(x) \equiv \left(\frac{\partial^2}{\partial t^2} - \nabla^2\right)\phi(x) = -m^2\phi(x) \tag{6.3}$$

where m is the pion mass (with the same equation applying for the complex conjugate field ϕ^*, which together with ϕ can be used as the two independent states instead of $\phi_1(x)$ and $\phi_2(x)$). This equation is invariant under Lorentz transformations since both the d'Alembertian operator $\Box = \partial^2/\partial t^2 - \nabla^2 = \partial_\mu\partial^\mu$ and m^2 are invariant. It is in fact the simplest possible relativistic wave equation for a free scalar field (that is, without interactions).

Equation (6.3) can be derived as the Euler Lagrange equation (see (B.27))

$$\frac{\partial\mathcal{L}}{\partial\phi^*} - \frac{\partial}{\partial x^\mu}\left(\frac{\partial\mathcal{L}}{\partial\phi^*_{,\mu}}\right) = 0 \tag{6.4}$$

if we choose the Lorentz-invariant Lagrangian density $\mathcal{L}(x)$ to be

$$\mathcal{L}(x) = \frac{1}{2}\left(\partial_\mu\phi(x)\right)^*\left(\partial^\mu\phi(x)\right) - \frac{m^2}{2}\phi^*(x)\phi(x) \tag{6.5}$$

We see that (6.5) remains unchanged if we change the phase of ϕ by the same amount everywhere, $\phi \rightarrow e^{i\alpha}\phi$ with α constant. This is called a *global invariance* of the Lagrangian. Suppose that we now demand that the theory should also be invariant if we change the phase by a different amount at different space-time points, that is, if we let $\alpha = \alpha(x)$. This is called local invariance or *gauge invariance*. Then the derivatives in (6.5) spoil the invariance, unless we add another set of fields $A_\mu(x)$ to the derivative ∂_μ,

$$\partial_\mu \rightarrow D_\mu \equiv \partial_\mu - iA_\mu(x) \tag{6.6}$$

and let

$$A_\mu(x) \rightarrow A_\mu(x) + \partial_\mu\alpha(x) \tag{6.7}$$

when $\phi(x) \rightarrow e^{i\alpha(x)}\phi(x)$. The field A_μ that we introduced in this way is the electromagnetic potential!

Equation (6.6) defines the *gauge covariant* derivative, and has obvious similarities to the covariant derivative we introduced in general relativity (see (3.36)). The equation of motion for ϕ becomes

$$\left(D_\mu D^\mu + m^2\right)\phi(x) = 0 \tag{6.8}$$

which describes the electromagnetic properties of a charged scalar field.

A very common and practical pictorial way to describe interactions through particle exchange is by means of Feynman diagrams, named after the late American physicist Richard P. Feynman, one of the inventors of QED. In Fig. 6.2, we show an example of such a diagram representing the scattering of two electrons by photon exchange.

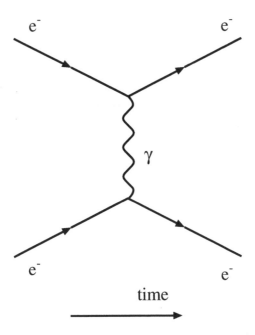

Fig. 6.2. A Feynman diagram representing the scattering of two electrons through photon exchange. Time runs from left to right: that is, the initial state is to the left and the final state to the right.

It was known from studies of neutron decay that there has to exist another interaction, much weaker than electromagnetism, therefore called the *weak interaction*. This is also the interaction that makes neutrinos experimentally detectable. However, a neutrino of 1 MeV energy interacts so weakly that the mean free path in ordinary matter is of the order of a light year!

6.7 Massive Gauge Bosons and the Higgs Mechanism

Since QED had been so successful, it was natural to generalize it to other interactions as well. In the 1970s, a gauge theory of the weak interactions was worked out, and indeed it is so intimately related to QED that the two interactions are nowadays considered as one *electroweak* interaction. Of course,

for the interaction between, for example, neutrinos there are corresponding
exchange particles, the $W^{\pm}$ and Z^0 bosons. In 1982 these were experimentally discovered at CERN, the European Laboratory for Particle Physics in
Geneva, Switzerland. In Fig. 6.3 the Feynman diagram describing neutron
β decay through the exchange of a W boson is shown. As can be seen in
this figure, the quark flavour can change through the so-called weak charged
current mediated by $W^{\pm}$-bosons.

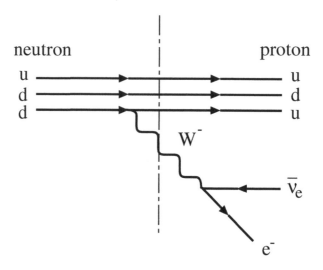

Fig. 6.3. Feynman diagram showing the decay of a neutron into a proton, an
electron and a neutrino. To be precise, an antineutrino is produced, which is why
the arrow for that particle is drawn in the backward direction. This is a general
convention in Feynman diagrams. The vertical line shows an intermediate state at
one particular time.

The masses of these particles were found to be very high, 80-90 GeV, in
distinct contrast to the photon which is believed to be strictly massless. This
asymmetry of mass of the respective exchange particles in the electroweak
theory necessitates a mechanism for symmetry breaking, called the *Higgs
mechanism*. It is one final part of the electroweak theory which is not yet
proven experimentally. If it is correct, there has to exist a new field, the
Higgs field, with its corresponding particle, which has the property of giving
mass to the W and Z bosons. It is also needed to give quarks and leptons
their mass.

The Higgs field in the Standard Model is a scalar field which means that it
has a Lagrangian density as given by (6.5). An important difference is, however, that it also has a self-interaction which can be described by a potential
$V(\phi)$ of the form

$$V(\phi) = b|\phi|^2 + \lambda|\phi|^4 + \text{const} \tag{6.9}$$

where λ must be positive to have a stable theory (V should be bounded from below). However, b can have either sign. If it is positive, we see that the minimum is given by $\phi = 0$; this is the unique symmetric vacuum state. However, if b is negative we can write

$$V(\phi) = \lambda \left(|\phi|^2 - v^2\right)^2 + const \tag{6.10}$$

with

$$v = \sqrt{\frac{-b}{2\lambda}} \tag{6.11}$$

which means that the ground state or vacuum state – that is, the state of lowest energy – is given not by a vanishing field $\phi = 0$, but rather $|\phi| = v$. The form of the Higgs potential (6.10), shown in Fig. 6.4, is typical of *spontaneous symmetry breaking*, and it is the non-vanishing vacuum expectation value of the Higgs field that can be shown (see, for example [31]) to give rise to masses of fermions and the W and Z bosons.

The Higgs mechanism is the only one known that can give mass to the gauge bosons. Gauge invariance means, as it does for the photon, that the exchange particle has to be massless. If one would try to introduce mass 'by hand', that is introducing a quadratic term in the boson fields, gauge invariance is spoiled which gives rise to a theory that cannot be treated (in particular, renormalized) using any known methods. Arguably, the Higgs mechanism seems a bit *ad hoc*, but it works and the hunt for the predicted Higgs particle has top priority at the biggest accelerators, at Fermilab and CERN. Results from the now discontinued Large Electron Positron collider (LEP) at CERN have now pushed the lower limit of the Higgs mass to above 115 GeV.

Without going too deep into field theory, it is useful for cosmological applications to display the expression for the energy-momentum tensor of a scalar field with a Lagrangian of the form

$$\mathcal{L} = \frac{1}{2}\partial^\mu\phi\partial_\mu\phi - V(\phi) \tag{6.12}$$

It is given by

$$T^{\mu\nu} = \partial^\mu\phi\partial^\nu\phi - \mathcal{L}g^{\mu\nu} \tag{6.13}$$

an expression that we shall use when discussing inflation.

The large mass scale of the weak bosons explains very nicely the feebleness of the weak interaction. This is not difficult to understand qualitatively. If we look at an 'intermediate state' in Fig. 6.3 (indicated by the vertical line in the figure), we see that it looks as though the neutron has decayed into a proton and a W boson. However, since the rest energy of the neutron is 0.94 GeV, and that of the W boson 80 GeV, this is not energetically possible. Have we done something wrong when drawing the diagram? No, we have to

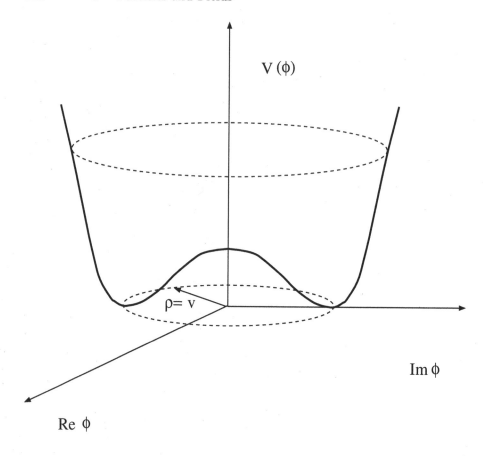

Fig. 6.4. The form of the Higgs potential that causes spontaneous symmetry breaking.

remember that in quantum mechanics energy need not be conserved during short periods of time. According to Heisenberg, we can 'borrow' energy ΔE during a time Δt as long as the uncertainty relation $\Delta E \Delta t \sim \hbar$ is fulfilled. However, such a large energy fluctuation is quite improbable, which explains why weak interactions are rare, meaning weak. A particle of mass M in a Feynman diagram which cannot exist as a real particle due to energy and momentum conservation (a so-called *virtual particle*) causes a suppression in the transition amplitude by a factor $1/M^2$, and therefore a factor $1/M^4$ in the interaction probability (the cross-section of a scattering process or decay rate for a decaying particle) since according to the laws of quantum mechanics this probability is proportional to the square of the amplitude.

6.8 Gluons and Gravitons

As a final triumph for the idea of a quantized field theory of basic interactions came the fundamental theory of the strong interactions, Quantum Chromo-Dynamics, QCD. According to QCD, the interactions between quarks are mediated by gauge bosons called *gluons*. Unlike the photons in QED (but similar to the W and Z bosons of the weak interactions), the gluons also interact directly with each other (see Fig. 6.5). This is because the gluons themselves carry the 'colour' charges of the strong interactions. In contrast, the photon couples to electrically charged particles, but is itself electrically neutral.

In fact, there are eight differently coloured charged gluons. The aforemen-

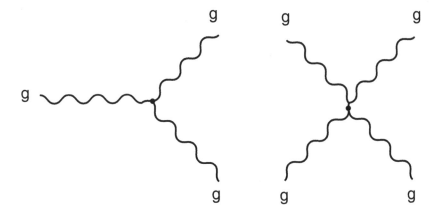

Fig. 6.5. Feynman diagrams showing self-interacting gluons.

tioned property of asymptotic freedom can be proven to exist in QCD, which explains why quarks and gluons appear almost as if they were free particles when probed during short times, despite the fact that they cannot survive as free particles during long periods of time. Also confinement is believed to be due to the self-interacting property of the gluons, but this has not yet been rigorously proven due to the complicated analytical structure of the theory. There exist, however, very strong indications from numerical calculations in so-called lattice gauge theory that this is indeed the case.

The gauge particles of the Standard Model are displayed in Table 6.2. In addition to these particles, the *graviton* is believed to exist. This is the mediator of the gravitational interaction, but since no consistent quantum theory of gravitation has yet been constructed its existence is still hypothetical. Einstein's classical theory of gravitation, general relativity, indeed has a structure that makes it very similar to a gauge theory with the gauge transformations corresponding to general coordinate transformations. The expression for the

Riemann tensor in general relativity looks very similar to the field strength of a non-abelian gauge theory like QCD, with the metric connections playing the role of the gauge potential.

We may mention that there have recently been very exciting developments in certain classes of gauge theory, which show the promise of elucidating the mechanism behind confinement (so-called *duality* properties). The reader is again directed to the electronic bulletin boards for the latest updates.

Table 6.2. The gauge particles of the Standard Model. Notice that since gluons are confined, the gluon mass is not a uniquely defined parameter.

Particle	Interaction	Mass	Electric charge
Photon	Electromagnetic	0	0
Z^0 boson	Weak (neutral current)	91 GeV	0
$W^\pm$ bosons	Weak (charged current)	80 GeV	± 1
g_i, i=1,2,..8 (gluons)	Strong	0	0

Let us continue our counting of the number of helicity states in the Standard Model. To the 90 states we found for fermions (including antiparticles) we add two for the photon and two for each of the eight gluons. They have spin 1 which means that they should each have $2s + 1$ helicity states with $s = 1$ giving three states, but since they are massless the only allowed helicities are +1 and −1, which means two states each. The W^+, W^- and Z^0 bosons are massive spin one particles, and therefore each have 3 helicity states. The physical Higgs boson is spinless and electrically neutral, with just one state. The total count thus becomes $g_{tot} = 90 + 18 + 9 + 1 = 118$. It is one of the most important tasks of modern particle theory to produce an explanation of the existence of these 118 states, Maybe there exists a grander structure which will unify these states into some more fundamental object, perhaps also having degrees of freedom in other dimensions.

6.9 Beyond the Standard Model

The Standard Model has been extremely successful in explaining all experiments and observations concerning the three fundamental forces: electromagnetism, weak interactions and strong interactions. Still, there are many unsolved fundamental problems which hint at a more complete theory, yet to be found:

- What determines the masses and couplings of all the particles of the Standard Model? Does there exist some unifying principle?
- There are indications that the coupling constants of the electromagnetic, weak and strong interactions are energy-dependent in

such a way that they would all be equal when extrapolating to an extremely high energy of around $10^{15} - 10^{16}$ GeV. Is this a coincidence, or is there a Grand Unification Theory (GUT) at this energy scale which unifies them all into one force?

- Why is the GUT scale and the mass scale of gravity, the Planck mass $1.2 \cdot 10^{19}$ GeV, so much higher than all the masses of the Standard Model? In particular, if one calculates the contribution to the Higgs boson mass that would come from quantum corrections (virtual particles) at the GUT scale, the Higgs mass, and therefore also the Z^0 and $W^\pm$ masses become huge. What is it that prevents these masses from becoming large?

- How can one describe quantum gravity? How does one unify gravity with the other three fundamental forces?

- Why is there more matter than antimatter in the Universe? In the Standard Model, there is a very tiny difference between particles and antiparticles (so-called CP violation), but this small difference does not seem to be enough to explain the baryon asymmetry of the Universe (see Section 1.3).

These questions are still unanswered, but there are great expectations that string theory and its extensions could provide some or all of the answers. While waiting for a complete theory to be constructed, there are some features which one already can deduce that such a theory should have. In particular, the problem of the disparate mass scales of Grand Unification and gravity on one hand and the Standard Model particles on the other, can be solved if there exists a whole new type of symmetry, called *supersymmetry*, in the new theory. Interestingly enough, supersymmetry also seems essential to unify gravity with the other forces.

6.9.1 Supersymmetry

Supersymmetry is a symmetry between fermions and bosons. In a supersymmetric theory there should first of all be an equal number of helicity states of fermions and bosons. Thus, to the spin-1 photon there should correspond a spin-1/2 particle called the *photino*. To the spin-1/2 fermions there should correspond spin-0 *sfermions* (*squarks* and *sleptons*). The supersymmetric partners of the Z^0, $W^\pm$, the gluons and Higgs bosons, are the spin-1/2 *zino*, *wino*, *gluino* and *higgsino*. The neutral particles of these are Majorana particles (see Section C.14) which means that they are their own antiparticles.

Secondly, if supersymmetry is unbroken the particles should have the same mass as their respective superpartners. This last property is obviously wrong (a selectron of 511 keV mass is definitely ruled out – present lower bounds from accelerator experiments are of the order of 100 GeV!). It is interesting, however, that it is possible to break supersymmetry in such a way that the Higgs mass problem is solved but the other attractive features of Grand Unification and so on are retained. For this to work, however, the lightest supersymmetric particle should not weigh more than a few hundred GeV, a fact that has initiated a dedicated search for supersymmetric particles in present and planned accelerators.

In many supersymmetric models there emerges a conserved multiplicative quantum number (called R-parity) which has the value $+1$ for ordinary particles and -1 for supersymmetric particles. This means that supersymmetric particles can only be created or annihilated in pairs. This implies in turn that a single supersymmetric particle cannot disappear by decaying into ordinary particles only. If it is heavy, it can decay into a lighter supersymmetric particle plus ordinary particles. From this it follows that the lightest supersymmetric particle is stable, since it has no allowed state to decay into.

If supersymmetry exists, it could have important consequences for cosmology. In the early Universe, the contribution from the supersymmetric fields to the effective potential energy could have been very important, perhaps driving inflation and other phenomena. Shortly after the Big Bang, when thermal energies were high compared to the supersymmetric particle masses, these particles should have been pair produced at a large rate. When the Universe then cooled and expanded, most of them decayed except the stable lightest supersymmetric particle. These particles could today still exist in the Universe as relics of the Big Bang. If they are electrically neutral they would have very weak interactions (like neutrinos), but if massive they could contribute to the *dark matter* of the Universe. Later we will show how to compute the relic density of such particles.

In supersymmetric theories, the most likely dark matter candidate is a quantum mechanical superposition, called the *neutralino χ* of electrically neutral supersymmetric fermions:

$$\chi = N_1 \tilde{\gamma} + N_2 \tilde{Z}^0 + N_3 \tilde{H}_1^0 + N_4 \tilde{H}_2^0 \tag{6.14}$$

where $\tilde{\gamma}$ is the photino, $\tilde{Z}^0$ is the zino and $\tilde{H}_1^0$ and $\tilde{H}_2^0$ are the superpartners of the two different neutral scalar Higgs particles which can be shown to be needed in supersymmetric theories. The coefficients N_i are normalized such that

$$\sum_{i=1}^{4} |N_i|^2 = 1 \tag{6.15}$$

Sometimes one defines the gaugino parameter

$$Z_g = |N_1|^2 + |N_2|^2 \tag{6.16}$$

and the higgsino parameter

$$Z_h = |N_3|^2 + |N_4|^2 \tag{6.17}$$

The mass and composition of the lightest neutralino depends on several presently unknown parameters of the supersymmetric models. The approach usually taken is to make large scans in that parameter space and to compute all the relevant quantities for each set of chosen parameters. The cosmologically interesting models generally have the lightest neutralino as a higgsino if the mass is high (above several hundred GeV up to a few TeV), and a gaugino or full mix of gaugino and higgsino for low-mass models (the lower mass limit consistent with accelerator data is somewhere between 20 and 30 GeV). At present, the neutralino is considered to be one of the most plausible candidates for the dark matter, and there are many experimental searches going on around the world to try to detect it. We shall return to this later.

6.10 Some Particle Phenomenology

In the previous sections, we encountered various particles like protons, neutrons, pions and their constituents, the quarks. The leptons: electrons, muons and τ-leptons and their respective neutrinos are as far as we know elementary. All of them played important roles in the early Universe when the temperature was very high, and many of them are still important today in various astrophysical processes.

Although the quarks and leptons enter the Standard Model on equal footing, the particle phenomenology generated by the quarks is quite a lot richer than that of the leptons. This is of course due to the fact that quarks are forced by the strong interactions to form bound states.

Since the u, d and s quarks are so much lighter than the other three, the hadrons (baryons and mesons) they form are most easily studied in accelerators, and were the first to be investigated experimentally. Before 1973, they were the only known quarks.

Since the strong interactions (in contrast to the electroweak ones) are the same for all quarks irrespective of their 'flavour', this should reflect itself in the properties of the hadrons. Indeed, the neutron and the proton are very similar particles. They have the same spin ($\hbar/2$), very similar strong interactions and the same mass within a fraction of one per cent. The only big difference is the electric charge, which we saw could be explained by the charges of the quarks. Actually, the small mass difference is also believed to be a consequence of the mass difference of the u and d quarks, plus perhaps an electromagnetic contribution generated by the difference in charges. Thus, it seems that hadrons should reflect a symmetry, meaning that exchanging u and d quarks should produce very similar hadrons (in particular, the mass should not change much).

In quantum mechanics, symmetries like the (approximate) one of exchange of quark flavours, are generated by so-called unitary operators. If we restrict ourselves to u and d quarks only, it seems that we can replace the doublet (u, d) by a linear combination

$$\begin{pmatrix} u' \\ d' \end{pmatrix} = \begin{pmatrix} \cos\theta & \sin\theta \\ -\sin\theta & \cos\theta \end{pmatrix} \begin{pmatrix} u \\ d \end{pmatrix} \tag{6.18}$$

without changing the strong interactions. This implies a symmetry, called *isospin invariance*, for strong interactions at low energy. However, the other quarks are much more massive and therefore higher flavour symmetries are not so good.

6.10.1 Estimates of Cross-Sections

The calculation of collision and annihilation cross-sections, and decay rates of particles, is an important task in particle physics. Here we will present only a brief outline of how this is done, and focus on 'quick-and-dirty' estimates which may be very useful in cosmology and astrophysics. For the local microphysics in the FLRW model, only three interactions – electromagnetic, weak and strong – between particles need to be considered. The gravitational force is completely negligible between individual elementary particles – for instance, the gravitational force between the proton and the electron in a hydrogen atom is around 10^{40} times weaker than the electromagnetic force. However, gravity, due to its coherence over long range, still needs to be taken into account through its influence on the metric. This means that the dilution of number densities due to the time dependence of the scale factor $a(t)$ has to be taken into account. In the next chapter we will see how this is done.

Let us begin with the interaction strengths. The strength of the electromagnetic interaction is governed by the electromagnetic coupling constant g_{em}, which is simply the electric charge. As usual, we take the proton charge e as the basic unit and can thus write

$$g_{em} = Qe \tag{6.19}$$

where Q is the charge of the particle in units of the proton charge (for a u-quark, for example, $Q_u = +2/3$). In our system of units,

$$\frac{e^2}{4\pi} \equiv \alpha_{em} \tag{6.20}$$

where α_{em} is the so-called fine structure constant which has the value of around $1/137$ at low energies.[4] (Usually, it is denoted just α without the subscript.) The weak coupling constant is of similar magnitude:

$$g_{weak} = \frac{e}{\sin \theta_W} \tag{6.21}$$

with θ_W the weak interaction (or Weinberg) angle, which has the numerical value $\sin^2 \theta_W \sim 0.23$. The fact that the weak and electromagnetic coupling constants are of the same order of magnitude is of course related to the fact that they are unified in the Standard Model to the 'electroweak' interaction.

---

[4] This coupling constant, as all others, depends on the energy scale, for example, the energy transfer, of the process – at 100 GeV energy $\alpha_{em} \sim 1/128$.

The coupling constant of the strong interaction, g_s, is somewhat higher. Also, it runs faster (it decreases) with energy than the electromagnetic coupling. At energies of a few GeV,

$$\alpha_s \equiv \frac{g_s^2}{4\pi} \sim 0.3 \qquad (6.22)$$

Let us look at the Feynman diagram for a simple process like $e^+e^- \rightarrow \mu^+\mu^-$ (Fig. 6.6). The amplitude will be proportional to the coupling constants at both vertices, which in this case are both equal to e. The cross-section, being proportional to the square of the amplitude, is thus proportional to $e^4 \propto (\alpha/4\pi)^2$.

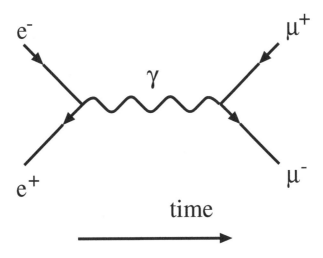

Fig. 6.6. A Feynman diagram representing the annihilation of an electron and a positron to a muon pair.

If we look at the total energy of the e^+e^- pair in the centre of momentum frame, we saw in Chapter 2 that it can be expressed as $E_{cm}(e^+)+E_{cm}(e^-) = \sqrt{s}$. Since the total momentum in this frame is zero, the four-momentum $p^\mu = (\sqrt{s},0,0,0)$ is identical to that of a massive particle of mass $M = \sqrt{s}$ which is at rest. Energy and momentum conservation then tells us that the photon in the intermediate state has this four-momentum. However, a freely propagating photon is massless, which means that the intermediate photon is virtual by a large amount. In quantum field theory one can show that the appearance of an intermediate state of virtual mass $\sqrt{s}$ for a particle with real rest mass M_i is suppressed in amplitude by a factor (called the *propagator* factor)

$$P(s) = 1/(s - m_i^2) \qquad (6.23)$$

In this case ($m_\gamma = 0$), which means a factor of $1/s$. The outgoing particles (in this case the muons) have a large number of possible final states to enter (for example, all different scattering angles in the centre of momentum frame). This is accounted for by the so-called phase space factor ϕ, which generally grows as s for large energies. For the cross-section σ

$$\sigma(e^+e^- \rightarrow \mu^+\mu^-) = const \cdot \phi \cdot \left(\frac{\alpha^2}{s^2}\right) \tag{6.24}$$

with ϕ the phase space factor. If s is large compared to m_e^2 and m_μ^2, $\phi \propto s$, and

$$\sigma(e^+e^- \rightarrow \mu^+\mu^-) \sim \frac{\alpha^2}{s} \tag{6.25}$$

This is not an exact expression. A careful calculation (see next section and Appendix D.2) gives $4\pi\alpha^2/(3s)$), but it is surprisingly accurate and often precise enough for the estimates we need in Big Bang cosmology.

Since the weak coupling strength is similar to the electromagnetic strength, the same formula is valid for, for example, $\nu_e + e \rightarrow \nu_\mu + \mu$ which goes through W exchange (see Fig. 6.7). The only replacement we need is $1/s \rightarrow 1/(s - m_W^2)$ for the propagator, thus

$$\sigma(\bar\nu_e + e^- \rightarrow \bar\nu_\mu + \mu^-) \sim \frac{\alpha^2 s}{(s - m_W^2)^2} \tag{6.26}$$

When $s \ll m_W^2$, this gives $\sigma_{weak} \sim \alpha^2 s/m_W^4$, which is a very small cross-section for, as an example, MeV energies (but notice the fast rise with energy due to the factor s). This is the historical reason for the name 'weak interaction', which as we see is really not appropriate at high energies (much larger than m_W), where the two types of cross-sections become of similar size.

Note that once one remembers the factors of coupling constants and the propagators, the magnitude of cross-sections can often be estimated by simple dimensional analysis. A cross-section has the dimension of area, which in our units means (mass)$^{-2}$. It is very useful to check that the expressions (6.25) and (6.26) have the correct dimensions.

A fermion has a propagator that behaves as $1/m$ (instead of $1/m^2$) at low energies. This means that the Thomson cross-section $\sigma(\gamma e \rightarrow \gamma e)$ at low energies $E_\gamma \ll m_e$ can be estimated to be (see Fig. 6.8)

$$\sigma_T \equiv \sigma(\gamma e \rightarrow \gamma e) \sim \frac{\alpha^2}{m_e^2} \tag{6.27}$$

6.11 Examples of Cross-Section Calculations

The foregoing simple estimates are in many cases sufficient for cosmological and astrophysical applications. However, there are cases when one would like

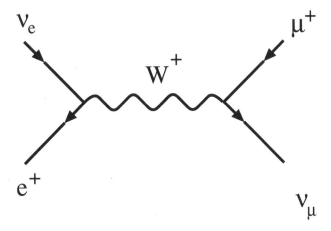

Fig. 6.7. A Feynman diagram representing the annihilation of an electron neutrino and a positron to a muon neutrino and a muon.

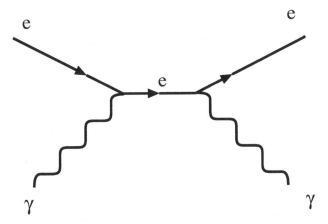

Fig. 6.8. A Feynman diagram representing the $\gamma e \rightarrow \gamma e$ process (Thomson scattering).

to have a more accurate formula (or when there are ambiguities concerning which mass scales to use for the simple estimates). Here we will provide only a couple of examples, relevant to later applications in the book. The detailed calculations require knowledge of the Dirac equation and are discussed in Appendix D. We summarize here the general framework and the main results.

6.11.1 Definition of the Cross-Section

In Appendix D, it is shown that the differential cross-section $d\sigma/dt$ for $2 \to 2$ scattering $a + b \to c + d$ is given by the expression

$$\frac{d\sigma}{dt} = \frac{|\widetilde{T}|^2}{16\pi\lambda\left(s, m_a^2, m_b^2\right)} \tag{6.28}$$

where (see Section 2.4.3) $t = (p_a - p_c)^2$ and $s = (p_a + p_b)^2$. Here $|\widetilde{T}|^2$ is the polarization-summed and squared quantum mechanical transition amplitude. The integration limits for the t variable were given in (2.60). A typical calculation (Appendix D) involves computing the matrix element in terms of s and t and carrying out the t integration to obtain the total cross-section.

In the one-photon exchange approximation, the cross-section for $e^+e^- \to \mu^+\mu^-$ is

$$\sigma(e^+e^- \to \mu^+\mu^-) = \frac{2\pi\alpha^2}{s} v \left(1 - \frac{v^2}{3}\right) \tag{6.29}$$

where the only approximation made is to neglect m_e (this is allowed, since $m_e^2/m_\mu^2 \ll 1$). Here v is the velocity of one of the outgoing muons in the centre-of-momentum frame, $v = \sqrt{1 - 4m_\mu^2/s}$. In the relativistic limit of $s \gg m_\mu^2$, $(v \to 1)$, this becomes

$$\sigma\left(e^+e^- \to \mu^+\mu^-\right)_{\text{large } s} = \frac{4\pi\alpha^2}{3s} \tag{6.30}$$

as noted previously.

6.11.2 Neutrino Interactions

For the neutrino process $\bar{\nu}_e e^- \to \bar{\nu}_\mu \mu^-$ the cross-section is found to be

$$\sigma\left(\bar{\nu}_e e^- \to \bar{\nu}_\mu \mu^-\right)_{m_\mu^2 \ll s \ll m_W^2} = \frac{g_{weak}^4 s}{96\pi m_W^4} \tag{6.31}$$

Before it was known that W bosons existed, Fermi had written a phenomenological theory for weak interactions with a dimensioned constant (the Fermi constant) G_F. The relation is

$$\frac{G_F}{\sqrt{2}} = \frac{g_{weak}^2}{8m_W^2} \simeq 1.166 \cdot 10^{-5} \text{ GeV}^{-2} \tag{6.32}$$

Using this, the cross-section can simply be written as

$$\sigma\left(\bar{\nu}_e e^- \to \bar{\nu}_\mu \mu^-\right)_{m_\mu^2 \ll s \ll m_W^2} = \frac{G_F^2 s}{3\pi} \tag{6.33}$$

We note that the cross-section rises with $s \simeq 2E_\nu m_e$ (c. f. Example 2.4.5) and thus linearly with neutrino energy. When s starts to approach m_W, the W

propagator $1/(s - m_W^2)$ has to be treated more carefully. It can be improved by writing it in the so-called Breit Wigner form

$$\frac{1}{s - m_W^2} \rightarrow \frac{1}{s - m_w^2 + i\Gamma m_W} \qquad (6.34)$$

where Γ is the total decay width (around 2 GeV) of the W. We see from this that a substantial enhancement of the cross-section is possible for $s \simeq m_W^2$. This is an example of a resonant enhancement in the s-channel. For a target electron at rest, this resonance occurs at around 6.3 PeV (the so-called Glashow resonance) If there exist astrophysical sources which produce electron antineutrinos with such high energies, the experimental prospects of detecting them would be correspondingly enhanced. Well above the resonance, this particular cross-section will again start to decrease like $1/s$, just as the electromagnetic $e^+e^- \rightarrow \mu^+\mu^-$ one does.

We should remark that the latter process, $e^+e^- \rightarrow \mu^+\mu^-$, also receives a contribution from an intermediate Z boson. At low energies this is completely negligible, but due to the resonant enhancement it will dominate near $s \simeq m_Z^2$. This is the principle behind the Z studies performed at the LEP accelerator at CERN (where all other fermion-antifermion pairs of the Standard Model can also be produced except $t\bar{t}$, which is not kinematically allowed). In a full calculation, the two contributions have to be added coherently and may in fact interfere in interesting ways, producing for example, a backward forward asymmetry between the outgoing muons.

6.11.3 The $\gamma\gamma ee$ System

By different permutations of the incoming and outgoing particles, the basic $\gamma\gamma ee$ interaction (shown in Fig. 6.8) can describe $\gamma\gamma \rightarrow e^+e^-$, $e^+e^- \rightarrow \gamma\gamma$, and $\gamma e^\pm \rightarrow \gamma e^\pm$. For $\gamma\gamma \rightarrow e^+e^-$ the result is

$$\sigma\left(\gamma\gamma \rightarrow e^+e^-\right) =$$
$$\frac{\pi\alpha^2}{2m_e^2}\left(1 - v^2\right)\left[\left(3 - v^4\right)\ln\left(\frac{1 + v}{1 - v}\right) + 2v\left(v^2 - 2\right)\right] \qquad (6.35)$$

where v now is the velocity of one of the produced electrons in the centre-of-momentum frame, $v = \sqrt{1 - 4m_e^2/s}$. Near threshold, that is, for small v the expression in square brackets can be series expanded to $2v + \mathcal{O}(v^2)$, and thus

$$\sigma\left(\gamma\gamma \rightarrow e^+e^-\right)_{\text{small } v} \simeq \frac{\pi\alpha^2}{m_e^2} \qquad (6.36)$$

In the other extreme, $v \rightarrow 1$,

$$\sigma\left(\gamma\gamma \rightarrow e^+e^-\right)_{s \gg 4m_e^2} \simeq \frac{4\pi\alpha^2}{s}\left[\ln\left(\frac{\sqrt{s}}{m_e}\right) - 1\right] \qquad (6.37)$$

Again we notice with some satisfaction that we could have guessed most of this to a fair degree of precision by the simple dimensional and vertex-counting rules. At low energy, the only available mass scale is m_e, so the factor α^2/m_e^2 could have been guessed for that reason. The factor v could also have been guessed with some more knowledge of non-relativistic partial wave amplitudes. At low energy, the $\ell = 0$ (S-wave) amplitude should dominate, and this contributes to the cross-section proportionally to v. A partial wave ℓ contributes to the total cross-section with a term proportional to $v^{2\ell+1}$.[5] At high energy, when m_e can be neglected, the dimensions have to be carried by s. Only the logarithmic correction factor in (6.37) could not have been very easily guessed.

We see from these formulae that the $\gamma\gamma \to e^+e^-$ cross-section rises from threshold to a maximum at intermediate energies and then drops roughly as $1/s$ at higher energy (see Fig. 6.9).

The results for the reverse process $e^+e^- \to \gamma\gamma$ are of course extremely similar. Now, the process is automatically always above threshold. For $v \to 0$ (with v now the velocity of one of the incoming particles in the cm-system, still given by the formula $v = \sqrt{1 - 4m_e^2/s}$), the flux factor $\sim 1/v$ in (D.7) diverges. Since the outgoing photons move away with $v = c = 1$ there is no partial-wave suppression factor, and we can thus expect the cross-section at low energy to behave as

$$\sigma \left(e^+e^- \to \gamma\gamma\right)_{\text{low energy}} \sim \frac{\alpha^2}{vm_e^2} \tag{6.38}$$

and the high-energy behaviour by the same formula, with m_e^2 replaced by s (and possibly a logarithmic factor). These expectations are borne out by the actual calculation, which gives

$$\sigma \left(e^+e^- \to \gamma\gamma\right) = \frac{\pi\alpha^2 \left(1 - v^2\right)}{2vm_e^2} \left[\frac{3 - v^4}{2v} \ln \left(\frac{1+v}{1-v}\right) - 2 + v^2\right] \tag{6.39}$$

Note the similarity with (6.35)

As the final example, we consider Compton scattering $\gamma + e^- \to \gamma + e^-$. Usually, one then has an incoming beam of photons of energy ω which hit electrons at rest. For scattering by an angle θ with respect to the incident beam, the outgoing photon energy ω' is given by energy-momentum conservation to be

$$\omega' = \frac{m_e\omega}{m_e + \omega \left(1 - \cos\theta\right)} \tag{6.40}$$

In this frame, the unpolarized differential cross-section, as first computed by Klein and Nishina, is

$$\frac{d\sigma}{d\Omega} = \frac{\alpha^2}{2m_e^2} \left(\frac{\omega'}{\omega}\right)^2 \left[\frac{\omega'}{\omega} + \frac{\omega}{\omega'} - \sin^2\theta\right] \tag{6.41}$$

Integrated over all possible scattering angles this gives the total cross-section

---

[5] We see from (6.29) that in the case of $e^+e^- \to \mu^+\mu^-$ the S-wave dominates at low energy, but when $v \to 1$, the P-wave contribution is $1/3$.

$$\sigma(\gamma + e \rightarrow \gamma + e) = \frac{\pi\alpha^2 \left(1 - v\right)}{m_e^2 v^3} \times$$

$$\left[\frac{4v}{1+v} + \left(v^2 + 2v - 2\right) \ln\left(\frac{1+v}{1-v}\right) - \frac{2v^3 \left(1+2v\right)}{\left(1+v\right)^2}\right] \tag{6.42}$$

where v is now the incoming electron velocity in the centre of momentum frame, $v = (s - m_e^2)/(s + m_e^2)$. If one expands this result around $v = 0$, one recovers the Thomson scattering result

$$\sigma_{\text{Thomson}} = \frac{8\pi\alpha^2}{3m_e^2} \tag{6.43}$$

and the large-s, so-called Klein Nishina regime gives

$$\sigma_{\text{KN}} = \frac{2\pi\alpha^2}{s} \left[\ln\left(\frac{s}{m_e^2}\right) + \frac{1}{2}\right] \tag{6.44}$$

We see that for photon energies much larger than m_e – that is, in the Klein Nishina regime – the Compton cross-section falls quite rapidly.

These formulae have many applications. In the classical Compton scattering situation, the outgoing photon energy is always less than the incoming one. Thus, energetic photons travelling through a gas of cold electrons will be 'cooled' by Compton scattering. In other cases (for example for the cosmic microwave background radiation passing through a galaxy cluster with hot gas) energetic electrons may transfer energy to photons, thereby 'heating' them. This is sometimes called *inverse Compton scattering* in the literature (although if we just express the energies in terms of s, it is the same formula which applies in the two cases).

When computing actual numbers for the cross-sections (which should have the dimensions of area) in our units, a useful conversion factor is

$$1 \text{ GeV}^{-2} = 0.389 \cdot 10^{-27} \text{ cm}^2 \tag{6.45}$$

In Fig. 6.9 the numerical results are summarized. The cross-sections are shown (in cm^2) for $\gamma\gamma \rightarrow ee$, $ee \rightarrow \gamma\gamma$ and $\gamma e \rightarrow \gamma e$ as a function of the available energy in the c.m.s., in units of the electron mass:

$$y = \frac{\sqrt{s} - \sqrt{s_{\text{min}}}}{m_e} \tag{6.46}$$

where $\sqrt{s_{\text{min}}} = 2m_e$ for $\gamma\gamma \rightarrow ee$ and $ee \rightarrow \gamma\gamma$, and $\sqrt{s_{\text{min}}} = m_e$ for Compton scattering $\gamma e \rightarrow \gamma e$. We see in this figure the different behaviours at low energy (small y) already discussed, but that they show a similar decrease at high energy.

Another process of great astrophysical importance is that of *bremsstrahlung*. This is the emission of photons from charged particles when they are accelerated or decelerated. If this acceleration is due to circular motion in a magnetic field, the term *synchrotron radiation* is used. Through these processes (unlike Compton scattering) the number of photons can change. This

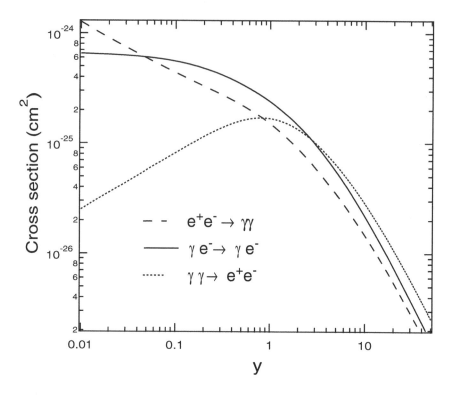

Fig. 6.9. The cross-sections (in cm^2) for photon-photon annihilation, $e^+e^- \to \gamma\gamma$ and Compton scattering as a function of the scaled available centre of momentum energy y.

is needed if thermal equilibrium is to be maintained, since the number density of photons depends strongly on temperature. Most of the produced photons have very low energy (long wavelength). If fast electrons pass through a region where synchrotron radiation and bremsstrahlung occur, these low-energy photons may have energy transferred to them through the inverse Compton process. This may explain the observations of very high-energy photons in active galactic nuclei (see Chapter 13).

(For a detailed discussion of these and other QED processes, see standard textbooks on quantum field theory, for example, [22].)

6.12 Processes Involving Hadrons

Since protons and neutrons are among the most common particles in the Universe, it is of course of great interest to compute processes where these and other hadrons (such as pions) are involved. This is, however, not easy to do from first principles. The reason that in the previous section we could compute weak and electromagnetic processes so accurately is that we could use perturbation theory (as summarized, for example, in the Feynman diagrams). The expansion parameter, the electroweak gauge coupling constant g or rather $\alpha_{ew} = g^2/(4\pi) \sim 10^{-2}$, is small enough such that a lowest-order calculation is sufficient to obtain very accurate results.

In QCD, we also have a coupling constant α_s, but it is large (~ 0.2 at scales of order 10 GeV). The 'running' of the coupling constant with energy scale means that it is smaller than this at high energies but larger at small energies. The energy scale is set, for example, by the energy or momentum transfer Q ($Q^2 \equiv -t$ with t the Mandelstam variable introduced in Section 2.4) in the process. So, for processes with large Q^2, we should be able to use perturbative QCD, although the results may not be as accurate as those of QED due to the importance of higher-order corrections. At low energies, say 1 GeV and below, the QCD coupling becomes of the order unity and perturbation theory breaks down. In this nonperturbative regime we have to rely on empirical methods or eventually on large computer simulations, where QCD is solved as accurately as possible by being formulated as a field theory on a lattice. Despite some early optimism, progress in this field has, however, been slow during the last few years.

For processes such as proton-proton scattering at low energies, the picture of strong interactions being due to the exchange of gluons breaks down. Instead one may approximate the exchange force as being due to pions and other mesons with surprisingly good results (this is in fact what motivated Yukawa to predict the existence of pions). If one wants to make crude approximations of the strong interaction cross-section in this regime, $\sigma_{strong} \sim 1/m_\pi^2$ is not a bad estimate.

In the perturbative regime at high Q^2, the scattering, for example, of an electron by a proton ('deep inelastic scattering') may be treated by the successful so-called parton model. According to this description, the momentum of a hadron at high energy is shared between its constituents. The constituents are of course the quarks that make up the hadron (two u and one d quarks in the case of the proton). These are called the *valence quarks*. In addition, there may be pairs of quarks and antiquarks created through quantum fluctuations at any given time. The incoming exchange photon sent out from an electron may happen to hit one of these *sea quarks*, which will therefore contribute to the scattering process (see Fig. 6.10).

Since the partons interact with each other, they can share the momentum of the proton in many ways. Thus, there will be a probability distribution, $f_i(x, Q^2)$, for a parton of type i (where i denotes any quark, antiquark or gluon) to carry a fraction x of the proton momentum. These functions cannot be calculated from first principles. However, if they are determined (by guess or by experimental information from various processes) at a particular value of Q_0^2, then the evolution of the structure functions with Q^2 can be predicted. This analysis, first performed by Altarelli and Parisi, gives rise to a predicted

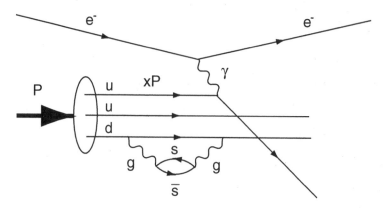

Fig. 6.10. Deep-inelastic scattering of an electron by a proton in the parton picture of the proton, valid at high energy and momentum transfers. The exchange photon can hit any of the constituents, and also one of the virtual sea quarks (here an $s\bar{s}$ quark pair created by a virtual gluon). The total scattering cross-section is a sum over all constituents and an integral over x, where x is the fraction of the proton momentum P carried by an individual constituent. The hit parton (and also the 'spectator quarks') develop a cascade which ultimately ends up in hadrons. This 'jet' of particles is a distinct signature of perturbative QCD processes which has been experimentally verified to exist, and agree extremely well with the QCD predictions.

variation of the deep inelastic scattering probability with Q^2 (so-called scaling violations) which has been successfully compared with experimental data.

With the QCD parton model, we can now compute many electromagnetic and weak processes, including when hadrons are involved. For instance, the neutrino proton scattering cross-section will be given by the scattering of a neutrino on a single quark and antiquark. This calculation is easily done in a way similar to how we computed the $\bar{\nu}_e + e^- \rightarrow \bar{\nu}_\mu + \mu^-$ cross-section. The only change is that the contributions from all partons have to be summed over, and an integral of x performed.

Example 6.12.1 Give an expression for the electromagnetic cross-section $p + p \rightarrow \mu^+\mu^-$ in the QCD parton model. (This is called the Drell-Yan process.)

Answer: The fundamental process must involve charged partons (since we neglect the weak contributions), $q + \bar{q} \rightarrow \gamma^* \rightarrow \mu^+\mu^-$, with a quark taken from one of the protons and an antiquark from the other proton. The momentum transfer in the process can be taken to be $Q^2 = \hat{s}$, where $\hat{s} = (p_{\mu^+} + p_{\mu^-})^2$. We know from (6.30) that the parton level cross-section is $4\pi\alpha e_q^2/3\hat{s}$ (where we have taken into account that the quark charge e_q is not the unit charge). Since the parton from proton 1 carries the fraction x_1 and that from proton 2 x_2 of the respective parent proton, $\hat{s} = x_1 x_2 s$, with $s = (p_1 + p_2)^2$. The total cross-section for producing a muon pair of momentum transfer $\hat{s}$ is thus

$$\frac{d\sigma}{d\hat{s}} = \frac{4\pi\alpha^2}{3\hat{s}} k_c \times$$

$$\sum_q e_i^2 \int_0^1 dx_1 \int_0^1 dx_2 \left[f_q(x_1, \hat{s}) f_{\bar{q}}(x_2, \hat{s}) + f_{\bar{q}}(x_1, \hat{s}) f_q(x_2, \hat{s}) \right] \delta(\hat{s} - x_1 x_2 s)$$

Here k_c is a colour factor, which takes into account that for a given quark of a given colour, the probability to find in the other proton an antiquark with a matching (anti-) colour is $1/3$. Thus, in this case $k_c = 1/3$. In the reverse process, $\mu^+ + \mu^- \rightarrow q + \bar{q}$, all the quark colours in the final state have to be summed over (each contributes to the cross-section), so in that case $k_c = 3$.

6.13 Vacuum Energy Density

The absolute energy concept is not relevant in classical mechanics as the dynamics of a system depends only on energy differences, e.g. $F = -\frac{dU}{dx}$, which remain unchanged for the transformation $U \rightarrow U + U_0$. However, quantum mechanical systems are subject to a lowest possible energy level: the *vacuum energy* . For example, the energy levels in a quantum harmonic oscillator are $E_n = (n + \frac{1}{2})\hbar\omega$. Thus, the zero-point energy of the $n = 0$ ground state is finite, i.e. the system cannot be completely at rest. This is a consequence of the uncertainty principle which states that the location and momentum of a particle cannot be known simultaneously. The generalization to quantum field theory implies that a relativistic field may be viewed as a collection of harmonic oscillators. Thus, a massless scalar field has a vacuum energy which is given by the sum of all the frequency modes:

$$E_0 = \sum_i \frac{1}{2}\hbar\omega_i \tag{6.47}$$

The sum is performed by putting the system in a box of volume $V = L^3$. Boundary conditions are imposed such that $L = \lambda_i \cdot n_i$, for some integers n_i, or in terms of the wave vector $\mathbf{k}$, we can write $\mathbf{n} = \frac{L\mathbf{k}}{2\pi}$. The number of frequency modes in the interval $(\mathbf{k}, \mathbf{k} + d\mathbf{k})$ is therefore $\frac{L^3 d^3\mathbf{k}}{(2\pi)^3}$, and the sum of the energy modes becomes:

$$E_0 = \frac{1}{2}\frac{\hbar L^3}{(2\pi)^3} \int \omega_k d^3\mathbf{k} \tag{6.48}$$

Letting $L \rightarrow \infty$, the vacuum energy density is obtained by performing the integral including all the energy modes for which we are confident in our theory, i.e. up to some value k_{max}:

$$\rho_{vac} = \lim_{L\to\infty} \frac{E_0}{L^3} = \frac{1}{2}\frac{\hbar}{(2\pi)^3} \int_0^{k_{max}} k \cdot 4\pi k^2 dk = \frac{\hbar k_{max}^4}{16\pi^2}, \tag{6.49}$$

which diverges for $k_{max} \to \infty$. However, it is believed that beyond the Planck energy, $E_{Pl} = 1.2 \cdot 10^{19}$ GeV, conventional field theory breaks down, i.e. $k_{max} \sim E_{Pl}/\hbar$. Hence, $\rho_{vac} \approx 10^{74}$ GeV$\hbar^{-3} = 10^{92}$ g/cm^3. In Section 4.1, it was shown that the configuration with no matter or radiation contribution to the the energy momentum tensor, the vacuum energy density, could be described with an equation of state parameter $\alpha_{vac} = -1$, i.e. $p_{vac} = -\rho_{vac}$, hence the association with the cosmological constant, $\rho_{vac} = \rho_\Lambda \equiv \Lambda/8\pi G$. Our mere existence rules out such a large vacuum energy density at the present epoch, since it would induce an expansion rate of the Universe inconsistent with the existence of any structure. This is known as the cosmological constant problem. Astronomical observations suggest that there is component of the Universe with negative pressure but then the cut-off energy is only at the 10^{-3} eV, more than 30 orders of magnitude below the Planck energy! We see that the effect of the vacuum energy is proportional to $\hbar$, i.e., it is not a problem for classical physics, where $\hbar \to 0$. Probably a solution to the cosmological constant problem has to come from a theory that can unify classical general relativity with quantum mechanics, such as string theory.

In the 1980s when supersymmetry was proposed, it was argued that a cancellation of the cosmological constant was possible as bosons and fermions would yield contributions with opposite signs to the integral in (6.49). However, as supersymmetry is clearly broken in the Universe of today, no theoretical explanation exists for the measured value of the cosmological constant.

6.14 Summary

- The Standard Model contains quarks and leptons.
- Quarks are confined to live within hadrons. The colour charge being non-zero, they cannot be free outside. However, at large momentum transfers, quarks effectively behave as free particles.
- Quarks have electric charges equal to $+2/3$ (for the three up-type quarks) and $-1/3$ (for the three down-type quarks). Leptons have charge -1 (for all the three electron-like leptons) and 0 (for the three neutrinos).
- Fundamental quark states are either composed of three quarks (baryon) or a $q\bar{q}$ (meson) states.
- The forces of the Standard Model can be obtained by postulating certain space-time dependent gauge symmetries.
- The potential of the fields of the Standard Model is such that $W^\pm$ and Z^0 become massive.
- The supersymmetric extension of the Standard Model is a plausible new framework, which will be testable at the new accelerator at CERN.

- To compute the order of magnitude of a cross-section for a pair of Standard Model particles, it is usually a good approximation to use the momenta and charges of the external particles.
- One useful quantity when computing cross-sections is $1 \text{ GeV}^{-2} = 0.389 \cdot 10^{-27} \text{ cm}^2$.
- Our present field theory does not tell us anything about the value of the cosmological constant.

6.15 Problems

6.1 Show that with ∂_μ replaced by the covariant derivative $D_\mu \equiv \partial_\mu - iA_\mu(x)$ and the rule $A_\mu(x) \to A_\mu(x) + \partial_\mu \alpha(x)$, the Lagrangian (6.5) is invariant under local gauge transformations $\phi(x) \to e^{i\alpha(x)}\phi(x)$.

6.2 In the minimal supersymmetric extension of the Standard Model (called the MSSM), the 'ordinary' particles are the same as in the Standard Model, with the exception of five Higgs particles (three neutral and two charged) instead of only one (plus of course the supersymmetric partners of all ordinary particles). Compute the total number of helicity states in the MSSM.

6.3 Give a quick estimate of the size of the following cross-sections: (a) $e^+ + e^- \to \gamma + \gamma$, (b) $u + \bar{u} \to g + g$, (c) $u + \bar{d} \to \nu_\mu + \mu^+$, (d) $u + \bar{u} \to d + \bar{d}$.

6.4 Derive (D.9).

7 Phase Transitions

7.1 Introduction

With the advent of relativistic quantum field theory as the most accurate description of the fundamental particles and their interactions at energy scales all the way up to the Grand Unification scale of around 10^{15} GeV, a number of interesting consequences for cosmology were soon identified. One of them, which we shall treat in Chapter 10, is the possibility that the vacuum energy of some fields generated an extremely fast expansion (inflation) of the very early Universe. Another is the observation that since the Universe started out from a very hot state and subsequently cooled, there could have been a whole series of phase transitions in the primordial 'soup', just like water will successively exist in the states of vapour, liquid and solid when cooled from a temperature above the boiling point to below the freezing point under standard conditions of temperature and pressure.

Although some 'defects' formed during phase transitions such as monopoles and domain walls are excluded because they would overclose the Universe, there are others – strings and textures – that may have formed after inflation (if there was inflation – this is an attractive but not compulsory possibility). They could have had striking effects on structure formation, and have been searched for in balloon-borne and satellite experiments on the microwave background. No signal has been found so far, which means that if cosmic defects exist, they must play a subdominant role. Anyway, the study of cosmological phase transitions is a fascinating branch of cosmology which has strong ties both to particle and condensed matter physics.

7.2 Phase Transitions in Condensed Matter

In statistical physics, one generally distinguishes between *first order* or *discontinuous* and *second order* or *continuous* phase transitions. In a first order phase transition (such as the boiling of water), there is latent heat involved, and the phase transition often proceeds through the nucleation of bubbles. In continuous phase transitions the change of phase is less dramatic, and when it takes place the regions of the new phase become larger and larger, as quantified by the so-called *correlation length*.

In any type of phase transition one can, for the other external parameters fixed (pressure, volume, magnetic field...), identify a *critical temperature T_c*. Near the critical temperature many of the quantities can be expanded in the reduced temperature

$$\epsilon = \frac{T - T_c}{T_c} \tag{7.1}$$

where, for example, the behaviour of the correlation length ξ as one approaches T_c from above in a second order phase transition is

$$\xi = \xi_0 \epsilon^{-\nu} \tag{7.2}$$

where $\nu > 0$ and ξ_0 is a constant that depends on the details of the interactions.

Obviously, it would be useful to be able to estimate ν and other *critical exponents*. In fact, very powerful theoretical methods have been developed to analyse the behaviour of the system as the length scale is successively changed: the *renormalization group* approach (see [16]). A more intuitive, but less powerful, method was developed by Landau in the 1950s. To analyse a system, Landau's prescription was to write down an *effective Hamiltonian* that should reflect the symmetries and essential dynamics of the system.

7.2.1 The Landau Description of Phase Transitions

We shall give an example of Landau's method that will also prepare us for the description of cosmic string formation in the early Universe. The first task is to identify some quantity that can describe the state of the system we are considering, and that can tell in which phase a particular portion of the system resides. A quantity with these properties is called an *order parameter*. For example, in liquid ^{4}He a phase transition to a superfluid state can occur at very low temperatures (of the order of 2 K). In the low-temperature phase the helium atoms form a Bose Einstein condensate which can be described by a coherent, macroscopic 'wave function' $\Psi(\mathbf{r}) = \eta e^{i\phi(\mathbf{r})}$, where η^2 is related to the density of the superfluid. (It is different from an ordinary Schrödinger wave function since the superposition principle is not valid.) At high temperature the motion of the atoms is random, so that $\eta = 0$. At lower temperatures, below the critical temperature T_c, the condensate develops, and $\eta \neq 0$.

To describe the dynamics of this simple system, we choose $\Psi(\mathbf{r})$ as the order parameter, and notice that since Ψ has the expectation value zero in the high-temperature phase and non-zero below the critical temperature, the effective Hamiltonian near T_c should be very similar to that of the Higgs field that we encountered in Section 6.6. Ψ is a complex field, and we want the system to be described by a real-valued, globally $U(1)$-invariant Hamiltonian ($U(1)$ being the group of multiplications of complex numbers $e^{i\alpha}$ on the unit circle which we encountered in Section 6.6, $\Psi \rightarrow e^{i\alpha}\Psi$). In other words, the free energy of the system should not depend on the overall phase of the wave function. The simplest possible terms are then uniquely given by

$$\mathcal{H}_{eff} = |\nabla\Psi|^2 + b(T)|\Psi|^2 + c(T)|\Psi|^4 \tag{7.3}$$

As we saw in Section 6.7, spontaneous symmetry breaking occurs if the co-efficient b of the second-order term is negative. Symmetry breaking at $T=T_c$ is thus guaranteed by the assumption

$$b(T) = b' \cdot (T - T_c) \tag{7.4}$$

with $b' > 0$.

Above the phase transition temperature, the minimum of the effective potential is achieved by $|\nabla\Psi| = |\Psi| = 0$, whereas below T_c it is given by $|\nabla\Psi| = 0$, $|\Psi|^2 = v^2$, with $v = \sqrt{-b/2c}$. Thus the state of lowest energy, the *vacuum state* per definition, has both constant phase θ and magnitude v. However, any value $\theta = $ const is as good as any other; the vacuum state is not unique, in contrast to the unique vacuum state $\Psi = 0$ above the phase transition.

Here we note a very important property of a system described by (7.3). It has a correlation length ξ associated with it, which can be shown to vary with temperature in accordance with the rule $\xi(T) = \xi_0\epsilon^{-\nu}$, where $\nu \sim 0.5$. The significance of the correlation length is that the field is not correlated for larger distances than ξ, so that when the system cools and the vacuum expectation value of Ψ develops, regions separated by a distance larger than ξ will obtain different values of θ. Loosely speaking, there is no way for one corner of the system to know what is going on at another corner, if the two are separated by more than a correlation length. Fig. 7.1 shows what can happen in this case as the system cools and the regions of different phase come close together. There can be a 'clash' of phases, or a *defect*, which can be stable over long periods of time if the structure of the effective potential has certain properties.

If we imagine that we have a region such as the one shown in Fig. 7.1, and we compute the change of phase $\Delta\theta$ of Ψ as we go one turn around the point C in a closed loop, we find it to be an integer multiple of 2π (this is required to obtain a single-valued Ψ). Thus, $\Delta\theta = 2n\pi$ with n an integer. Let us say that we have one unit of winding number (that is, $n = 1$). If we now gradually shrink the size of the loop, the winding number will not change unless we cross a singularity of undefined phase. This is so because the winding number can be written as a line integral of the gradient of the phase, and is thus an analytic mapping of the phase field $\theta(\mathbf{r})$ onto the integers. A continuous change of line contour cannot generate a discontinuous change of the winding number unless a singularity is crossed. However, it must then be possible to shrink the size of the loop to zero, without changing the winding number. In this way, we have located a point in space where the phase has to be undefined, and the only possibility is that $v = 0$ at this point and

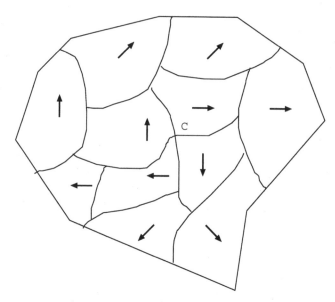

Fig. 7.1. The phase in each region is represented by the direction of the arrow. A topological defect forms when there is a 'clash' of directions in neighbouring regions, here most apparent at the point C.

therefore the phase is undefined. Thus, at this point the field sits in the 'wrong', symmetric, vacuum at $\Psi = 0$.

By the same kind of reasoning, we realise that there has to be a line-like set of singular points, a *topological string*, because otherwise we could continuously deform the loop to pass beside the singular point. $U(1)$ strings like these can only be closed or infinite (that is, in a real condensed matter system, ending at a boundary). They are very stable, since the 'unwrapping' of the string would mean that the field has to coherently rearrange itself on a macroscopic scale. The energetics of a string can be easily investigated by dimensional arguments. Since the core of the string represents the wrong vacuum with $v = 0$, having a large core increases the energy. On the other hand, if the core is made smaller the gradients of the field become larger, which also increases the energy (according to (7.3)), so it is not difficult to understand that there is a core radius which minimizes the energy per unit length.

When strings form during a continuous phase transition, one may expect a density of the order of one string per coherence volume from this reordering mechanism. One point may need clarification here. We have said that the correlation length goes to infinity at the critical temperature T_c. Then the strings would be separated by the infinitely large correlation length: that is, formed at a negligible density! In fact this is not so, for several reasons. First, for already formed strings to unwind, thermal energy is needed to overcome the potential barrier. When the temperature is below the so-called Ginzburg temperature, defects are 'frozen in' and do not evolve further. Secondly, in a real situation the cooling time past the transition may be shorter than the dynamical time scale needed for the string network to rearrange itself according to the new temperature (so there is a 'freeze-out' of the network strings).

The formation of strings has been experimentally verified in superfluid ^{3}He (the subject of the 1996 Nobel prize in physics) where the microscopic description is more complicated (pairs of ^{3}He atoms bind together similar to Cooper pairs in superconductivity) but the effective Hamiltonian contains a vacuum manifold with the same $U(1)$ structure as discussed above. It is reassuring that the experimental results verify the existence of string-like defects and the formation mechanism we have just presented. This research produces very interesting analogies between condensed matter physics and cosmology.

7.3 Domain Walls, Strings and other Defects

The idea that the vacuum structure of the fundamental fields in nature depends on temperature and maay therefore give rise to a series of phase transitions in the evolution of the Universe has been very fruitful. Not only can one obtain mechanisms for generating structure in the Universe – the absence of certain types of topological defects also places constraints on the underlying theory. Let us remind ourselves of the Lagrangian that we used before to illustrate the Higgs mechanism in particle physics as well as the formation of strings in condensed matter systems.

To make the discussion rather more general, let us introduce an arbitrary number of real components of the field ϕ. Another way to express this is that we have a whole set of real-valued fields $\phi_i(\mathbf{x})$, $i = 1 \ldots N$ which interact in such a way that it is natural to group them all together in a column vector

$$\phi = \begin{pmatrix} \phi_1(\mathbf{x}) \\ \cdot \\ \cdot \\ \phi_N(\mathbf{x}) \end{pmatrix}$$

The $O(N)$ model is defined as being invariant under orthogonal rotations of the 'coordinates' $\phi_i(\mathbf{x})$, for fixed $\mathbf{x}$. A simple way to guarantee this rotation invariance (in 'internal', that is, field space) is to write the potential energy density for the fields as functions of the 'squared length'

$$|\phi|^2 = \phi^T \phi$$

$$V(\phi) = -\mu^2 \phi^T \phi + \frac{\lambda}{2} |\phi|^4 + \text{const.} \tag{7.5}$$

Here we only consider **x**-independent (global) transformations of the fields, so for the kinetic term we just choose $\partial_\mu \phi^T \partial^\mu \phi$. The Lagrangian for our $O(N)$ model is thus

$$\mathcal{L}(x) = \partial_\mu \phi^T \partial^\mu \phi + \mu^2 \phi^T \phi - \frac{\lambda}{2}(\phi^T \phi)^2 + \text{const} \qquad (7.6)$$

The sign of μ, which we saw can be temperature-dependent, determines whether the vacuum is trivial ($\phi_i = 0$ for all i), or given by the manifold in the space spanned by the ϕ_i (field space) defined by the equation

$$\phi^T \phi = v^2$$

with

$$v^2 = \frac{\mu^2}{\lambda}$$

For the simplest case $N = 1$, the equation for the vacuum state (denoting the only component ϕ_i by ϕ) is

$$\phi^2_{vac} = v^2 \qquad (7.7)$$

which has the two solutions $\phi_{vac} = \phi_+ = v$ and $\phi_{vac} = \phi_- = -v$. Suppose now that after the phase transition, one region of the Universe went into the state ϕ_+ and a causally disconnected region into the state ϕ_-. As the horizon expanded with the Hubble expansion, the two regions would have met with a 'mismatch' between the two values of ϕ_{vac}. A discontinuous change from ϕ_+ to ϕ_- would require infinite energy (because the gradient part of the energy density would diverge). It would also require (almost) infinite energy to transform one state into the other (because of the energy barrier separating the two states). Therefore the configuration of lowest energy will involve a continuous change of the field from ϕ_+ to ϕ_- over a length scale δ. This configuration is called a *domain wall*. Suppose we have a wall in the yz plane. Then the equations of motion (6.4) for the Lagrangian (7.6) take the form

$$\frac{\partial^2 \phi}{\partial x^2} - \lambda \phi(\phi^2 - v^2) = 0 \qquad (7.8)$$

with boundary conditions $\phi(x = -\infty) = \phi_- = -v$, $\phi(x = +\infty) = \phi_+ = v$. This equation has the 'kink' solution

$$\phi_{kink} = v \tanh\left(\frac{x}{\delta}\right) \qquad (7.9)$$

with the width parameter δ given by

$$\delta = \sqrt{\frac{2}{\lambda v^2}} \qquad (7.10)$$

Since the field ϕ leaves the vacuum manifold in the region characterized by δ, the surface energy density σ in such a domain wall is of the order of λv^4 (see (7.6) for $\phi = 0$) times the width δ, thus $\sigma \sim \sqrt{\lambda} v^3$. A wall dividing the present Universe in half: that is, of area proportional to H_0^{-2}, would give a contribution to Ω of around 10^{13} for $v \sim 1$ TeV! This is, of course, in gross violation of observational bounds $\Omega \sim 1$, which puts constraints on models of, for example, supersymmetry breaking at these mass scales. One cannot allow models where the scalar fields have a potential corresponding to (7.6) with $N = 1$.

For $N = 2$, the situation is a little more interesting. As we noted in Section 6.6, two scalar fields coupled in this way can be equally described as one complex field. Since electromagnetism involves this type of complex field (with somewhat different, local, that is, space-time dependent $U(1)$ transformations), it can be expected to appear generically when gauge symmetries are broken. This is of course the model we have already encountered in Section 7.2.1, which we saw allows string solutions.

The equation for the vacuum manifold when $N = 2$ is

$$\phi_1^2 + \phi_2^2 = v^2 \qquad (7.11)$$

which defines a circle in the (ϕ_1, ϕ_2) field space. When one turn around this circle in field space is performed as one goes around a closed loop in real space, a string is present. The mass density of such a string is not at all catastrophically large (especially not for local $U(1)$ strings – gauge strings). The mass density μ per unit length of string is of the order of $\sqrt{\lambda}v^2$, which is large enough to seed structure in the Universe but small enough not to violate bounds on the total energy density. For $v \sim E_{GUT} \sim 10^{15}$ GeV, this means a mass density of 10^{16} kg/cm.

Strings have very interesting gravitational properties. The Einstein equation for a straight infinite string in the z direction with energy momentum tensor

$$T^{\mu\nu} = \mu\delta(x)\delta(y) \begin{pmatrix} 1 & 0 & 0 & 0 \\ 0 & 0 & 0 & 0 \\ 0 & 0 & 0 & 0 \\ 0 & 0 & 0 & -1 \end{pmatrix} \qquad (7.12)$$

can be solved, and gives the metric (in cylindrical coordinates (ρ, ϕ, z))

$$ds^2 = dt^2 - dz^2 - d\rho^2 - (1 - 4G\mu)^2 \rho^2 d\phi^2 \qquad (7.13)$$

By making the variable substitution $\phi \to (1 - 4G\mu)\phi$ this becomes the Minkowski metric, but with the angle ϕ only running from 0 to $2\pi(1 - 4G\mu)$. This is called a conical space, and leads to various interesting effects such as double images of quasars behind a string, fluctuations in the microwave background, and a compression of matter where a moving string is passing.

$N = 3$ can be shown to correspond to monopole defects, which we said are more or less excluded because of their large energy density. $N = 4$ and higher gives a vacuum manifold that has a non-trivial structure, but no stable defects are formed. (The 'knots' that can appear, have the possibility of unwinding because of the extra internal field dimensions.) However, even these unstable *textures* can have cosmologically interesting effects on, for example, structure formation, and will be searched for in the next generation of cosmic microwave background experiments.

7.4 Summary

- The Universe may have gone through a series of phase transitions.
- The Landau description of phase transitions utilizes an order parameter with an effective Hamiltonian similar to that of the Higgs field. The coefficients are temperature-dependent, which can explain the occurrence of different phases.
- Cosmic strings and textures are defects associated with a nontrivial vacuum manifold of some set of scalar fields. They may play a role for structure formation in the Universe. Domain walls and monopoles generally contribute too highly to the energy density of the Universe, and are thus excluded (or diluted to negligible density by inflation).

8 Thermodynamics in the Early Universe

8.1 Introduction

We shall now explore the consequences of the Hot Big Bang model based on the FLRW metric according to which the Universe started out from a much hotter and more compressed state than that which we observe today. In particular, we shall see how we can follow the evolution of the number densities of different kinds of particles and radiation throughout the history of the early Universe. This will enable us to derive certain predictions for the abundance of various light elements, photons and neutrinos that should be present in the Universe we observe today. The agreement of the calculations with observations of the concentrations of helium, deuterium and lithium is one of the cornerstones of the Big Bang scenario.

We shall again choose units in such a way that the equations become as simple as possible. In particular, as usual we set

$$c = \hbar = 1 \tag{8.1}$$

meaning, as we saw in Section 2.4.1, that all quantities with dimension can be expressed in terms of mass energy, eV (or, more often, MeV or GeV). To obtain results expressed in the usual SI units one then has to reinsert appropriate powers of $\hbar$ and c in the final expressions. Which powers of $\hbar$ and c to use can, for example, be determined by dimensional analysis (Section 2.4.2).

For instance, to write Newton's gravitational constant in terms of the Planck mass, we first look at the SI value

$$G = 6.673 \cdot 10^{-11} \ \mathrm{m^3 kg^{-1} s^{-2}} \tag{8.2}$$

We then try to write this as a product $\hbar^\alpha c^\beta m_{Pl}^\gamma$, where the constants α, β and γ can be determined by demanding that this combination has the same dimension as the expression for G. (Problem 8.1.) This then defines a mass, the Planck mass, such that

$$G = \frac{\hbar c}{m_{Pl}^2} \tag{8.3}$$

with the numerical value

$$m_{Pl} = \sqrt{\frac{\hbar c}{G}} = 1.221 \cdot 10^{19} \text{ GeV}/c^2 \tag{8.4}$$

or, again within the convention of putting $c = 1$,

$$m_{Pl} = 1.221 \cdot 10^{19} \text{ GeV} \tag{8.5}$$

This is a huge mass scale (we saw in section 6.2 that the heaviest elementary particle found today, the t quark, has a mass of 'only' 175 GeV). When we discuss thermodynamics in the early Universe, it is also convenient to measure temperature in units of energy or mass, which means that we set the Boltzmann constant $k_B = 1$ (a useful conversion factor is then 1 MeV $=$ $1.1605 \cdot 10^{10}$ K).

We consider only the Friedmann-Lemaître-Robertson-Walker metric, from which the Friedmann equation (4.16) followed from the time-time component of the Einstein equation:

$$H^2(t) + \frac{k}{a^2} = \frac{8\pi G\rho}{3} \tag{8.6}$$

where, as usual, the Hubble parameter is $H = H(t) = \dot{a}(t)/a(t)$, with $H(now) \equiv H_0 = h \cdot 100 \text{ km s}^{-1} \text{ Mpc}^{-1}$, and where according to present observational data $h \sim 0.7 \pm 0.1$.

As we saw in Section 4.5, there can be various different contributions to $\Omega = \frac{\rho}{\rho_{crit}}$, such as radiation Ω_R, matter Ω_M and vacuum energy Ω_Λ. In this chapter we shall, however, keep the curvature term (proportional to k in (8.6)) explicit and not introduce Ω_K as was done in Section 4.5. When treating the earliest Universe, the curvature turns out to be less important, as shown below. The equations of motion for the matter in the Universe are given by the vanishing of the covariant divergence of the energy-momentum tensor

$$T^{\alpha\beta}{}_{;\beta} = 0 \tag{8.7}$$

This gave, for the FLRW metric, (see (4.21))

$$\frac{d}{dt}(\rho a^3) = -p\frac{d}{dt}a^3 \tag{8.8}$$

which shows that the change of energy in a comoving volume element is equal to minus the pressure times the change in volume. (We also saw in Section 4.2 that this can be rewritten as

$$a^3\frac{dp}{dt} = \frac{d}{dt}[a^3(\rho + p)] \tag{8.9}$$

which we shall see can be interpreted as a conservation law for the entropy in a volume $a^3(T)$.) For radiation, where $p = \rho/3$, (8.8) gives $\rho \sim a^{-4}$. Note that all particle species which are light enough such that their average thermal kinetic energies at a certain temperature are larger than the rest mass have the equation of state of radiation.

For matter domination, where $p = 0$, $\rho \sim a^{-3}$. In either case, we see that in the early Universe (that is, for small a) the curvature term $\sim k/a^2$ was much less important than the energy density term $\sim \rho$. Of course, the cosmological constant which today gives a contribution of at most ρ_{crit}^0, was completely negligible then (except in a hypothetical early period of inflation when the value of Λ must have been huge (see Chapter 10)).

The Friedmann equation for the early Universe is thus simplified to

$$H^2(t) = \frac{8\pi G\rho}{3} \tag{8.10}$$

where as a good approximation only the relativistic species contribute appreciably to ρ. (We shall make this more quantitative in the following section.) Note that the Hubble parameter $H(t)$ has units of $1/(\text{time})$. This means in our units that it has dimensions of mass (see Problem 2.10). In Section 4.5.1 we saw that at any time the age of the Universe is of the order of H^{-1}, at least when the scale factor increases as a power of t.

We now want to treat the thermodynamics of an expanding Universe. The first question to ask is whether this is at all possible. A crucial point here is the microscopic understanding of thermodynamics in terms of the statistical mechanics of a large number of elementary particles or quanta. Generally, thermodynamic equilibrium requires very frequent interactions between the constituents of the system we are considering. If these interactions are frequent enough then the description of the Universe as evolving through a sequence of states of local thermal equilibrium is good; and we can use the thermodynamical quantities, temperature T, pressure p, entropy density s, and other quantities, at each time t to describe the state of the Universe. If the Universe has constituents with number density n and typical relative velocities v, and if they interact through scattering processes that have a cross-section σ, the interaction rate per particle Γ is given by $\Gamma = n\sigma v$. The condition that the interactions maintain equilibrium is that the interaction rate should be much larger than the expansion rate of the Universe:

$$\Gamma \gg H \tag{8.11}$$

Typically, the number density of particles decreases faster with temperature and therefore with time than does the Hubble parameter. This means that at certain epochs some particles will leave thermodynamic equilibrium. Their number density will then be 'frozen' at some particular value which then only changes through the general dilution due to the expansion. As we shall see, such 'freeze-out' of particles is a very important mechanism that can explain much of the particle content of the Universe we observe today.

8.2 Equilibrium Thermodynamics

We work in the dilute, weakly interacting gas approximation, where the distribution function $f_i(\mathbf{p})$ for particle species of type i is given by

$$f_i(\mathbf{p}) = \frac{1}{e^{\frac{(E_i - \mu_i)}{T}} \pm 1} \tag{8.12}$$

where $E_i = \sqrt{\mathbf{p}^2 + m_i^2}$ is the energy, μ_i is the chemical potential and T is the temperature (remember that we put $k_B = 1$). The minus sign is for particles that obey Bose Einstein statistics (bosons) and the plus sign is for particles obeying the exclusion principle and therefore Fermi Dirac statistics (fermions). It is usually assumed that the chemical potentials may be neglected in the very early Universe (see [51] for a full discussion).

Another important quantity is the number g_i of internal degrees of freedom of the particle, since each adds independently to the number and energy densities, pressure, etc. In the previous chapter we enumerated the degrees of freedom of the particles in the Standard Model. The photon has two polarization states and therefore $g_\gamma = 2$. The neutrinos only have one polarization state, giving $g_\nu = 1$, electrons and muons have $g_{e,\mu} = 2$ (and the same numbers for the antiparticles).

With these definitions, we can write for a species i its number density

$$n_i = \frac{g_i}{(2\pi)^3} \int f_i(\mathbf{p}) d^3 p \tag{8.13}$$

its energy density

$$\rho_i = \frac{g_i}{(2\pi)^3} \int E_i(\mathbf{p}) f_i(\mathbf{p}) d^3 p \tag{8.14}$$

and its pressure

$$p_i = \frac{g_i}{(2\pi)^3} \int \frac{|\mathbf{p}|^2}{3E_i(\mathbf{p})} f_i(\mathbf{p}) d^3 p \tag{8.15}$$

As the distribution functions only depend on $|\mathbf{p}|$, we let $d^3 p \to p^2 dp d\Omega$, (with $p \equiv |\mathbf{p}|$) where the integral over $d\Omega$ just gives a factor 4π. By differentiating the relation (see (2.45)) $E_i^2 = p^2 + m_i^2$, we obtain $pdp = E_i dE_i$, so that

$$d^3 p \to 4\pi \sqrt{E^2 - m_i^2} E dE \tag{8.16}$$

The resulting expressions for $n(T)$ and $\rho(T)$ are shown in Fig. 8.1 (a) and (b). Note that for small T/m – that is, when the thermal motion of the particles is non-relativistic – firstly there is no difference between bosons and fermions, and secondly the densities decrease very rapidly (exponentially) with decreasing temperature.

In the non-relativistic limit $T/m \ll 1$ one can solve the integrals analytically, with the result (both for Fermi Dirac and Bose Einstein particles)

$$n_{NoRe} = g_i \left(\frac{mT}{2\pi}\right)^{\frac{3}{2}} e^{-\frac{m}{T}} \tag{8.17}$$

$$\rho_{NoRe} = m \cdot n_{NoRe} \tag{8.18}$$

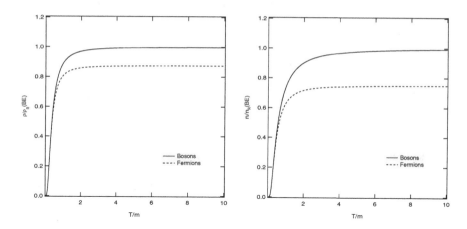

Fig. 8.1. The energy (a) and number density (b) for fermions and bosons, as a function of T/m. The quantities have been normalized to the relativistic expressions for bosons, $n_R(BE) = \frac{\zeta(3)}{\pi^2} g_i T^3$ and $\rho_R(BE) = \frac{\pi^2}{30} g_i T^4$, respectively.

$$p_{NoRe} = T \cdot n_{NoRe} \ll \rho_{NoRe} \tag{8.19}$$

Non-relativistically, $\langle E \rangle = m + 3T/2$. (If we re-insert units, this is written $mc^2 + 3k_B T/2$. This produces the well-known result that apart from the rest mass energy mc^2, the average thermal energy of a point particle is $3k_B T/2$.)

In the ultra-relativistic approximation, $T/m \gg 1$, the integrals can also be performed with the results

$$\rho_{Re} = \frac{g_i}{6\pi^2} \int_0^\infty \frac{E^3 dE}{e^{\frac{E}{T}} \pm 1} = \begin{cases} \frac{\pi^2}{30} g_i T^4, & \text{Bose Einstein} \\ \frac{7}{8}\left(\frac{\pi^2}{30} g_i T^4\right), & \text{Fermi Dirac,} \end{cases} \tag{8.20}$$

$$n_{Re} = \begin{cases} \frac{\zeta(3)}{\pi^2} g_i T^3, & \text{Bose Einstein} \\ \frac{3}{4}\left(\frac{\zeta(3)}{\pi^2} g_i T^3\right), & \text{Fermi Dirac,} \end{cases} \tag{8.21}$$

where $\zeta(x)$ is the Riemann zeta function, $\zeta(3) = 1.20206...$ The average energy for a relativistic particle is obtained by forming the ratio ρ/n, which gives

$$\langle E \rangle_{BE} \sim 2.70 \cdot T \tag{8.22}$$

and

$$\langle E \rangle_{FD} \sim 3.15 \cdot T \tag{8.23}$$

For photons, with the mass $m_\gamma = 0$, and $g_\gamma = 2$, the expression for $\rho_\gamma(T) \sim T^4$ is the famous Stefan Boltzmann law for electromagnetic black-body radiation.

We now come to the problem of calculating the total contribution to the energy and number density of all kinds of particles in the early Universe. Since we have seen that the energy and number density of a non-relativistic species is exponentially suppressed compared to a relativistic species, it is often a good approximation at a given temperature to sum only over the relativistic particles which are in equilibrium at that temperature. This means that the energy density will be of the form of the Stefan Boltzmann law,

$$\rho_{Re}(T) = \frac{\pi^2}{30} g_{\text{eff}}(T) T^4 \tag{8.24}$$

$$p_{Re}(T) = \frac{1}{3}\rho_{Re}(T) = \frac{\pi^2}{90} g_{\text{eff}}(T) T^4 \tag{8.25}$$

where the effective degeneracy factor $g_{\text{eff}}(T)$ counts the total number of internal degrees of freedom (such as spin, colour, etc.) of the particles that are relativistic and in thermal equilibrium at temperature T (that is, whose mass $m_i \ll T$). The expression for $g_{\text{eff}}(T)$ also contains the factor 7/8 for fermions that enters formula (8.20) for $\rho(T)$ (see (8.27) below).

It may be instructive to calculate $g_{\text{eff}}(T)$ for a temperature of, say, 1 TeV when all the particles of the Standard Model were relativistic and in thermal equilibrium. The total number of internal degrees of freedom of the fermions is 90 and for the gauge and Higgs bosons 28, so the total expression for g_{eff} is

$$g_{\text{eff}}(T = 1\,\text{TeV}) = 28 + \frac{7}{8} \cdot 90 = 106.75 \tag{8.26}$$

If it happens (as we shall see it does for neutrinos) that the interaction rate becomes smaller than the expansion rate, then those particles will have a lower temperature than the photons, but will still be relativistic (the neutrinos will be unaffected by heating that takes place for photons after the neutrinos have decoupled). This can be handled by introducing a specific temperature T_i for each kind of relativistic particle, which can be included in the effective g_i.

The total number of effective degrees of freedom g_{eff} is then

$$g_{\text{eff}} = \sum_{i=bosons} g_i \left(\frac{T_i}{T}\right)^4 + \frac{7}{8} \sum_{j=fermions} g_j \left(\frac{T_j}{T}\right)^4 \tag{8.27}$$

If we insert this expression into the Friedmann equation (8.10) we get for the radiation-dominated epoch in the early Universe

$$H^2 = \frac{8\pi G}{3}\rho_{Re} = \frac{8\pi G}{3}\frac{\pi^2}{30}g_{\text{eff}}T^4 = 2.76\frac{g_{\text{eff}}T^4}{m_{Pl}^2} \tag{8.28}$$

or

$$H = 1.66\sqrt{g_{\text{eff}}}\frac{T^2}{m_{Pl}} \tag{8.29}$$

This is one of the most important formulae of the physics of the early Universe.

From the relations between the scale factor a and the time t derived in Section 4.2 we obtained (see (4.29))

$$a(t) \sim \sqrt{t} \quad \text{radiation domination} \tag{8.30}$$

that is, for the equation of state $p = \rho/3$. For matter domination, that is, for $p \sim 0$, one finds

$$a(t) \sim t^{\frac{2}{3}} \quad \text{matter domination.} \tag{8.31}$$

Thus, for radiation domination,

$$H = \dot{a}/a = \frac{1}{2t} \quad \text{radiation domination} \tag{8.32}$$

and the time temperature relation becomes

$$t = 0.30 \frac{m_{Pl}}{\sqrt{g_{\text{eff}}} T^2} \sim \left(\frac{1 \text{ MeV}}{T}\right)^2 \text{sec} \tag{8.33}$$

This is a convenient formula to memorize, valid during the important temperatures around 1 MeV, when most of nucleosynthesis and neutrino decoupling occurred, as we shall see.

8.3 Entropy

We now have to determine which particles are in thermal equilibrium at a given temperature, so that we can calculate $g_{\text{eff}}(T)$. It is useful to first go through some basic thermodynamic relations. In particular, we will show that in the case of thermal equilibrium the entropy within a volume $a^3(t)$ is conserved.

The *entropy* $S(V,T)$ is introduced in one of the central equations of thermodynamics through the definition of its differential

$$dS(V,T) = \frac{1}{T} [d(\rho(T)V) + p(T)dV] \tag{8.34}$$

Identifying the coefficient functions multiplying dT and dV in this expression with the general form of the differential

$$dS(V,T) = \frac{\partial S(V,T)}{\partial V} dV + \frac{\partial S(V,T)}{\partial T} dT \tag{8.35}$$

we find

$$\frac{\partial S(V,T)}{\partial V} = \frac{1}{T} (\rho(T) + p(T)) \tag{8.36}$$

and

$$\frac{\partial S(V,T)}{\partial T} = \frac{V}{T} \frac{d\rho(T)}{dT} \tag{8.37}$$

Equality of the mixed derivatives,

$$\frac{\partial^2 S(V,T)}{\partial V \partial T} = \frac{\partial^2 S(V,T)}{\partial T \partial V} \tag{8.38}$$

then gives

$$\frac{\partial}{\partial T}\left(\frac{1}{T}(\rho(T)+p(T))\right) = \frac{\partial}{\partial V}\left(\frac{V}{T}\frac{d\rho(T)}{dT}\right) \tag{8.39}$$

This can be simplified to

$$\frac{dp(T)}{dT} = \frac{1}{T}(\rho(T)+p(T)) \tag{8.40}$$

which could alternatively have been derived (Problem 8.5) directly from equations (8.14) and (8.15). Inserting this into (8.34) gives

$$dS(V,T) = \frac{1}{T}d[(\rho(T)+p(T))V] - \frac{V}{T^2}(\rho(T)+p(T))dT \tag{8.41}$$

which can be integrated to show that the entropy $S(V,T)$ is, up to an integration constant, given by

$$S(V,T) = \frac{V}{T}(\rho(T)+p(T)) \tag{8.42}$$

We now recall (8.9):

$$a^3\frac{dp(T)}{dt} = \frac{d}{dt}[a^3(\rho(T)+p(T))] \tag{8.43}$$

which combined with (8.40) can be written

$$\frac{d}{dt}\left(\frac{a^3}{T}[\rho(T)+p(T)]\right) = 0 \tag{8.44}$$

Identifying the volume V with $a^3(t)$ and comparing with (8.42) we then finally have the advertised law of conservation of entropy in the volume $a^3(t)$. Sometimes it is more useful to work with the entropy density $s(T)$ rather than the total entropy $S(V,T)$ within the volume V. The definition is thus:

$$s(T) \equiv \frac{S(V,T)}{V} = \frac{\rho(T)+p(T)}{T} \tag{8.45}$$

(This section has been rather formal. The important things to remember are the expressions (8.42) for the entropy, (8.45) for the entropy density, and the conservation equation (8.44).)

In the early Universe, both the energy density ρ and the pressure p were dominated by relativistic particles with the equation of state $p = \rho/3$. Using (8.45) and the relativistic expressions (8.24, 8.25) for the energy density and the pressure, we find for the entropy density s

$$s = \frac{2\pi^2}{45}g_{\text{eff}}^s T^3 \tag{8.46}$$

where g_{eff}^s is defined in a similar way to g_{eff}:

$$g_{\text{eff}}^s = \sum_{i=bosons} g_i \left(\frac{T_i}{T}\right)^3 + \frac{7}{8} \sum_{j=fermions} g_j \left(\frac{T_j}{T}\right)^3 \tag{8.47}$$

Since s and n_γ both vary as T^3 there is a simple relationship between them. With

$$n_\gamma = \frac{2\zeta(3)}{\pi^2} T^3 \tag{8.48}$$

one finds

$$s = \frac{\pi^4}{45\zeta(3)} g_{\text{eff}}^s n_\gamma \sim 1.8 g_{\text{eff}}^s n_\gamma \tag{8.49}$$

We now turn to the question of the decoupling of neutrinos from the thermal 'bath' in the early Universe. As previously mentioned, weakly interacting particles like neutrinos decouple below some temperature T_{dec} when the rate of interaction between particles is not fast enough to keep pace with the Hubble expansion. The weak interactions are mediated by the W and Z particles that are massive, $m_W \sim 80$ GeV and $m_Z \sim 91$ GeV. At temperatures much smaller than 80–90 GeV the exchanged W and Z bosons are virtual so that their propagators behave like $1/m_W^2$ (see Section 6.10.1).

The cross-section for the weak interactions was found in Section 6.10.1 to be proportional to $\alpha^2 s/m_W^4$. For relativistic neutrinos (in the probably excellent approximation that neutrinos are massless this is always the case, since massless particles always have to move at the speed of light; in general, the requirement is $m_\nu \ll T$) and for relativistic charged leptons a typical process maintaining thermal equilibrium like $\nu_e + e^+ \to \nu_\mu + \mu^+$ will have the cross-section $\sigma_{weak} \sim \alpha^2 T^2/m_W^4$. This is because s is given by the energy squared of the reacting particles, and the average energy is proportional to T. The interaction rate $\Gamma = \sigma|\mathbf{v}|n$ is thus, since $|\mathbf{v}| = c = 1$ and $n \sim T^3$

$$\Gamma_{weak} \sim \frac{\alpha^2 T^5}{m_W^4} \tag{8.50}$$

This now has to be compared with the expansion rate H. As we saw in (8.29), $H \sim T^2/m_{Pl}$, so the ratio becomes

$$\frac{\Gamma}{H} \sim \frac{\alpha^2 m_{Pl} T^3}{m_W^4} \tag{8.51}$$

Decoupling occurs when this ratio drops below unity, meaning

$$T_{dec} \sim \left(\frac{m_W^4}{\alpha^2 m_{Pl}}\right)^{\frac{1}{3}} \sim 4 \text{ MeV} \tag{8.52}$$

What happens after the neutrinos have decoupled? All neutrinos will move as free particles following the general Hubble expansion. This means, just as for photons, that their energies redshift by the factor a/a_{dec}. They will thus remain in a thermal (Fermi Dirac) distribution with an effective temperature given by

$$T_\nu = T_{dec} \frac{a_{dec}}{a} \sim a^{-1} \tag{8.53}$$

If we recall the conservation of entropy for particles in thermal equilibrium

$$g_{\text{eff}}^s (aT)^3 = const$$

we see that

$$T \sim (g_{\text{eff}}^s)^{-1/3} a^{-1}$$

Thus, the neutrino distribution after decoupling *will look as if it is still in thermal equilibrium* as long as g_{eff}^s does not change. However, g_{eff}^s will change when the electrons and positrons become non-relativistic and annihilate through $e^+e^- \to \gamma\gamma$. This happens at a temperature around 1 MeV, since below that temperature the inverse process $\gamma\gamma \to e^+e^-$ is no longer kinematically possible (the rest mass of an electron positron pair is 1.02 MeV). We thus calculate the number of degrees of freedom before and after e^+e^- annihilation. The neutrinos have already decoupled, so at temperatures somewhat higher than 1 MeV the relativistic species in thermal equilibrium are e^+, e^- and γ, which gives $(g_{\text{eff}}^s)^{before} = 2 \cdot 2 \cdot 7/8 + 2 = 11/2$ and below 1 MeV only the photons are in equilibrium giving $(g_{\text{eff}}^s)^{after} = 2$. Since the total entropy for the equilibrium particles is conserved,

$$\left(g_{\text{eff}}^s (aT)^3\right)_{before} = \left(g_{\text{eff}}^s (aT)^3\right)_{after} \tag{8.54}$$

meaning

$$(aT)_{T \leq 1 \ MeV} = \left(\frac{11}{4}\right)^{\frac{1}{3}} (aT)_{T \geq 1 \ MeV} \sim 1.4 \, (aT)_{T \geq 1 \ MeV} \tag{8.55}$$

There is an entropy transfer from the decoupling e^+e^- particles to the photons, usually called reheating (although the temperature actually does not rise, it only decreases less rapidly for the photons due to the entropy transfer). The already decoupled neutrinos, on the other hand, do not benefit from this reheating. They just follow the redshift relation $(aT_\nu)_{before} = (aT_\nu)_{after}$. This can be interpreted as saying that the neutrino entropy is separately conserved after decoupling. It means that there will be a difference in temperature for neutrinos and photons after e^+e^- decoupling, namely

$$T_\nu = \left(\frac{4}{11}\right)^{\frac{1}{3}} T_\gamma \tag{8.56}$$

Since the cosmic microwave background photons now have a temperature of 2.73 K, there should also exist a *cosmic neutrino background* having a Fermi Dirac energy spectrum with a temperature of $T_\nu = 1.95$ K. As neutrinos of such low energies interact extremely weakly with matter, it will be a tremendous challenge for experimental physicists of the coming century to detect this cosmic neutrino background directly. Indirectly, neutrinos of non-zero mass could leave an imprint on the large scale structure, in particular suppressing structure on small scales.

What is the total entropy and radiation energy density today? It is given by the contributions from the photons and the three species of neutrinos (ν_e, ν_μ and ν_τ, assumed massless), thus

$$g_{\text{eff}}^{tot}(today) = 2 + \frac{7}{8} \cdot 2 \cdot 3 \cdot \left(\frac{4}{11}\right)^{\frac{4}{3}} = 3.36 \tag{8.57}$$

and

$$(g_{\text{eff}}^{s})^{tot}(today) = 2 + \frac{7}{8} \cdot 2 \cdot 3 \cdot \left(\frac{4}{11}\right) = 3.91 \tag{8.58}$$

In the case that neutrinos have mass (as we shall see in Chapter 14 is most likely) they would most probably be nonrelativistic by now.

The total (current) radiation energy density in the form of photons is

$$\rho_{R_\gamma} = \frac{\pi^2}{30} g_{\text{eff}}^{tot} T^4 = 4.8 \cdot 10^{-34} \text{ g/cm}^3 \tag{8.59}$$

which corresponds to a contribution to $\Omega = \rho/\rho_{crit}$ of

$$\Omega_{R_\gamma} h^2 = 2.6 \cdot 10^{-5} \tag{8.60}$$

The number density of microwave background photons is

$$n_\gamma = \frac{2\zeta(3)}{\pi^2} T^3 \sim 410 \text{ cm}^{-3} \tag{8.61}$$

for $T = T_0 = 2.725$ K. Despite contributing numerically a very small fraction of Ω today, the CMBR at the time of its emission was much more important dynamically. However, the most important use of the CMBR today is that it essentially provides a 'snapshot' of the Universe at a redshift of around 1100. In Section 11.2 we shall see how tiny differences of CMBR temperature in various directions on the sky provides a clue as to how the first structures formed in the Universe.

Could there be any other type of radiation present as a relic of the early Universe? Yes, there are strong reasons to believe that *gravitons*, the gauge particles of gravitation, exist. Since they are connected to gravity, the mass scale for their interaction is the Planck mass, $\sigma_{grav} \sim T^2/m_{Pl}^4$, and $\Gamma_{grav}/H \sim T^3/m_{Pl}^3$, so that the decoupling temperature was enormous, $T_{dec} \sim m_{Pl} \sim 10^{19}$ GeV. We saw before that the number of degrees of freedom of the Standard Model was very large (106.75) at high temperature. At the Planck mass it was most probably much larger due to many additional heavy particles connected with supersymmetry and Grand Unification. This means that

$$T_{grav} = \left(\frac{(g_{\text{eff}}^{s})^{now}}{(g_{\text{eff}}^{s})^{Planck}}\right)^{\frac{1}{3}} \cdot T_0 \leq \left(\frac{3.9}{106.75}\right)^{\frac{1}{3}} \cdot 2.73 \text{ K} = 0.9 \text{ K} \tag{8.62}$$

The contribution to the present energy density is then, since $\rho \sim T^4$, $\rho_{grav} \leq 0.012\rho_\gamma$.

After relativistic particles have decoupled, their contribution to the energy density goes down as $1/a^4$ (since $\rho \sim T^4$ and $aT \sim const$). For non-relativistic matter, however, the energy density is given by

$$\rho_{matter} \equiv \rho_m = m_{NoRe} \cdot n_{NoRe} \tag{8.63}$$

and for a stable particle (such as a baryon) $n_{NoRe} \sim 1/a^3 \sim T^3$. Therefore, eventually the Universe became matter dominated. When did that occur? With ρ_M denoting the *present* matter density, the matter contribution to the energy density is now written as

$$\rho_M = 1.9 \cdot 10^{-29} \, \Omega_M h^2 \, \text{g/cm}^3 \tag{8.64}$$

Then, for arbitrary times,

$$\rho_m = \rho_M \left(\frac{a_0}{a}\right)^3 = \rho_M (1+z)^3 \tag{8.65}$$

with the usual expression for the redshift factor $1 + z = a_0/a$. Similarly (ρ_R being the present value of the radiation energy density ρ_r),

$$\rho_r = \rho_R \left(\frac{g_{\text{eff}}}{g_{\text{eff}}^{now}}\right)(1+z)^4 \tag{8.66}$$

Equating (8.65) and (8.66) gives

$$(1 + z_{eq}) \sim \frac{\rho_M}{\rho_R} = 2.3 \cdot 10^4 \Omega_M h^2 \tag{8.67}$$

$$T_{eq} = T_0(1 + z_{eq}) = 5.5\Omega_M h^2 \, \text{eV} \tag{8.68}$$

and (see 4.79))

$$t_{eq} \sim \frac{2}{3} H_0^{-1} \Omega_M^{-1/2} (1 + z_{eq})^{-\frac{3}{2}} \sim 1.9 \cdot 10^3 / (\Omega_M h^2)^2 \, \text{years} \tag{8.69}$$

As will be explained in the section on structure formation, the time of the onset of matter domination is very important, since it was only then that structures could begin to grow.

8.4 Summary

- In the earliest Universe, only relativistic particles were important for the cosmic expansion rate. The contribution to the energy density by relativistic particles is

$$\rho_{Re}(T) = \frac{\pi^2}{30} g_{\text{eff}}(T) T^4$$

where

$$g_{\text{eff}} = \sum_{i=bosons} g_i \left(\frac{T_i}{T}\right)^4 + \frac{7}{8} \sum_{j=fermions} g_j \left(\frac{T_j}{T}\right)^4$$

Here the possibility of particles having different effective temperatures has been taken into account.

- In the radiation-dominated epoch (the first few hundred thousand years), the expansion rate is given by

$$H = 1.66\sqrt{g_{\text{eff}}}\,\frac{T^2}{m_{Pl}}$$

and the relation between temperature and time around 1 MeV was

$$t = 0.30\frac{m_{Pl}}{\sqrt{g_{\text{eff}}}T^2} \sim \left(\frac{1 \text{ MeV}}{T}\right)^2 \text{ sec}$$

- The total entropy $S(V,T)$ in a region of the Universe was conserved during periods of thermal equilibrium,

$$\frac{dS}{dt} = \frac{d}{dt}\left(\frac{a^3}{T}[\rho(T) + p(T)]\right) = 0$$

The entropy density is given by

$$s = \frac{2\pi^2}{45}g^s_{\text{eff}}T^3$$

with

$$g^s_{\text{eff}} = \sum_{i=bosons} g_i \left(\frac{T_i}{T}\right)^3 + \frac{7}{8} \sum_{j=fermions} g_j \left(\frac{T_j}{T}\right)^3$$

- Neutrinos decoupled at a temperature around 4 MeV, but their distribution functions were still of the thermal form, only redshifted with the cosmic expansion. However, they did not participate in the reheating when electrons and positrons became non-relativistic. The 'cosmic neutrino background' therefore after this epoch had a temperature which is somewhat lower than that of the microwave background,

$$T_\nu = \left(\frac{4}{11}\right)^{\frac{1}{3}}T_\gamma \sim 0.71T_\gamma$$

- The contribution of the cosmic microwave background to the present energy density of the Universe is just

$$\Omega_{R_\gamma}h^2 = 2.5 \cdot 10^{-5}$$

and the number density is

$$n_\gamma = \frac{2\zeta(3)}{\pi^2}T^3 \sim 410 \text{ cm}^{-3}$$

8.5 Problems

8.1 Perform the dimensional analysis leading to (8.3).

8.2 Use dimensional analysis to show that the energy density of a photon gas has to be proportional to T^4 and the number density proportional to T^3.

8.3 Estimate roughly at what temperature strong interaction processes involving quarks and gluons would leave thermal equilibrium if they were to be unconfined and interacting with strength $\alpha_s \sim 0.3$ at all temperatures.

8.4 Suppose one finds a theory for massless gravitons where they interact by exchanging particles of mass m_{Pl} with electromagnetic interaction strength. Would they be in thermal equilibrium at temperatures $T < m_{Pl}$?

8.5 Derive (8.40) directly from (8.14) and (8.15).

8.6 (a) What was the energy density in radiation (photons and neutrinos) expressed in units of the critical density, at the time when the first quasars formed, at $z \sim 5$? Use $h = 0.65$.
(b) Suppose there now exists a homogeneous magnetic field B_0 in the Universe. How strong (measured in Gauss) would it have to be, if the energy density is the same as that in the cosmic microwave background? (Hint: $\rho_B = const \cdot B_0^2$; you may have to refer to a book on electromagnetism to find the constant in suitable units.)

8.7 Estimate the effective number of degrees of freedom g_{eff} in the Standard Model in the early Universe when the temperature was 50 GeV.

9 Thermal Relics from the Big Bang

The Universe has gone through a sequence of important events, where traces of its history, so called 'relics', have been left behind. In this chapter we shall look at the mechanisms involved in the freeze-out of hypothetical non-baryonic dark matter particles, that contribute 20 to 30 % of the matter in the Universe today. We shall also go through the basic processes that converted the neutrons and protons to light elements such as ^{4}He and deuterium. One of the most important epochs in the early Universe was when the Universe suddenly became transparent to optical photons. This happened when neutral hydrogen and helium gas formed, since a neutral gas is transparent, as compared with an ionized plasma. The radiation which could start to propagate freely at that time is still travelling through the Universe in all directions, and is now redshifted to the microwave region. This is of course the CMBR, which has been and will be an extremely important relic from a few hundred thousand years after the Big Bang. In this chapter we shall also study this process, leaving the details to Chapter 11.

9.1 Matter Antimatter Asymmetry

After the annihilation of electrons and positrons (that is, at temperatures much lower than 1 MeV), the number of positrons in the Universe was extremely small since they became non-relativistic and their number density was exponentially suppressed. However, there was an excess of electrons – just enough to balance the electric charge of the protons. The exact origin of the asymmetry between matter and antimatter in the Universe is still unknown. However, in the modern theories of particle physics, several mechanisms exist that may create such an asymmetry. The Russian physicist Andrei Sakharov showed that to obtain a matter antimatter asymmetry in the Universe, even from an initial state that was symmetric, three conditions are necessary:

- There have to exist CP violating processes. C is the charge conjugation operator and P is the parity operator. After the discovery of P violation in the weak interactions (as we remember from Section 6.2 only left-handed neutrinos seem to exist in nature), it was realised that the transformation which takes one from a particle state to an antiparticle state is not only C, but the combined action CP. (A CP transformation on a left-handed neutrino gives a right-handed antineutrino, which also exists in nature.) If there were to be no CP violation, particles and antiparticles would always interact in the same way, and an asymmetry could not arise. In the Standard Model, CP violation exists because of interference between the three families of quarks and leptons – only two families would not be enough! In the K^0-meson system, (a bound state of a d quark and an s antiquark) CP violation was discovered experimentally by Fitch and Cronin in the 1960s. Recently, CP violating decay have also been found in mesons containing a b quark.

- Baryon number violating processes. Since today there is an asymmetry between baryons and antibaryons (that is, a net baryon number $N_B - N_{\bar{B}} > 0$), some interactions must have taken place that violated baryon number conservation. In Grand Unified Theories (GUT) aimed at unifying the weak, strong and electromagnetic interactions there exist such interactions. A prediction, not yet experimentally verified, from these theories is that the proton should also decay. However, the average lifetime of a proton has to be longer than 10^{33} years to be consistent with present experimental bounds (compared with the age of the Universe, 10^{10} years!). Also in the minimal Standard Model there turns out to exist baryon number violating processes. These are nonperturbative so-called instanton processes found by G. 't Hooft. Today they are even smaller than the GUT processes because of a large energy barrier which makes quantum mechanical tunneling difficult, but in the early Universe thermal energies could have helped pass the barrier. This is at present an intensive field of theoretical research, and the details have not yet been worked out. It seems that the CP violation present in the minimal Standard Model is not enough for the mechanism to work, but it may be sufficient to introduce an extra doublet of Higgs fields, as is required in supersymmetric theories. In most of the models, the baryon asymmetry is exactly balanced by a lepton asymmetry. In fact, in some of the most promising models the primary source is lepton asymmetry, which then gives the baryon asymmetry through decay.

- Deviation from strict thermal equilibrium. This is necessary, because CPT symmetry, believed to be exact in nature (T is the time reversal operator) requires the masses of particles and antiparticles to be the same, and the abundance of a particle species in thermal equilibrium at a constant temperature depends only on the mass. In the Big Bang model this is not a problem, since as we have seen the steady decrease in temperature means that many species of particles successively leave thermal equilibrium. As we shall see, the matter antimatter asymmetry need only be of order 10^{-10} to explain observations.

9.2 Freeze-Out and Dark Matter

There are several important examples of freeze-out in the early Universe, for instance at the synthesis of light elements one second to a few minutes after the Big Bang, and the microwave photons from the surface of last scattering several hundred thousand years later. Before we calculate these processes, it is convenient to introduce a formalism which considers freeze-out in general: that is, what happens when a particle species goes out of equilibrium. A rigorous treatment has to be based on the Boltzmann transport equation in an expanding background, but here we give a simplified treatment (see, for example, the book by Kolb and Turner [26] for a more complete discussion).

We first consider a case of great interest for the dark matter problem. Suppose that there exists some kind of unknown particle χ, with antiparticle $\bar{\chi}$, that can annihilate each other and be pair created through processes $\chi + \bar{\chi} \leftrightarrow X + \bar{X}$, where X stands for any type of particle to which the χs can annihilate.[1] We further assume that the X particles have zero chemical potential and that they are kept in thermal equilibrium with the photons and the other light particles in the early Universe (the X particles can be quarks, leptons etc.)

How will the number density n_χ evolve with time (and therefore with temperature)? It is clear that in exact thermal equilibrium the number of χ particles in a comoving volume $N_\chi = a^3 n_\chi$ will be given by the equilibrium value $n_\chi^{EQ}(T)$ (see (8.13)). (In exact thermal equilibrium the rate for the process $\chi + \bar{\chi} \leftrightarrow X + \bar{X}$ is the same in both directions.) If the actual number density $n_\chi(T)$ is larger than the equilibrium density the reaction will go faster to the right: that is, the χ particles will annihilate faster than they are created. The depletion rate of χ should be proportional to $\sigma_{\chi\bar{\chi}\to X\bar{X}}|\mathbf{v}|n_\chi^2$ (quadratic in the density, since it should be proportional to the product of n_χ and $n_{\bar{\chi}}$, and these are equal). However, χ particles are also created by the inverse process, with a rate proportional to $(n_\chi^{EQ})^2$. We have thus 'derived' the basic equation that governs the departure from equilibrium for the species χ:

$$\frac{dn_\chi}{dt} + 3Hn_\chi = -\langle\sigma_{\chi\bar{\chi}\to X\bar{X}}|\mathbf{v}|\rangle[n_\chi^2 - (n_\chi^{EQ})^2] \tag{9.1}$$

---

[1] The supersymmetric neutralino is actually its own antiparticle (just as the photon is its own antiparticle). The formalism is very similar in this case. In particular, a neutralino can annihilate with another neutralino giving other, non-supersymmetric particles in the final state.

The left-hand side comes from $\frac{1}{a^3}\frac{d}{dt}[n_\chi a^3]$; the term proportional to $3H$ just expresses the dilution that automatically comes from the Hubble expansion. The expression $\langle \sigma_{\chi\bar{\chi}\to X\bar{X}}|\mathbf{v}|\rangle$ stands for the thermally averaged cross-section times velocity. This averaging is necessary, since the annihilating particles have random thermal velocities and directions. Summing over all possible annihilation channels gives

$$\frac{dn_\chi}{dt} + 3Hn_\chi = -\langle \sigma_A|\mathbf{v}|\rangle[n_\chi^2 - (n_\chi^{EQ})^2] \tag{9.2}$$

where σ_A is the total annihilation cross-section.

Using the time-temperature relation (8.33) (for radiation dominance)

$$t = 0.30\frac{m_{Pl}}{T^2\sqrt{g_{\text{eff}}}} \tag{9.3}$$

this can be converted to an evolution equation for n_χ as a function of temperature. Introducing the dimensionless variable $x = m_\chi/T$, and normalizing n_χ to the entropy density:

$$Y_\chi = \frac{n_\chi}{s} \tag{9.4}$$

gives after some intermediate steps (Problem 9.1)

$$\frac{dY}{dx} = -\frac{m_\chi m_{Pl}c_{\text{eff}}}{x^2}\sqrt{\frac{\pi}{45}}\langle \sigma_A|\mathbf{v}|\rangle(Y_\chi^2 - (Y_\chi^{EQ})^2) \tag{9.5}$$

where

$$c_{\text{eff}} = \frac{g_{\text{eff}}^s}{\sqrt{g_{\text{eff}}}} \tag{9.6}$$

or after some rearrangement

$$\frac{x}{Y_\chi^{EQ}}\frac{dY}{dx} = -\frac{\Gamma_A}{H}\left[\left(\frac{Y_\chi}{Y_\chi^{EQ}}\right)^2 - 1\right] \tag{9.7}$$

where $\Gamma_A = n_\chi^{EQ}\langle \sigma_A|\mathbf{v}|\rangle$. This equation can be solved numerically with the boundary condition that for small x, $Y_\chi \sim Y_\chi^{EQ}$ (since at high temperature the χ particles were in thermal equilibrium with the other particles). We see from (9.7) that, as expected, the evolution is governed by the factor Γ_A/H, the interaction rate divided by the Hubble expansion rate.

The solutions to these equations have to be obtained numerically in the general case to find the temperature T_f and therefore the value of x_f of freeze-out and the asymptotic value $Y_\chi(\infty)$ of the relic abundance of the species χ. There are, however, some simple limiting cases. If the species χ is relativistic at freeze-out ($x_f \leq 3$, say), then Y_χ^{EQ} is not changing with time during the period of freeze-out, and the resulting $Y_\chi(\infty)$ is just the equilibrium value at freeze-out,

$$Y_\chi(\infty) = Y_\chi^{EQ}(x_f) = \frac{45\zeta(3)}{2\pi^4}\frac{g_{eff}}{g_{\text{eff}}^s(x_f)} \tag{9.8}$$

where $g_{eff} = g$ for bosons and $3g/4$ for fermions. A particle that was rela-tivistic at freeze-out is called a *hot* relic. A typical example is the neutrino. The present mass density of a hot relic with mass m is

$$\Omega_\chi h^2 = 7.8 \cdot 10^{-2} \frac{g_{eff}}{g_{\text{eff}}^s(x_f)} \left(\frac{m_\chi}{1 \text{ eV}} \right) \tag{9.9}$$

Note that *today* the motion of a particle with mass greater than the small number $T_0 = 2.73$ K $= 2.4 \cdot 10^{-4}$ eV is of course non-relativistic and therefore the contribution to the energy density is dominated by its rest mass energy. An ordinary neutrino has $g_{eff} = 2 \cdot 3/4 = 1.5$ and decouples at a few MeV (see (8.52)), when $g_{\text{eff}}^s = g_{\text{eff}} = 10.75$. Demanding that the neutrinos do not overclose the Universe ($\Omega_{\nu\bar{\nu}} h^2 < 1$; this can alternatively be stated as a condition on the age of the Universe; see Section 4.5.1) produces a famous condition on the neutrino mass (or, really, the sum of the masses of the three different kinds of neutrino in nature):

$$\sum_i m_{\nu_i} < (94 \text{ eV}) \cdot \Omega_M h^2 \tag{9.10}$$

This analysis has been valid for hot relics, or *hot dark matter*. For *cold* relics (particles that were non-relativistic at freeze-out) (9.7) has to be found numerically. Solutions to this equation are shown in Fig. 9.1, for different values of $\langle \sigma_A |\mathbf{v}| \rangle$.

As can be seen, the value of x_f (when Y_χ leaves the equilibrium curve) is lower for a smaller cross-section σ_A: that is, more weakly interacting parti-cles decouple earlier. Since the equilibrium curve for a non-relativistic species drops fast with increasing x, this means that the more weakly coupled parti-cles will have a higher relic abundance.

Going through the numerical analysis one finds that a hypothetical neu-trino with mass $m_\nu \sim 3$ GeV would also have about the right mass to close the Universe. On the other hand, the range between 90 eV and 3 GeV is cosmologically disallowed for a stable neutrino. There are arguments from large-scale structure formation that favour cold relics over hot relics, so such a neutrino would be a good dark matter candidate. Data from the LEP ac-celerator at CERN have, however, excluded any ordinary neutrino with a mass in the GeV range.

So, what could the dark matter be? It turns out that in particle physics, there are hypothetical particles, like supersymmetric partners of ordinary particles discussed in Section 6.9.1, that have the right interaction strength and mass range to be promising dark matter candidates. In particular, the neutralino (6.14), has all the properties of a good dark matter candidate. Since it is electrically neutral it does not emit or absorb radiation which makes it 'dark' (invisible matter is thus a better term than dark matter). The couplings of neutralinos are generally of weak interaction strength, but the large number of possible annihilation channels, which depends on the unknown supersymmetry breaking parameters, makes an exact prediction of

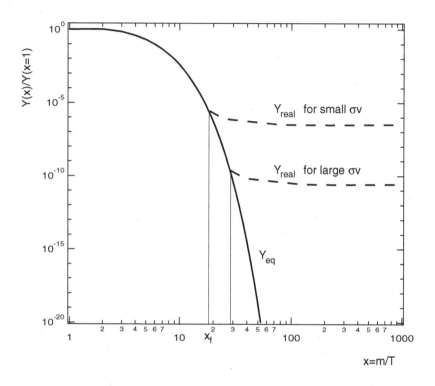

Fig. 9.1. The freeze-out of a massive particle. At a certain value $x_f = m_\chi/T_f$ the number density Y (normalized to the entropy density s, and in the figure arbitrarily normalized to the value at $x = 1$) leaves the equilibrium abundance curve Y_{eq} (the solid line) and gives an actual abundance Y_{real} shown by the dashed lines. As can be seen, a higher annihilation rate σv means a smaller relic abundance, since the actual curve tracks the equilibrium curve to smaller temperatures. For weakly interacting massive particles, x_f is of the order of 20. Adapted from [26].

mass and relic abundance uncertain. Scans of parameter space show, however, that a neutralino in the mass range between 30 GeV and a few TeV could give a relic density close to the critical density. This is currently an active area of research, and there are several experiments being conducted around the world to try to detect supersymmetric dark matter, if it exists.

Another candidate for the dark matter is provided by the axion, a hypothetical light boson which was introduced for theoretical reasons to explain the absence of CP violation in the strong interactions (as far as we know, CP violations only take place in the weak interactions). It turns out that for a mass range between 10^{-6} and 10^{-3} eV, the axion could make a sizeable contribution to Ω_M. It couples very weakly to ordinary matter, but it may be converted into a photon in a cavity containing a strong magnetic field (the basic coupling is to two photons, but here the magnetic field takes the role of one photon). Experiments in the USA and Japan are currently probing parts of the interesting mass region.

9.3 Nucleosynthesis

One of the cornerstones of Big Bang cosmology is the observational fact that the mass fraction of ^{4}He is about 24 per cent, whereas hydrogen makes up the remaining part of nuclei except for very small abundances of ^{3}He, ^{2}H (deuterium, D), and ^{7}Li. Heavier elements, which are common in our immediate surroundings, only make up a small fraction of the baryonic matter in the Universe as a whole. The generally accepted picture is that all elements heavier than ^{7}Li have been produced in the interior of stars or in other astrophysical processes (supernova explosions, spallation by cosmic rays, etc.). Evidence for this general picture comes from many places. Firstly, the amount of heavy elements is consistent with estimates based on known star-formation rates and inferred star-formation history. Secondly, the large amount of helium is impossible to explain by such stellar processing (see Example 1.2.1). Recent observations (for example, from the Hubble Space Telescope) of 'unprocessed' gas in the form of clouds at high redshift, show abundances of helium and deuterium that agree with Big Bang nucleosynthesis predictions. Similarly, when looking at old, metal-poor stars one finds the abundance of ^{4}He and ^{7}Li to reach 'plateaux': that is, values which are independent of the abundance of heavy elements. The idea is that stellar production of ^{4}He and ^{7}Li inevitably also leads to production of heavier elements such as oxygen. (The jargon is such that anything heavier than helium is called a 'metal', and the abundance of such elements is called 'metallicity'). If lithium were only synthesized in stars, one would thus expect its abundance to decrease with the metallicity of the observed stars. The only natural interpretation of the observed plateaus is that they represent the primordial abundances of these elements.

Big Bang nucleosynthesis is a mature field which has been treated very carefully using numerical methods which enable one to follow the number density of various nuclei with time during the first few seconds after the Big Bang when the light elements were synthesized. A very important and non-trivial test is provided by the agreement between calculations and observations for all the four elements ^{4}He, ^{3}He, D and ^{7}Li. This agreement works for a small range of the baryon-to-photon ratio η_B. Since the number density of photons is dominated and experimentally fixed by the cosmic microwave background radiation (CMBR), the upper limit of η_B from nucleosynthesis gives the maximum allowed contribution Ω_B (in fact, $\Omega_B h^2$) to the total energy density of the Universe. Since this appears to be smaller than dynamical estimates of the total mass density of, for example, galaxy clusters, it gives an indication that dark matter may be needed. (The computation of the dark matter density for some hypothetical particles will be carried out in the next section.)

We shall not go into the details of nucleosynthesis here – the equations are non-linear and can in practice only be solved numerically. It is, however, not so difficult to give a schematic outline of how the most important abundance, that of ^{4}He, can be estimated.

At the earliest times, ($t \ll 1$ s) neutrinos, electrons and positrons were still in equilibrium through weak interactions such as

$$n \leftrightarrow p + e^- + \bar{\nu}_e$$
$$\nu_e + n \leftrightarrow p + e^-$$
$$e^+ + n \leftrightarrow p + \bar{\nu}_e \tag{9.11}$$

These reactions are all governed by the weak interaction, and correspond to typical cross-sections of order $\sigma_{weak} = \alpha^2 s/m_W^4$. When the temperature of the Universe was much larger than the mass difference between the proton and the neutron, $\Delta m \equiv m_n - m_p = 1.29$ MeV, the reactions went equally fast in both directions, and there were equally many neutrons and protons. When the temperature decreased towards 1 MeV, the suppression of the number density of neutrons because of their higher mass started to become important. Since protons and neutrons were non-relativistic at these temperatures ($T \sim 1$ MeV $\ll m_n, m_p \sim 940$ MeV) we can use (8.17) to compute the ratio:

$$\frac{n_n}{n_p} = e^{-\Delta m/T} = e^{-(1.29 \text{ MeV})/T} \tag{9.12}$$

(since the number of helicity states $g_p = g_n = 2$). This shows that at high temperature the ratio is close to unity. If equilibrium were to be maintained, the ratio would decrease to a very small value at low temperature. However, we know from the recombination calculation that what determines the abundance is often the 'freeze-out' of the abundance due to the fact that the reaction rate such as

$$\Gamma(\nu_e + n \leftrightarrow p + e^-) \sim 2.1 \left(\frac{T}{1 \text{ MeV}}\right)^5 \text{ sec}^{-1} \tag{9.13}$$

falls below the expansion rate when $T < 0.8$ MeV. Therefore, neutrons are not destroyed (or created) by the two last reactions in (9.11), but may still be destroyed by neutron decay (the first reaction in (9.11)). However, this is governed by the same process that causes a free neutron to decay, and its average lifetime is measured in laboratory experiments to be rather long (around 890 seconds). The neutron abundance is therefore 'frozen in' at the value around the temperature of 0.8 MeV, which gives

$$\frac{n_n}{n_p} \sim e^{-1.29/0.8} \sim 0.2 \tag{9.14}$$

Before having time to decay, most neutrons will end up in helium nuclei through either of the two chains

$$p + n \leftrightarrow d + \gamma$$
$$d + d \leftrightarrow {}^3\text{He} + n$$
$${}^3\text{He} + d \leftrightarrow {}^4\text{He} + p \tag{9.15}$$

or

$$d + d \leftrightarrow {}^3\text{H} + p$$
$${}^3\text{H} + d \leftrightarrow {}^4\text{He} + n \tag{9.16}$$

The ratio of the rate for $p + n \leftrightarrow d + \gamma$ to the expansion rate is given by

$$\frac{\Gamma_{pn}}{H} \sim 2 \cdot 10^3 \left(\frac{T}{0.1 \text{ MeV}}\right)^5 \frac{n_p}{n_p + n_n} \Omega_B h^2 \tag{9.17}$$

which is a large ratio for $T \gg 0.1$ MeV. Since the number density of photons was so high, photodisintegration of deuterium was very efficient, and the deuterium abundance was held below 10^{-10} during equilibrium. This caused the $d + d$ reactions to go very slowly (the rate was proportional to the square of the small deuterium abundance), so that not much helium was produced for $T > 0.1$ MeV. Below this temperature, photodisintegration became inefficient, and the deuterium abundance rose to $\sim 10^{-5} - 10^{-3}$, which led to rapid $d + d$ fusion of helium. This consumed most of the neutrons, so a good estimate of the ^{4}He abundance is given by

$$Y({}^4\text{He}) \equiv \frac{4n_{He}}{n_{tot}} = \frac{4(n_n/2)}{n_n + n_p} = \frac{2n_n/n_p}{1 + n_n/n_p} \tag{9.18}$$

The ratio n_n/n_p was found to be around 0.2 at $T \sim 0.8$ MeV. After that time, some neutrons decay, so the ratio at the end of nucleosynthesis ($T \sim 0.01$ MeV) was around 0.13. This gives an abundance of around $Y({}^4\text{He}) \sim 0.24$, which also emerges after a full calculation. This is precisely the value that measurements of the helium abundance in stars and metal-poor gas clouds give within measurement uncertainties.

The agreement between calculations and observations of the helium abundance, besides being a strong piece of evidence in favour of the Big Bang model, can also be used to constrain the laws of physics that were valid during this early epoch of the Universe. An important effect for the value of $Y({}^4\text{He})$ was the depletion of the neutron-to-proton ratio due to neutron decay. If the expansion of the Universe was faster than the standard analysis gives, fewer neutrons would have had time to decay before being 'saved' into the stable existence inside helium nuclei, and the helium abundance would have increased. Since according to the Friedmann equation the expansion rate $H^2 \propto \rho$, and ρ is dominated by relativistic species, additional neutrinos (besides the three of the Standard Model) would speed up the expansion to make the helium abundance higher than allowed by observations. This fact was used before the LEP accelerator at CERN (near Geneva) went into operation to limit the number of neutrinos to less than or equal to four. (LEP subsequently determined the number to be $N_\nu = 3$.)

Some traces of ^{3}He and D also remain (being frozen out when the reactions in (9.15) and (9.16) dropped out of equilibrium), at the level of $10^{-5} - 10^{-4}$. Recently, the deuterium abundance has been in focus thanks to new measurements by D. Tytler and collaborators. With the Keck Telescopes and the Hubble Space Telescope, absorption lines corresponding to deuterium have been detected at high redshift. The light from distant quasars occasionally passes through intervening clouds. If these are also at high redshift, the material in the clouds should closely reflect the primordial abundance. (At least one obtains a good lower bound on the abundance, since deuterium is very fragile and can, as far as is known, only be destroyed, not produced, in stellar processes.) Deuterium is a very sensitive probe of the baryon-to-photon ratio, since the curve is very steep as a function of η_B, and therefore to $\Omega_B h^2$ (see Fig. 9.2). The current measurements indicate a quite high value of η_B, corresponding to

$$\Omega_B h^2 \sim 0.02 \tag{9.19}$$

This is still far from what is needed to explain the observations of the total matter density, however.

An abundance of the order of $10^{-10} - 10^{-9}$ of ^{7}Li is also predicted through the reactions

$$^4\text{He} + {}^3\text{H} \leftrightarrow {}^7\text{Li} + \gamma$$
$$^4\text{He} + {}^3\text{He} \leftrightarrow {}^7\text{Be} + \gamma$$
$$^7\text{Be} + n \leftrightarrow {}^7\text{Li} + p \tag{9.20}$$

Heavier elements are produced in truly negligible quantities, since there are no stable elements with $A = 5$ and $A = 8$ that could serve as intermediate steps. (Also, for nuclei with more than three protons, the Coulomb repulsion is too strong to be overcome by the thermal energies at the time of nucleosynthesis.)

9.4 Photon Recombination and Decoupling

9.4.1 Ionization Fraction – the Saha Equation

Nucleosynthesis took place during the first few minutes after the Big Bang, at temperatures between 1 MeV and 0.01 MeV. We now look at what happened long after that, when the temperature had fallen to the eV scale. To a good approximation we may, incorporating the asymmetry between matter and antimatter, set $n_{e^+} = 0$, $n_{\bar{p}} = 0$ and $n_{e^-} = n_p$. The electrons and the photons were thermally coupled to each other until their interaction rate fell below the expansion rate of the Universe.

The basic mechanism for scattering low-energy ($E_\gamma \ll m_e$) photons on electrons is Thomson scattering $\gamma + e^- \to \gamma + e^-$, which according to the estimate (6.27) has a cross-section of

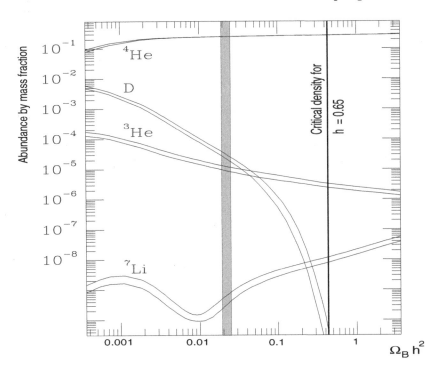

Fig. 9.2. The Big Bang nucleosynthesis predictions for the abundances of the light elements, as a function of the quantity $\Omega_B h^2$, which is the baryonic contribution to the present matter density. The vertical band indicates where the primordial deuterium abundance measurements lie. The observed helium and lithium abundances are in accordance with this range, given the measurement uncertainties. See [44] for further details.

$$\sigma_T = \frac{\alpha^2}{m_e^2} \qquad (9.21)$$

The total interaction rate per photon is then (since $v = c = 1$ for photons)

$$\Gamma_\gamma = n_e \sigma_T \qquad (9.22)$$

and when $\Gamma_\gamma / H < 1$ the photons decouple. This takes place at a temperature somewhere between 1 and 10 eV.

When we try to calculate n_e as a function of temperature a new feature appears. Electrons may 'disappear' by combining with protons to give atoms of neutral hydrogen plus photons. The hydrogen thus formed could then be reionized by photons. We thus have to consider reactions of the type $p + e^- \leftrightarrow H + \gamma$ in the primordial plasma. This means that the chemical potentials fulfil

$$\mu_p + \mu_e = \mu_H \qquad (9.23)$$

in equilibrium (photons have zero chemical potential). Since the baryons (protons) can either be free or bound as hydrogen, we introduce the total baryon density n_B

$$n_B = n_p + n_H \tag{9.24}$$

Charge neutrality also implies that

$$n_p = n_e \tag{9.25}$$

Since we are interested in processes that appear at temperatures of at most 10 eV (compare the binding energy of the ground state of hydrogen $B_1 = 13.6$ eV) e, p, and H are all non-relativistic to a very good approximation. We can therefore use (8.17) in the integrals defining n_i to find, in the non-relativistic limit,

$$n_i = g_i \left(\frac{m_i T}{2\pi} \right)^{\frac{3}{2}} e^{\frac{\mu_i - m_i}{T}} \tag{9.26}$$

for $i = e, p, H$. Using (9.23) and the relation $m_H = m_e + m_p - B$ which defines the binding energy B, we can solve for n_H to find

$$n_H = \frac{g_H}{g_e g_p} n_e n_p \left(\frac{m_e m_p T}{2\pi m_H} \right)^{-\frac{3}{2}} e^{\frac{B}{T}} \tag{9.27}$$

This can be simplified further by noting that $n_e = n_p$ and to a good approximation $m_p/m_H = 1$. Defining the ionization fraction X_e as

$$X_e \equiv \frac{n_p}{n_B} = \frac{n_p}{n_p + n_H} \tag{9.28}$$

and the baryon to photon ratio η_B

$$\eta_B \equiv \frac{n_B}{n_\gamma} \tag{9.29}$$

one obtains (Problem 9.5)

$$\frac{1 - X_e}{X_e^2} = \frac{4\sqrt{2}\zeta(3)}{\sqrt{\pi}} \eta_B \left(\frac{T}{m_e} \right)^{\frac{3}{2}} e^{B/T} \tag{9.30}$$

This is the so-called Saha equation for the fractional ionization at equilibrium. It can be solved to express X_e as a function of T and therefore as a function of redshift z (since $T = 2.73(1 + z)$ K). The baryon-to-photon ratio η_B is related to the baryon contribution Ω_B to the density of the Universe by (see (8.59) and (8.64))

$$\eta_B \equiv \frac{n_B}{n_\gamma} = 2.7 \cdot 10^{-8} \Omega_B h^2 \tag{9.31}$$

We shall see that Big Bang nucleosynthesis (the synthesis of the light nuclei helium, deuterium and lithium) constrains η_B to be in the interval $(5 - 6) \cdot 10^{-10}$, corresponding to $\Omega_B h^2 \sim 0.02$.

As shown in Fig. 9.3, the ionization fraction drops below 10 per cent at a redshift somewhere in the range 1200 - 1300. The process when the electrons are captured by the protons to form a hydrogen bound state is called recombination.

We have just shown that the redshift z_{rec} of the recombination is

$$z_{rec} \sim 1300 \tag{9.32}$$

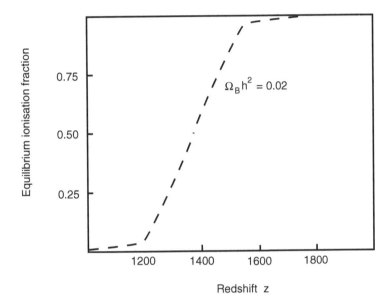

Fig. 9.3. The equilibrium fractional ionization as a function of redshift z, for a value of $\Omega_B h^2 = 0.02$.

This corresponds to a temperature at recombination of

$$T_{rec} = T_0(1 + z_{rec}) \sim 2.7 \cdot 1300 \text{ K} = 3500 \text{ K} \sim 0.3 \text{ eV} \tag{9.33}$$

and a time at recombination of (see (8.69))

$$t_{rec} = \frac{2}{3} H_0^{-1}(1 + z_{rec})^{-\frac{3}{2}} \sim 1.4 \cdot 10^5 / h \text{ years} \tag{9.34}$$

(If $\Omega_M \neq 1$, the right-hand side should be divided by $\sqrt{\Omega_M}$.)

Naively, one would expect that recombination occurs around $T = 13$ eV, since this corresponds to the binding energy of hydrogen. However, the Bose Einstein distribution of photons produces a long tail of energies higher than T. Since there are so many photons compared to the number of baryons (see equation (9.31) which is one of the most remarkable numbers in cosmology) they were efficient in ionizing hydrogen down to a temperature of 0.3 eV.

Our treatment has been based on the Saha equation, which is valid for processes that are in thermal equilibrium. Thus, we have to demand that the rate for the process $p + e^- \leftrightarrow H + \gamma$ is faster than the Hubble expansion rate. We shall soon see that this is the case down to redshifts around 1100. After that, equilibrium could not be maintained and the ionization fraction was frozen at the value it had around $z = 1100$. Also, the number density of photons per comoving volume was fixed, and it is the redshifted population of photons from this epoch that we can observe today as the cosmic microwave background radiation.

The residual ionization determines the mean free path of photons after recombination (since the dominant mechanism is Thomson scattering), and one can calculate that the mean free path becomes larger than the radius of the observable Universe at redshifts less than around 1050. The region around $z_{dec} \sim 1100$ is therefore sometimes referred to as the surface of last scattering of the cosmic microwave background photons.

We can finally address the question of the photon decoupling: that is, the 'freeze-out' of the ionization fraction. By comparing the formalism of $\chi + \bar{\chi} \leftrightarrow X + \bar{X}$ with $e + p \leftrightarrow H + \gamma$ we can immediately write down the simplified Boltzmann equation for n_e:

$$\dot{n}_e + 3Hn_e = -\langle\sigma_{rec}|\mathbf{v}|\rangle[n_e^2 - (n_e^{EQ})^2] \tag{9.35}$$

with $\langle\sigma_{rec}|\mathbf{v}|\rangle$ the thermally averaged recombination cross-section which can be calculated to be

$$\langle\sigma_{rec}|\mathbf{v}|\rangle = 4.7\cdot 10^{-24}\left(\frac{1\text{ eV}}{T}\right)^{\frac{1}{2}}\text{cm}^2 \tag{9.36}$$

The numerical analysis of solving (9.35) gives

$$T_f \sim 0.25\text{ eV} \tag{9.37}$$

and the residual ionization fraction

$$X_e(\infty) \sim 2.7\cdot 10^{-5}(\Omega_B h)^{-1} \tag{9.38}$$

It is quite plausible that this value increased when the first epoch of star-formation occurred, due to the injection into the primordial gas of radiation with high enough energy to re-ionize the gas.

Recombination is a very important process that took place a few hundred thousand years after the Big Bang. The details of the recombination history were worked out by Peebles [34]. The most important change from the very simplified calculation given here involves a more realistic treatment of the photon emission for electron capture. For instance, capture directly to the ground state immediately causes emission of a very energetic photon that directly can ionize another atom, leaving no net change. Also, capture to a highly excited state causes Lyman series photons to be produced in the allowed decays to the ground state, and these Lyman photons excite other atoms to states where they are photoionized again.

The production of atomic hydrogen rather occurs by two-photon decay from the metastable $2s$ level to the ground state. The loss of Lyman-α resonance photons by the cosmological redshift is also important, and has to be included in a full analysis [34].

We have seen that there exist many interesting relics from the Big Bang which appeared (froze out) at epochs with very different temperature, and where the effective number of relativistic species g_{eff} varied by two orders of magnitude. In Fig. 9.4 this number is shown as a function of time and temperature, and three very important epochs are indicated (we shall treat the cosmic microwave background in Chapter 11).

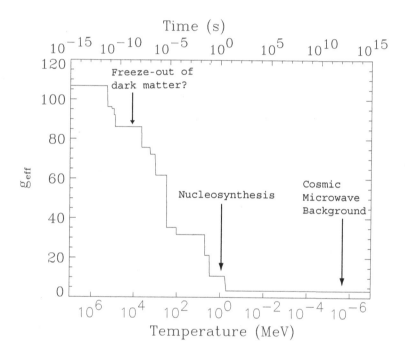

Fig. 9.4. The behaviour of the effective number of relativistic species g_{eff} computed according to the Standard Model, as a function of time and temperature. The three important epochs of the freeze-out of dark matter, synthesis of light elements, and release of the cosmic microwave background, are indicated. The numbers at the earliest epochs may be different, depending on the particle spectrum of the sector that accounts for the dark matter.

9.5 Summary

- The matter antimatter asymmetry of the Universe is of order 10^{-10}. Although the ingredients needed to generate such an asymmetry in the Universe exist in particle physics models, a detailed understanding of its origin is still lacking.

- Any stable neutral particle with weak interactions should have been produced in large quantities in the early Universe and would make a non-negligible contribution to the matter density in the Universe today, perhaps explaining the dark matter problem.
- Stable neutrinos in the mass range between 90 eV and 3 GeV are excluded for cosmological reasons.
- Big Bang nucleosynthesis took place during the first few minutes after the Big Bang. The observed abundances of helium, deuterium and lithium agree well with the predictions of the Big Bang model.
- Photons decoupled from thermal equilibrium at a redshift around 1100. Since that time of last scattering they have been redshifted by a corresponding factor and are observed today as the cosmic microwave background, one of the cornerstones of the Big Bang model.

9.6 Problems

9.1 Derive equation (9.7). (Hint: Use that $\dot{n}_\chi + 3Hn_\chi = s\dot{Y}_\chi$ due to the conservation of entropy, $sa^3 = const.$)

9.2 Suppose there exists a very light stable fermion ($g = 2$) which only had interactions with ordinary matter in the early Universe caused by the exchange (with the same coupling strength as the ordinary W, Z bosons) of a very heavy mediator (gauge boson) B, with $m_B \sim 10$ TeV.
(a) Estimate the decoupling temperature of the fermion.
(b) How heavy can the fermion maximally be not to overclose the Universe (that is, we demand $\Omega_M h^2 < 1$), if it was relativistic at freeze-out?

9.3 Assume that at $T_1 = 150$ MeV the scale factor was a_1 and the relativistic particles in thermal equilibrium were μ^+, μ^-, e^+, e^-, neutrinos and photons. At $T_2 = 10$ MeV, the μ^+ and μ^- particles had all annihilated or decayed, and the scale factor was a_2. Compute a_2/a_1.

9.4 Estimate the change in ^{4}He abundance that would be caused by a 10 per cent increase of the neutron proton mass difference.

9.5 Derive (9.30).

10 The Accelerating Universe

10.1 Problems of the Standard Big Bang Model

As the historic measurements by the COBE (Cosmic Background Explorer) satellite, beginning in 1991, have shown, the Universe is extremely isotropic on large scales (of the order of 1000 Mpc). The first anisotropy one sees when analyzing the data is a dipole pattern – the microwave background is somewhat colder than the average 2.73 K in one direction and warmer by the same fraction (of order 10^{-3}) in the diametrically opposite direction – explained by our peculiar motion with respect to the cosmic rest frame. Subtracting this dipole component, the temperature fluctuations that have been measured by COBE correspond to

$$\frac{\Delta T}{T} \sim 2 \cdot 10^{-5} \qquad \text{at an angular scale of 10 degrees} \qquad (10.1)$$

This is one of the motivations for using, as we have done, the Friedmann-Lemaître-Robertson-Walker metric for the expanding Universe.

We have seen that local thermal equilibrium could be maintained as long as the interaction rates were greater than the Hubble expansion rate. However, as discussed in Section 4.3.3, there are causal horizons which prevent regions that are far apart from interacting with each other. For matter domination in a flat ($k = 0$) Universe, the horizon distance grows as $d_H(t) = 3t$, whereas for radiation domination $d_H(t) = 2t$. Since $a(t) \sim t^{2/3}$ (matter domination) or $a(t) \sim t^{1/2}$ (radiation domination), we see that $d_H/a \sim t^{1/3}$ or $\sim t^{1/2}$: that is, in the past a much smaller fraction of the Universe was causally connected than it is today. (Note that as an estimate for both the matter and radiation eras one may take horizon size to be the Hubble time $H^{-1}(t) \sim t$.) How then is it possible that in well-separated regions in the sky the temperature is equal to such a high accuracy?

We can make this problem more quantitative by introducing the total entropy as a measure of the size of the causally connected region of the Universe. During radiation domination,

$$s = \frac{2\pi^2}{45} g_{\text{eff}}^s T^3 \qquad (10.2)$$

so that

$$S_{Hor}^{aD} \sim \frac{4\pi}{3} d_H^3 s \sim 0.1 \frac{1}{\sqrt{g_{eff}^s}} \left(\frac{m_{Pl}}{T}\right)^3 \tag{10.3}$$

whereas during matter domination $s = s_0(1+z)^3 \sim 3000(1+z)^3$ cm^{-3} and

$$S_{Hor}^{MD} \sim \frac{4\pi}{3} d_H^3 s \sim 10^{88}(1+z)^{-\frac{3}{2}} \tag{10.4}$$

This means that at recombination the entropy within the horizon was about 10^{83}, a factor of 10^5 smaller than the entropy of the observable Universe today. It must be explained therefore, how 10^5 regions, which were causally disconnected when the light was emitted, can have the same temperature to such a high accuracy. This is sometimes called the horizon problem. The angle subtended today by the horizon at photon decoupling is only around 0.8 degrees, which is thus the largest scale on which there should be causal smoothing of the microwave background. Yet it is isotropic over all angular scales.

Another problem is the mismatch between the smoothness on the largest scales and the inhomogeneities observed on smaller scales, such as galaxies, clusters of galaxies, superclusters and perhaps even larger structures. What is the mechanism that can generate the seeds of the perturbations that evolve into these structures?

Perhaps the most difficult problem for standard cosmology is that we cannot determine whether the Universe is flat, open or closed although the Universe is very old. That is, we know from observation that Ω_T is, to be generous, somewhere in range 0.2–2. If we remember that Ω varies with time, this has striking consequences. We have already made use of the fact that in the early Universe the curvature term $\sim k/a^2$ in the Friedmann equation was less important than the energy density term $8\pi G\rho/3$. We can make this more quantitative by writing approximately

$$\Omega_t(t) - 1 = \delta(t) \tag{10.5}$$

with

$$\delta(t) = \frac{k/a^2}{8\pi G\rho/3} \sim \begin{cases} a(t), & \text{matter domination} \\ a^2(t), & \text{radiation domination} \end{cases} \tag{10.6}$$

This shows that for early times (small a) Ω_t must have been very close to unity. For example, at the time of nucleosynthesis (around 1 sec), $\Omega_t(1\text{sec}) = 1 \pm 10^{-16}$, and in the earliest Universe at the Planck time $t = 10^{-43}$ sec, $\Omega_t(10^{-43} \text{ sec}) = 1 \pm 10^{-60}$.

Another way to express this is that the Universe must have been extremely flat at early times. Defining the physical radius of curvature of the Universe by $a_{curv} = a(t)/\sqrt{|k|}$ (this means that for a closed Universe it is just the physical radius of the three-sphere), one can show (see Appendix A)

$$a_{curv} = \frac{1}{H}\sqrt{\frac{k}{\Omega_t - 1}}, \quad k \neq 0 \tag{10.7}$$

Note that although we have scaled the metric so that $k = 0, +1, -1$, the closed and open models represent infinite classes of models, where one particular representative is given by specifying the a_{curv} at some given epoch.

From (10.7) and (10.6), we see that a_{curv} must have been enormous compared to the Hubble radius H^{-1}:

$$a_{curv}(1 \text{ sec}) > 10^8 \ H^{-1} \tag{10.8}$$

and

$$a_{curv}(10^{-43} \text{ sec}) > 10^{60} \ H^{-1} \tag{10.9}$$

The difficulty in explaining this remarkable flatness of the early Universe is sometimes called the flatness problem.

One should note that the horizon and flatness problems are not in contradiction with the Big Bang model itself: it is just a problem of initial conditions. Since the initial conditions for our observable Universe were set up at an epoch where physics is unknown, it is not clear how serious these problems are. However, it would be attractive to have a physical mechanism that gives such smooth initial conditions generically, without having to fine-tune the parameters of the model. Inflation is such a mechanism.

Another problem that is rather more technical, but which is also solved by inflation, is the so-called monopole problem. There are strong reasons to believe (as G. 't Hooft and A. Polyakov showed) that the grand unification of forces in nature implies that superheavy magnetic monopoles should have been produced at $T \sim 10^{15}$ GeV. Calculating their relic abundance one finds that they would have overclosed the Universe by a large factor unless there is some mechanism to dilute their number density. Inflation provides such a dilution mechanism.

Finally, there exists another problem in cosmology that is related to inflation but is not solved by it. This has to do with the extremely small value of the cosmological constant. We go back to the full Friedmann equation (4.16), showing explicitly the contribution of the cosmological constant Λ:

$$\left(\frac{\dot{a}}{a}\right)^2 = \frac{8\pi G\rho}{3} - \frac{k}{a^2} + \frac{\Lambda}{3} \tag{10.10}$$

and we see that the relevant comparison between the matter term $\sim \rho$ and the cosmological constant (vacuum energy) term $\sim \Lambda$ is given by their ratio:

$$r_\Lambda(T) \equiv \frac{\Lambda}{8\pi G\rho(T)} \tag{10.11}$$

From observation, we know that $r_\Lambda^0 < 1$, which means that at the Planck epoch, $r_\Lambda^{Planck} < 10^{-122}$, one of the smallest dimensionless numbers encountered in physics! The cosmological constant represents vacuum energy, and since the energy scale for gravity is m_{Pl} one would expect $\Lambda/8\pi G \sim m_{Pl}^4$, which is thus wrong by 122 orders of magnitude (see the discussion in Section 4.7). The reason for the smallness of the cosmological constant is still unknown, and therefore it is customary to just assume that today it is zero or very close to zero. However, at phase transitions (as in the Higgs mechanism), vacuum energy may be released. So, if the vacuum energy is close to zero today it may have been different from zero in the early Universe, before the phase transition. This is what led A. Guth (1981) to consider what happens to a cosmological model during a phase transition when vacuum energy is released.

10.2 The Inflation Mechanism

The Einstein equation including a cosmological constant reads

$$R_{\mu\nu} - \frac{1}{2}g_{\mu\nu}\mathcal{R} = 8\pi G T_{\mu\nu} + \Lambda g_{\mu\nu} \tag{10.12}$$

which shows that a cosmological term acts as a stress-energy tensor with the unusual equation of state $p_{vac} = -\rho_{vac}$. (We have noted before that one may include vacuum energy in the term proportional to G, with $\rho_\Lambda = \Lambda/(8\pi G)$.) This means that the entropy density $s \sim \rho + p \sim 0$ – that is, when vacuum energy dominates – we have a vanishing entropy.[1] In a situation where a constant vacuum energy dominates the expansion the Friedmann equation (10.10) becomes very simple:

$$H^2 = \left(\frac{\dot{a}}{a}\right)^2 = \frac{\Lambda}{3} \tag{10.13}$$

or

$$H = \frac{\dot{a}}{a} = \sqrt{\frac{\Lambda}{3}} = \text{const} \tag{10.14}$$

with the solution

$$a \sim e^{Ht} \tag{10.15}$$

This is the meaning (also in economic theory!) of inflation: the expansion rate is constant, leading to an exponential growth of the scale factor (which in economics is the price of a given commodity).

In typical models for inflation, the phase transition of a scalar field (sometimes called the inflaton field) took place at temperatures around the Grand Unification scale $T_{GUT} \sim 10^{15}$ GeV, when the Hubble time $H^{-1} \sim 10^{-34}$ sec. Suppose that the Universe remained in the inflationary state for 10^{-32} sec. (This may appear to be a short time, but remember that the relevant

---

[1] This can be understood from statistical mechanics. Entropy is related to the total number of degrees of freedom, and the vacuum (at least if it is unique) is just one state, that is only one degree of freedom. Of course, the entropy that was in a patch before inflation will be there after inflation – but it will be diluted by an exponential factor.

timescale in cosmology is the Hubble time, so inflation then lasted for 100 Hubble times: that is, 100 times the age of the Universe at the time inflation started.) After inflation stopped, the vacuum energy of the inflaton field was transferred to ordinary particles, so a reheating of the Universe took place. (During inflation itself, the Universe supercooled, since the entropy was constant, and low, meaning aT was constant so that $T \sim e^{-Ht}$.) The reheating temperature is of the order of the temperature of the phase transition, so $T_{RH} \sim 10^{15}$ GeV (if the inflaton is strongly enough coupled to ordinary matter, as it is in successful models of inflation).

Consider a small region with radius around, say, 10^{-23} cm before inflation. The entropy within that volume is only around 10^{14}. After inflation, the volume of the region has increased by a factor of $(e^{100})^3 = 1.9 \cdot 10^{130}$, so after the entropy generation provided by reheating, the total entropy within the inflated region has grown to around 10^{144}. The entropy is generated because the equation of state changes from $p = -\rho$ to $p = \rho/3$, which means that the entropy density $s \sim p + \rho$ increases dramatically.

This huge entropy increase solves three of the four problems mentioned above. The horizon problem is solved since our whole observable Universe may have arisen from a very small volume that was in thermal contact before inflation. The smooth region after inflation has more than enough entropy to encompass our observable Universe many times over.

During inflation the energy density of the Universe is constant, whereas the scale factor a increases exponentially. This means (see (10.6)) that Ω_t after inflation must have been exponentially close to unity, and the present value $\Omega_t(t_0) \equiv \Omega_T = 1$ to an accuracy of many decimal places. Inflation thus solves the flatness problem. (Another way to see this is that all local curvature of the original volume is smoothed out by the expansion.) Even if $\Omega_T = 1$ is predicted to high accuracy, there is nothing a priori which tells us the subdivision of Ω_T into contributions from radiation, matter and vacuum energy. As we have noted, however, the 'natural' contribution of Ω_Λ is either extremely small or extremely large. Only during very brief epochs can Ω_Λ be of similar magnitude as the matter contribution Ω_M (see Section 4.3.2).

The monopole problem is also solved, since the number density of such objects is diluted by a factor of the order of e^{300} by the inflation. The same is true for cosmic strings and all other topological defects. (However, if there were other phase transitions after the epoch of inflation, defects could have been formed again.)

The period of inflation and reheating is strongly non-adiabatic, since there is an enormous generation of entropy at reheating. After the end of inflation, the Universe 'restarts' in an adiabatic phase with the standard conservation of aT, and it is because the Universe automatically restarts from very special initial conditions that the horizon, flatness and monopole problems are avoided. This is schematically shown in Fig. 10.1.

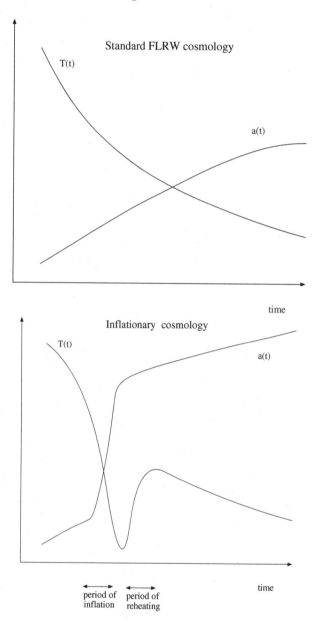

Fig. 10.1. The schematic evolution of a and T in inflationary cosmology. During the inflation epoch, T decreases exponentially as the scale factor a increases exponentially. At the end of inflation, an enormous amount of entropy is generated, and the Universe starts out in a phase which is similar to ordinary FLRW cosmology from then on.

10.3 Models for Inflation

Inflation is a very attractive scenario, and there is even some observational support for it (the pattern of fluctuations in the cosmic microwave background seem consistent with the predictions from inflation). However, it has proven to be quite difficult to construct theoretical models which give the right amount of inflation, and which end the inflationary epoch in the required way with huge entropy generation.

Remember (see (6.12)) that a Lagrangian density of the form

$$\mathcal{L} = \frac{1}{2}\partial^\mu\phi\partial_\mu\phi - V(\phi) \tag{10.16}$$

gives a contribution to the energy-momentum tensor $T^{\mu\nu}$ of the form

$$T^{\mu\nu} = \partial^\mu\phi\partial^\nu\phi - \mathcal{L}g^{\mu\nu} \tag{10.17}$$

For a homogeneous state, the spatial gradient terms vanish, and $T^{\mu\nu}$ becomes that of the perfect fluid type[2] with

$$\rho = \frac{\dot\phi^2}{2} + V(\phi) \tag{10.18}$$

and

$$p = \frac{\dot\phi^2}{2} - V(\phi) \tag{10.19}$$

as can be seen by inserting (10.16) and (10.17) and comparing with (3.53).

The equations of motion of ϕ can be derived from the condition of vanishing covariant divergence of the energy-momentum tensor, $T^{\mu\nu}_{\;;\nu} = 0$, which gives (Problem 10.2)

$$\ddot\phi + 3H\dot\phi + V'(\phi) = 0 \tag{10.20}$$

This is similar to the equation of motion of a ball in a potential well with 'Hubble friction' $\sim 3H\dot\phi$, and can be solved by elementary methods. We assume that at very high temperatures, $\phi = 0$ gives the minimum of the potential, but temperature-dependent terms in the effective potential generate another minimum for $\phi = \phi_{vac} \neq 0$ (spontaneous symmetry breakdown). To produce a long enough period of inflation and a rapid reheating after inflation, the potential $V(\phi)$ has to look something like the one shown in Fig. 10.2.

In the beginning, on the almost horizontal slow 'roll' towards the deep potential well, $\ddot\phi$ can be neglected, and the slow-roll equation of motion

$$3H\dot\phi + V'(\phi) = 0, \tag{10.21}$$

together with the Friedmann equation

---

[2] If one keeps the gradient terms, one sees that they are divided by $a(t)^2$, which means that after a short period of inflation they are exponentially suppressed.

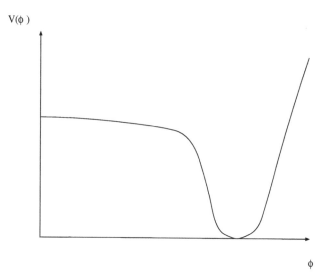

V(ϕ)

ϕ

Fig. 10.2. Approximate shape of the inflaton potential $V(\phi)$ needed to produce an acceptable time evolution during and after inflation.

$$H^2 = \frac{8\pi G}{3}\left[\frac{1}{2}\dot{\phi}^2 + V(\phi)\right], \tag{10.22}$$

which during slow roll can be approximated by

$$H^2 = \frac{8\pi G}{3}V(\phi), \tag{10.23}$$

gives (Problem 10.4) a convenient expression for the number N_ϕ of e-folds of the scale factor. Here we use the near constancy of H to write $a_2/a_1 = \exp(\int_{t_1}^{t_2} H(t))$: that is,

$$N_\phi \equiv \log\left(\frac{a_2}{a_1}\right) = \int H\,dt = -8\pi G \int_{\phi_1}^{\phi_2}\frac{V(\phi)}{V'(\phi)}d\phi \tag{10.24}$$

Thus, for a large growth of the scale factor, $V(\phi)$ has to be very flat ($V'(\phi) \sim 0$). At present, there is no natural explanation for such a potential except perhaps in some supersymmetric theories where 'flat directions' can occur because of the pattern of supersymmetry breaking. In a situation of such a slow roll of the inflaton field, the exact form of the potential does not matter so much, and the relevant physics can be expressed in terms of the so-called slow-roll parameters [28]

$$\varepsilon = -\frac{\dot{H}}{H^2} = 4\pi G\frac{\dot{\phi}^2}{H^2} = \frac{1}{16\pi G}\left(\frac{V'}{V}\right)^2 \tag{10.25}$$

$$\eta = \frac{1}{8\pi G}\left(\frac{V''}{V}\right) = \frac{V''}{3H^2} \tag{10.26}$$

where the second equation in (10.25) comes from taking the derivative of (10.22) and inserting into (10.20). The variable ε is a measure of the change of the Hubble expansion during inflation; for inflation to happen at all, $\varepsilon < 1$ is needed (Problem 10.3).

In the picture of the rolling ball, reheating corresponds to oscillations in the potential well. Thus, for enough entropy to be generated the well has to be rather steep. The problem of constructing a suitable potential is to simultaneously have it flat near $\phi = 0$ and steep near $\phi = \phi_{min}$.

A way to avoid a phase transition, and in fact the simplest model of inflation is the *chaotic inflation* model of Andrei Linde [29]. It relies on the fact that the key thing for inflation to occur is that the field is rolling slowly, so that the energy density is nearly constant while the scale factor grows exponentially. Since the rolling is damped by the presence of the term proportional to H in (10.20), and H according to the Friedmann equation is given by the height of the potential (if kinetic terms can be neglected), inflation will be possible for any positive, power-law potential $V(\phi)$, for example the simplest $V(\phi) = \frac{1}{2}m^2\phi^2$, as long as the field values start out large. As Linde has argued, this may not be unreasonable since these initial values may be given randomly ('chaotically') at the Planck epoch, and those regions where the field values are large start to inflate rapidly dominating the volume of the Universe.

Inflation is very close to becoming an ingredient in the Standard Model of cosmology. Active research is going on to solve the theoretical problem of generating a suitable potential, and to describe reheating in a more detailed way than has been sketched here.

If a value of the total energy density $\Omega_T = 1$ is found observationally in the future[3], the most natural explanation would be that the Universe has indeed gone through a period of inflation. Another indirect test of inflation may be produced by the upcoming measurements of the detailed pattern of temperature fluctuations in the cosmic microwave background radiation. We will see in Chapter 11 that inflation predicts a nearly scale-invariant spectrum of such fluctuations on large scales.

10.4 Dark Energy

Current observations from distant Type Ia supernovae as well as the combination of the sub-degree anisotropies of the CMBR (described in Chapter 11) and mass energy density estimates from galaxy clusters, weak lensing and large scale structure suggest that the Universe is also going through an inflationary phase at present. More specifically, the density in some sort of *dark energy*, ρ_Q, with a negative equation of state parameter, $\alpha_Q = p_Q/\rho_Q < 0$,

---

[3] Current measurements of the CMBR anisotropy yield $\Omega_T = 1.02 \pm 0.02$, as discussed in Chapter 11

should have overcome the ever diluting matter content at about a redshift $z \sim 1$. It is unclear if the dark energy is identical with the vacuum energy density related to the cosmological constant, ρ_Λ, or if it is in the form of some new type of matter: *'quintessence'*. The most popular models of quintessence invoke what we so far have considered for the inflation of the early Universe: a scalar field Q slowly rolling in a potential. Thus, quintessence models are dynamical as opposed to the case of the cosmological constant. This presents some advantages while trying to match the current observations. A dynamical energy density may evolve with time towards zero, and the current small values measured may be explainable in such scenarios. Moreover, a successful quintessence model could shed light on the coincidence problem related to Λ. While Ω_m and Ω_λ are very comparable *at present*, they were not in the early Universe $(a \to 0)$ nor in the future $(a \to \infty)$. As the Universe expands the relative abundances change as:

$$\frac{\rho_\lambda}{\rho_m} \propto a^3 \tag{10.27}$$

In the cosmological constant scenario, it would seem that there is something very special and possibly unnatural about the present time. An observer on one of the first galaxies, $z \sim 10$, would have probably been unable to detect the existence of a non-vanishing Λ. In the future, the Λ term would be the only thing that would be noticed. In fact, eventually will distant galaxies reach exponentially large redshifts and the Universe become cold and dark. Thus, the ability of an observer to deduce the 'true' content of the Universe would be optimal at about *this* time.

Following these considerations, it seems useful to explore the dynamical dark energy models, as these may ease the cosmological constant problems. The dark energy may have been decreasing since the start of time, $\rho_{de} \propto t^{-\beta}$, and could therefore be very small today.

Invoking the relations in (10.18) and (10.19) for a homogeneous scalar field fluid in a potential $V(Q)$, we find that the equation of state parameter for the fluid becomes:

$$\alpha_Q = \frac{\dot{Q}^2/2 - V(Q)}{\dot{Q}^2/2 + V(Q)}. \tag{10.28}$$

Thus, for $\dot{Q}^2 \ll V$, $\alpha_Q \to -1$, i.e. like α_Λ case. In general we find $-1 \leq \alpha_Q \leq 1$. Recalling (4.28) we notice that an accelerating Universe is only possible when the dominating fluid fulfills $\alpha_Q < -\frac{1}{3}$:

$$\frac{\ddot{a}}{a} = \frac{-4\pi G}{3} \sum_i \rho_i (1 + 3\alpha_i) \tag{10.29}$$

Some scalar field potentials are particularly interesting as they produce solutions where the dark energy scale with the matter or radiation energy independently of initial conditions. In addition, they may be derived from particle-physics models. Examples are $V(Q) \propto e^{-Q}$ and $V(Q) \propto Q^{-1}$. The

scalar field couples with the other fluids through the Friedmann equation
(4.16), the equation of motion of Q (10.20) and the energy conservation equation (4.24):

$$
\begin{cases}
H^2 = \frac{8\pi G}{3}\left(\rho + \frac{1}{2}\dot{Q}^2 + V\right) \\[2mm]
\ddot{Q} + 3H\dot{Q} + V'(Q) = 0 \\[2mm]
\dot{\rho} + 3(1+\alpha)H\rho = 0,
\end{cases}
\qquad (10.30)
$$

where ρ and α refer to the background dominant fluid of standard cosmology, i.e. radiation in the early Universe and later non-relativistic matter.
Solutions to the coupled equations in (10.30) for exponential potentials yield
$\rho_{de} = \left(V + \frac{1}{2}\dot{Q}^2\right) \propto t^{-2}$, just like ρ_m and ρ_{rad}. Thus, the fraction of dark
energy in the Universe is constant in time. For other potentials there are
solutions for which the dark energy density falls less rapidly with time than
the background density, $\rho_{de} \propto t^{-(2-\delta)}$, leading to a transition from matter to
dark energy domination. The exact epoch at which the dark energy overtakes
the background density is not predicted by the models, thus the coincidence
problem is not completely solved.

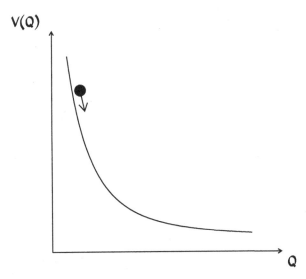

Fig. 10.3. Schematic view of a rolling quintessence scalar field.

Example 10.4.1 Show that $\rho_{de} \propto t^{-2}$ for the exponential potential, $V = e^{-Q}$.
Use the *ansatz* $a = t^\beta$ and $Q = Q_0 \ln t$.

Answer: With the *ansatz* on Q we find the derivatives Q to be $\dot{Q} = Q_0 t^{-1}$ and $\ddot{Q} = -Q_0 t^{-2}$. The potential and its gradient become $V' = dV/dQ = -t^{-Q_0} = -V$.

Inserting these expressions into the equation of motion with $H = \frac{\dot{a}}{a} = \beta t^{-1}$ we arrive at the equation:

$$-\frac{Q_0}{t^2} + 3\beta\frac{Q_0}{t^2} - \frac{1}{t^{Q_0}} = 0,$$

which is fulfilled for $\beta = \frac{1}{2}; Q_0 = 2$.

Inserted into the expression for the dark energy density, $\rho_{de} = V + \frac{1}{2}\dot{Q}^2$ we find the anticipated result, $\rho_{de} \propto t^{-2}$.

The behaviour of the slowly rolling 'quintessence' field shown in Fig. 10.3, approximately tracking the energy density of the dominant background fluid, may be qualitatively understood examining the coupled equations in (10.30). The size of the force term $V'(Q)$ scales with $-V$, or even a higher power of V. It can be shown that the friction is subdominant during this period. Thus, for a large value of V the dark energy decreases rapidly, faster than matter or radiation. On the other hand, if $\rho_m \gg \rho_{de}$ or $\rho_{rad} \gg \rho_{de}$, the force is small in comparison with the friction term $3H\dot{Q}$ and Q comes nearly to a halt until the radiation or matter density is small enough. Thus, stability is approached when the dark energy density is comparable to the leading term of the background fluid.

Unfortunately, the dynamical energy models introduce new problems. For example, in order for the scalar field Q to play a an important role at this time it should have a mass comparable to the current Hubble scale, $m_Q \sim H_0 \sim 10^{-33}$ eV. Such a light scalar appears extremely unnatural.

10.5 Summary

- Inflation is a generic mechanism which can solve many of the problems of standard cosmology. Among these are the smoothness, flatness and monopole problems.
- The most solid predictions from inflation are that $\Omega_T = 1$ to high accuracy, and that fluctuations in the microwave background should be nearly scale-invariant.
- The simplest particle physics models for inflation utilize a potential for a hypothetical inflaton field which is temperature-dependent and similar to a Higgs potential. However, to make inflation work quantitatively the potential must have unusual properties. No compelling explicit model has yet been found.
- Dark energy may be causing the present Universe to accelerate in a similar fashion as the inflation of the early Universe. This

behaviour may be due to a cosmological constant or a dynamical scalar field 'rolling' in a suitable potential.

10.6 Problems

10.1 Derive (A.58).

10.2 Derive the equation of motion $\ddot{\phi}+3H\dot{\phi}+V'(\phi)=0$ from the vanishing covariant divergence of the energy-momentum tensor.

10.3 Show that $\frac{\ddot{a}}{a}=\dot{H}+H^2$, and use this to show that inflation only takes place when the slow-roll parameter $\varepsilon<1$.

10.4 Derive (10.24).

10.5 Compute $N_\phi=\log(a_2/a_1)$ for a potential of the form $V(\phi)=\lambda\phi^4+c_1$, with λ small and c_1 a constant. Express the solution in terms of H and λ.

11 The Cosmic Microwave Background Radiation and Growth of Structure

11.1 The First Revolution: the 2.7 K Radiation

The detection of the cosmic microwave radiation was the most spectacular evidence supporting the Big Bang theory after Hubble's discovery of the expansion of the Universe. Two radio astronomers at the Bell Telephone Laboratories, Arno Penzias and Robert Wilson, submitted their revolutionary result for publication in May 1965. By then, they had 'failed' to eliminate an excess noise[1] at 4080 MHz in their 20-foot antenna at Holmdel, New Jersey (USA). The background 'noise' came from all directions, and the authors estimated that the excess signal corresponded to a residual temperature of 3.5 ± 1.0 Kelvin.[2]

The explanation for the source of the observed signal was published in an accompanying paper in the same volume of the *Astrophysical Journal* by the Princeton group consisting of R.H. Dicke, P.J.E. Peebles, P.G. Roll and D.T. Wilkinson.[3] Let us recap on the origin of the CMBR (cosmic microwave background radiation). In a hot and dense medium, such as the early Universe soon after the Big Bang, we have seen that thermal electromagnetic radiation was generated and kept in equilibrium through reactions of the type $\gamma + \gamma \leftrightarrow e^+ + e^-$ for temperatures $T \sim m_e$. In the subsequent expansion of the Universe the radiation cooled adiabatically. Below the threshold for pair-production the photons were kept in thermal equilibrium through processes such as Compton scattering on free electrons:

$$e^- + \gamma \to e^- + \gamma$$

The electrons, in turn, were thermally linked to the protons through electromagnetic interactions. Eventually, as the temperature fell well below the photoionization energy for hydrogen, the reaction

---

[1] Now recognized as a *signal*.

[2] Penzias and Wilson shared the 1978 Nobel prize in physics.

[3] The Princeton group had already started the construction of their own radiometer to look for the cosmic microwave background radiation when, by chance, they heard about the 'problems' that Penzias and Wilson were having with an unexplained source of isotropic noise in the telescope at Holmdel, at just the right temperature level to be consistent with the relic cosmic radiation.

$$H + \gamma \leftrightarrow p + e^-$$

was no longer in thermal equilibrium, and the photons 'decoupled' from matter and still move essentially unscattered through the entire Universe. We saw in (9.37) that this happened a few hundred thousand years after the Big Bang at the freeze-out temperature of $T_f \sim 0.25$ eV. These photons have thus been travelling on geodesics, without scattering, through the Universe for 13 - 14 billion years and have by now cooled to just below 3 K due to the expansion.

The existence of a residual electromagnetic radiation from the Big Bang in the microwave range had already been predicted in 1946 by George Gamow and co-workers. In spite of their assumptions being partially wrong, their estimate of the relic radiation temperature was very close to the temperature measured 19 years later. Gamow's result was a by-product as he tried to establish that all nuclei, including the heavy ones, were produced during the Big Bang.[4]

At present, the relic temperature of the Universe has been measured with amazing precision: 2.725 ± 0.004 K (95 per cent confidence level) [14], mainly through air-borne radiometers. Ground-based detectors are at a disadvantage as they have to subtract the ambient 300 K radiation from the environment. Fig. 11.1 shows the result of the compiled data from 1985 to 1996. The bulk of the data comes from the FIRAS instrument on board the COBE satellite launched in late 1989.

The cosmic microwave background radiation decoupled from matter at some value of $z \sim 1000$, when the scale factor of the Universe was about 1000 times smaller than its current size. Thus, the original wavelength of the radiation was 1000 times smaller, and the energy consequently 1000 times larger, than that observed today (see (4.57)).

11.1.1 Thermal Nature of the CMBR

In the hot Big Bang picture, photons in the early Universe were continuously created, absorbed or annihilated and re-emitted: that is, the Universe was an almost perfect black-body.

Under such physical conditions, where an ensemble of photons is maintained at thermal equilibrium with the environment at some temperature T, the mean number of photons per oscillation mode is given by the Planck distribution:

$$\bar{n} = \frac{1}{e^{2\pi/T\lambda} - 1} = \frac{1}{e^{\omega/T} - 1} \tag{11.1}$$

---

[4] As we saw in Section 9.3, it is now believed that only the light elements up to ^7Li were produced in the Big Bang. The heavier elements have been produced in stars.

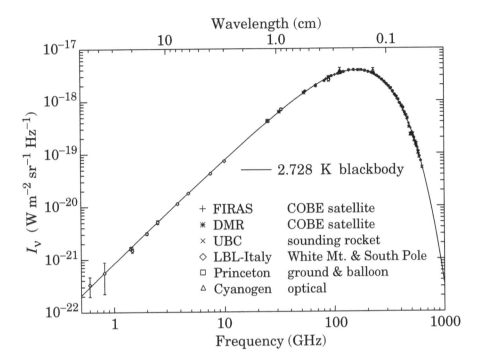

Fig. 11.1. Precise measurements of the CMBR spectrum. The line represents a 2.73 K black-body, which describes the spectrum very well, especially around the peak in intensity. The spectrum is less well constrained at frequencies of 3 GHz and below (10 cm and longer wavelengths). From G. Smoot, astro-ph/9705101 (1997), updated according to [14].

where $\omega = 2\pi/\lambda = 2\pi\nu$ is the angular frequency of the oscillation mode with wavelength $\lambda = 1/\nu$ (we put $c = \hbar = k_B = 1$, as usual).

We gave the expression for the well-known black-body radiation spectrum in section 8.2, but here we provide some more details to introduce a Fourier mode description which is useful when we later discuss fluctuations in the CMBR. To obtain the total spectrum we must multiply (11.1) with the number of oscillation modes per unit volume.

To obtain this we first, for a given cosmic time t, divide the entire space into an array of identical boxes with volume $V = L^3$, where the length of each side, L, is such that

$$L \gg \lambda \tag{11.2}$$

where $\lambda = 2\pi/k$ is the longest wavelength under consideration.

Next we impose the boundary condition that the plane wave of each single oscillation mode, $\psi = e^{i\mathbf{kr}} = e^{i(k_x x + k_y y + k_z z)}$, should be identical in each box:

$$\psi(x + L, y, z) = \psi(x, y, z)$$
$$\psi(x, y + L, z) = \psi(x, y, z) \tag{11.3}$$
$$\psi(z, y, z + L) = \psi(x, y, z)$$

Note that the requirement of periodicity does not affect the physics of interest as long as the box dimensions are large compared to the wavelength under consideration, as specified in (11.2).

By imposing the periodic boundary conditions in (11.3) the propagation vector becomes quantized:

$$\mathbf{k} = \frac{2\pi\mathbf{n}}{L} \tag{11.4}$$

where the components of $\mathbf{n}$, (n_x, n_y, n_z), are any set of integers: positive, negative or zero.

Thus, the number of possible states (plane waves) with propagation vector between $\mathbf{k}$ and $\mathbf{k} + d\mathbf{k}$ is

$$\begin{aligned}
\Delta N_k &= \Delta n_x \Delta n_y \Delta n_z \\
&= \left(\frac{L}{2\pi} dk_x\right)\left(\frac{L}{2\pi} dk_y\right)\left(\frac{L}{2\pi} dk_z\right) \\
&= \frac{V}{(2\pi)^3} d^3\mathbf{k} \\
&= \frac{V}{(2\pi)^3} 4\pi k^2 dk
\end{aligned} \tag{11.5}$$

where in the last step we have taken into account that the radiation is emitted in all directions and the integrated solid angles is therefore 4π.

We can thus conclude that there are $(2\pi)^{-3} 4\pi k^2 dk$ photon states with energy $E = k = \omega$ per unit volume. Since the mean number of photons with energy E is given by (11.1) we are ready to compute the mean number of photons per unit volume with angular frequency between ω and $\omega + d\omega$ in a photon gas at temperature T. Taking into account that there are *two* possible polarization states for each photon with angular frequency ω (that is, $g_\gamma = 2$) we find from (11.1) and (11.5), for completeness temporarily re-inserting factors of c, $\hbar$ and k_B:

$$n(\omega; T)d\omega = \frac{1}{\pi^2 c^3} \frac{\omega^2 d\omega}{e^{\hbar\omega/k_B T} - 1} \tag{11.6}$$

We have produced a full derivation of this result, but we could have taken it directly from (8.12) and (8.13).

What happens to the temperature of the photons as the Universe expands? In order to predict what can be observed *today*, billions of years after the decoupling of the radiation, we have only to carry out a small exercise. From (4.57) it follows that, at a later epoch, the angular frequency ω_0 is redshifted to

$$\omega_0 = \frac{\omega}{1 + z} \tag{11.7}$$

After decoupling, the Universe was essentially transparent to radiation. As the radiation is free of interactions the number of photons must be conserved.

To obtain the currently observable *density* of photons per unit frequency we introduce the variable transformation in (11.7) into the function in (11.6):

$$n(\omega; T)d\omega = \frac{1}{\pi^2 c^3} \frac{[\omega_0(1+z)]^2 d\omega_0 \cdot (1+z)}{e^{\frac{\hbar\omega_0(1+z)}{T}} - 1}$$

$$= n(\omega_0; T_0) \cdot (1+z)^3 \tag{11.8}$$

i.e, the density of photons preserves the black-body spectrum and scales proportionally to the inverse volume of the Universe. The black-body temperature decreases linearly with the radial scale; that is, with $(1+z)$

$$T_0 = T_e \frac{1}{1+z}. \tag{11.9}$$

The differential energy spectrum for the black-body radiation, $u(\omega)d\omega$, is obtained by multiplying the photon energy, $\hbar\omega$, in (11.6)

$$u(\omega; T)d\omega = \frac{\hbar}{\pi^2 c^3} \frac{\omega^3 d\omega}{e^{\hbar\omega/kT} - 1}. \tag{11.10}$$

Example 11.1.1 Estimate the fraction of the total energy density of the Universe, Ω_T, which is in the form of relic photons from the Big Bang, by integrating (11.10).

Answer:
The *total* energy density in the form black-body radiation, ρ_γ, in *all* frequencies is given by the integral over all frequencies:

$$\rho_\gamma = \int_0^\infty u(\omega; T)d\omega \tag{11.11}$$

To simplify this calculation we introduce a dimensionless parameter, ξ:

$$\xi = \frac{\hbar\omega}{kT} \tag{11.12}$$

Equation (11.11) can then be re-written as

$$\rho_\gamma = \frac{\hbar}{\pi^2 c^3} \left(\frac{kT}{\hbar}\right)^4 \int_0^\infty \frac{\xi^3 d\xi}{e^\xi - 1} \tag{11.13}$$

where the definite integral is just a numerical constant:

$$\int_0^\infty \frac{\xi^3 d\xi}{e^\xi - 1} = \frac{\pi^4}{15} \tag{11.14}$$

Thus we arrive at the familiar Stefan Boltzmann relation:

$$\rho_\gamma = \sigma T^4,$$
$$\sigma = 4.72 \cdot 10^{-3} \text{eV}/\text{cm}^3/\text{K}^4$$

We have seen in Section 4.3 that the critical density separating an open from a closed Universe is:

$$\rho_c = \frac{3H_0^2}{8\pi G} = 1.9 \cdot 10^{-29} h^2 \text{ g/cm}^3 = 1.1 \cdot 10^4 h^2 \text{ eV/cm}^3 \tag{11.15}$$

The fraction of total energy in the Universe in the form of relic radiation ($T = 2.73$ K) is thus (see (8.60)):

$$\Omega_\gamma h^2 \approx 4 \cdot 10^{-5}$$

The energy density at present is thus dominated by matter. The contribution from radiation is negligible.

11.2 The Second Revolution: the Anisotropy

Although the isotropic nature of the CMBR supports the Big Bang scenario, it poses difficulties in explaining the Universe at the present epoch. Clearly, if we look around us in the sky, what we see is far from homogeneous. The average temperature and density of galaxies differ dramatically from that of the space between them.[5]

11.2.1 Temperature Fluctuations and Density Perturbations

In the Big Bang model, the structure that we observe today is formed by gravitational instability. Small perturbations of density in the otherwise homogeneous matter distribution of the early Universe will, due to gravitational attraction, grow and eventually form the stars and galaxies known to exist today.

The observation of such *seeds* of density fluctuations by the COBE DMR instrument (colour Plate 5, in the middle of the book) thus propelled observational cosmology into a new era. Anisotropies at the 10^{-5} level in temperature were detected in the all-sky maps collected over several years, and were first announced in 1992. Temperature differences between patches of the sky appear when photons from a region close to a density enhancement get redshifted as they climb out of the gravitational potential well in the surface of last scattering.

On large scales, such as the ones probed by the DMR instrument on board the COBE satellite (with angular resolution about 10 degrees), deviations from temperature isotropy in the cosmic microwave background are due to the gravitational effects on the energy of the radiated photons: the *Sachs*

---

[5] A more subtle problem with the high degree of isotropy in the CMBR is that some of the patches of the sky that have the same temperature could not have been in causal contact at the time of decoupling. This problem was referred to as the *horizon problem* and was dealt with in the chapter about inflation.

Wolfe effect. Next we derive this relation between the observable temperature fluctuations in the sky and the underlying gravitational potential.[6]

Energy conservation implies that photons travelling through a changing gravitational potential ϕ lose energy as

$$\left(\frac{\Delta T}{T}\right)_0 = \left(\frac{\Delta T}{T}\right)_e + \phi_e \tag{11.16}$$

where the subscript 0 defines the observed temperature fluctuations, and the terms with subscript e describe the physical temperature fluctuation and gravitational field at the point of emission.

The first term on the right-hand side corresponds to an intrinsic temperature fluctuation in the early Universe, and the second is the potential well surrounding the photons.

For adiabatic fluctuations, the number of photons inside the potential well is expected to be larger than average: that is, one expects the temperature to be higher than in the surrounding regions. Therefore, one expects the two terms in (11.16) to partially cancel. In particular, because the over-density inside the potential well corresponds to a hot spot one expects the intrinsic temperature fluctuation to be proportional to the strength of the gravitational potential:

$$\left(\frac{\Delta T}{T}\right)_e \sim -\phi_e \tag{11.17}$$

We found in (11.9) that temperature is inversely proportional to the scale factor of the Universe, $aT = const$: that is, $\Delta(aT) = 0$. This in turn implies

$$\frac{\Delta T}{T} = -\frac{\Delta a}{a} \tag{11.18}$$

For an equation of state, $p = \alpha\rho$, it follows from Einstein's equations of the expansion of a flat Universe that the scale factor dependence with time is (see (4.29))

$$a(t) \sim t^{\frac{2}{3(1+\alpha)}} \tag{11.19}$$

which means

$$\frac{\Delta a}{a} = \frac{2}{3(1+\alpha)}\frac{\delta t}{t} \tag{11.20}$$

Recalling that in the neighbourhood of strong gravitational fields time is slowed according to (3.39):

$$d\tau = \sqrt{1 + 2\phi}\,dt \simeq (1 + \phi)dt \tag{11.21}$$

so that

---

[6] This simplified derivation, more intuitive than the one in Sachs and Wolfe (1967) was originally described by White and Hu (1996).

$$\frac{\delta t}{t} \simeq \phi \tag{11.22}$$

we find by insertion of (11.20) and (11.18) into (11.17):

$$\left(\frac{\Delta T}{T}\right)_e = -\frac{\Delta a}{a} = -\frac{2}{3(1+\alpha)}\frac{\delta t}{t} = -\frac{2}{3(1+\alpha)}\phi_e \tag{11.23}$$

Combined with (11.16) we reach the general result of this calculation:

$$\left(\frac{\Delta T}{T}\right)_0 = \frac{1+3\alpha}{3+3\alpha}\phi_e \tag{11.24}$$

For a matter dominated Universe, $\alpha = 0$, thus the measured temperature fluctuations correspond to the size of the potential well times a factor of $1/3$:

$$\left(\frac{\Delta T}{T}\right)_0 = \frac{1}{3}\phi_e \tag{11.25}$$

Note that there are two effects competing: the gravitational redshift which decreases the measured temperature of dense regions, and the heating caused by the local compression of the matter in dense regions. They partly cancel, but the net effect is that overdense regions appear cooler.

The DMR instrument measured temperature fluctuations of about $\frac{\Delta T}{T} \sim 10^{-5}$. These can be interpreted as being due to a spatially varying gravitational potential at the time the CMBR was emitted. We have seen that this happened at $z \sim 1000$, which corresponds to some 300,000 years after the Big Bang.

To summarize this section: the study of fluctuations in the CMBR allows us to determine the gravitational potential at the time the photons decoupled (the 'surface of last scattering'). The large-angle ($>$ a few degrees) observations of COBE represent truly primordial fluctuations in the gravitational potential, since the causal horizon at the time of decoupling subtends less than a degree on the sky today.

11.3 The New Generation of Observations

In spite of all the excitement generated by the 2.73 K spectrum and the temperature anisotropies measured by the COBE satellite, the best CMBR physics might be yet to come! While the temperature fluctuations between portions of the sky at separation of 10 degrees are crucial for the understanding of structure formation, they are not very sensitive to the parameters of the Standard Model of cosmology: the energy density due to the curvature parameter $\sim k/a^2$, the contribution from matter (also dark matter), Ω_M, the energy density contribution due to a non-zero cosmological constant, Ω_Λ and the Hubble parameter H_0. However, at much smaller angular separations, small enough to contain causally-connected regions at the surface of last scattering, the measurement of the power spectrum of temperature fluctuations might be sensitive to the Standard Model parameters with unprecedented precision.

As discussed in the previous section, fluctuations in density cause photons and matter to fall into potential wells. Within a horizon volume, the infall will be slowed and even reversed by radiation pressure. The baryon photon fluid starts making acoustic oscillations. Regions of compression (hot spots) and rarefaction (cold spots) will thus populate causally-connected volumes. As photons scattered for the last time at the decoupling redshift $z_{dec} \approx 1100$, a 'snap-shot' of the oscillating fluid can be deduced from CMBR measurements at angular scales below 1 degree.

So what is the connection with the cosmological parameters? The oscillations of the fluid are confined to the *Hubble radius* H_{dec}^{-1} at the redshift of recombination. Thus, in order to estimate the principal feature of the CMBR anisotropy spectrum at small scales we must compute the angle subtended today by $H(z = z_{rec})^{-1}$:

$$\Delta\theta(z_{dec}) = \frac{1}{d_A \cdot H} \tag{11.26}$$

with the angular size distance d_A in a Friedmann-Lemaître-Robertson-Walker Universe defined by (4.85) and (4.70). The position of this *First Acoustic Peak* of the CMBR anisotropy spectrum is expected to be at an angular scale $\Delta\theta$ which is sensitive to Ω_M and Ω_Λ, as shown in Fig. 11.2.

The amplitude of the peak (that is, the size of the anisotropy) increases with the baryon density fraction, Ω_B and depends also sensitively on H_0 and Ω_Λ. CMBR fluctuations at yet smaller angles are a result of higher frequency acoustic oscillations of the baryon and photon fluid. Their relative strength and angular position yield further information on the fractional contributions to the energy density of the Universe. In particular, the higher order acoustic peaks provide information about the nature of dark matter. In the next sections, we shall analyse this in more detail.

11.4 Fluid Equations

The analysis of how structure formed and grew in the early Universe is one of the most important and interesting activities in modern cosmology. The fact that the initial density perturbations as measured in the CMBR were so small simplifies things substantially. If we only give a zero-order (homogeneous) solution to the equations governing the cosmic 'fluid', we can use perturbation theory to write down and solve the hydrodynamical equations for the deviations from homogeneity. We shall solve a simplified model with only one species which makes up the cosmic fluid.

In the FLRW model the dynamics is essentially Newtonian for length scales much smaller than H^{-1}, if only the scale factor $a(t)$ is kept in the equations. The relevant equations are the continuity equation

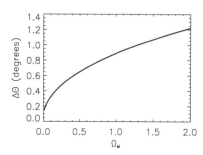
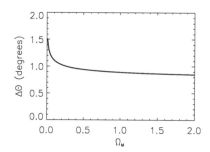

Fig. 11.2. Angular size of regions in causal contact ($r_H = H^{-1}$) at $z_{dec} = 1100$ as a function of Ω_M. In the left-hand plot $\Omega_\Lambda = 0$ and in the right-hand side for a flat Universe, $\Omega_\Lambda = 1 - \Omega_M$.

$$\dot{\rho} + \nabla \cdot (\rho \mathbf{v}) = 0 \qquad (11.27)$$

the Euler equation of hydrodynamics

$$\dot{\mathbf{v}} + (\mathbf{v} \cdot \nabla) \mathbf{v} = -\nabla \left(\Phi + \frac{p}{\rho} \right) \qquad (11.28)$$

and the Poisson equation of Newtonian gravity

$$\nabla^2 \Phi = 4\pi G \rho \qquad (11.29)$$

These equations have the zero-order solution

$$\rho_0(t, \mathbf{r}) = \frac{\rho_0}{a^3(t)}$$

$$\mathbf{v}_0(t, \mathbf{r}) = \frac{\dot{a}(t)}{a(t)} \mathbf{r}$$

$$\Phi_0(t, \mathbf{r}) = \frac{2\pi G \rho_0 r^2}{3} \qquad (11.30)$$

The first of these just shows the dilution of the mass density that is caused by the cosmic expansion, and the second shows again the homogeneous 'Hubble flow' according to the law $\mathbf{v}_0(t, \mathbf{r}) = H(t)\mathbf{r}$, valid for all observers locally at rest in the cosmic frame. The third simply follows from solving (11.29) with a constant ρ (see Problem 11.3).

Since we have spatially homogeneous equations, we try to find plane-wave solutions for the perturbations: that is, we take the Fourier transforms. It is convenient to 'take out' the expansion by using comoving space coordinates $\mathbf{x} = \mathbf{r}/a(t)$. Then, for instance, $\rho_0(t, \mathbf{r}) \to \rho_0(t, \mathbf{x}) = \rho_0(t)$: that is, in comoving coordinates the background density is independent of the scale factor $a(t)$.

We now write the first-order expanded quantities as

$$\rho(t, \mathbf{x}) = \rho_0(t) + \rho_1(t, \mathbf{x}) \equiv \rho_0(t) [1 + \delta(t, \mathbf{x})]$$

$$\mathbf{v}(t, \mathbf{x}) = \mathbf{v}_0(t, \mathbf{x}) + \mathbf{v}_1(t, \mathbf{x})$$
$$\Phi(t, \mathbf{x}) = \Phi_0(t, \mathbf{x}) + \Phi_1(t, \mathbf{x}) \tag{11.31}$$

where $|\delta|$, $|\mathbf{v}_1|$ and $|\Phi_1| \ll 1$. Taking the Fourier transform of all quantities, for example,

$$\delta(t, \mathbf{x}) = \frac{1}{(2\pi)^{\frac{3}{2}}} \int e^{i\mathbf{k}\cdot\mathbf{x}} \delta(t, \mathbf{k}) d^3 k \tag{11.32}$$

with the inversion formula

$$\delta(t, \mathbf{k}) = \frac{1}{(2\pi)^{\frac{3}{2}}} \int e^{-i\mathbf{k}\cdot\mathbf{x}} \delta(t, \mathbf{x}) d^3 x \tag{11.33}$$

one finds after some algebra (Problem 11.4)

$$\ddot{\delta}(t, \mathbf{k}) + 2\frac{\dot{a}(t)}{a(t)}\dot{\delta}(t, \mathbf{k}) + \left(\frac{k^2 v_s^2}{a^2(t)} - 4\pi G\rho_0\right)\delta(t, \mathbf{k}) = 0 \tag{11.34}$$

where v_s is the speed of sound,

$$v_s^2 = \left(\frac{\partial p}{\partial \rho}\right)_{\text{adiabatic}} \tag{11.35}$$

That the speed of sound enters the equations is natural, since it governs how fast perturbations in the fluid can propagate. For a relativistic fluid with equation of state $p = \rho/3$, we see that

$$v_s = \frac{1}{\sqrt{3}} \tag{11.36}$$

(that is, of the order of the velocity of light), whereas for a gas of neutral hydrogen

$$v_s = \sqrt{\frac{5T}{3m_p}} \tag{11.37}$$

which is much smaller (Problem 11.5).

We note that by using the Fourier transform of the linearly perturbed version of (11.29),

$$\nabla^2 \Phi_1 = 4\pi G\rho_0\delta, \tag{11.38}$$

using the first-order Friedmann equation (for a flat Universe),

$$H^2 = \frac{8\pi G}{3}\rho_0, \tag{11.39}$$

we find the relation

$$\delta(t, \mathbf{k}) = -\frac{2}{3}\left(\frac{k}{aH}\right)^2 \Phi_1(t, \mathbf{k}) \tag{11.40}$$

which is of use when relating a density perturbation to a perturbation of the Newtonian potential, as we will do in Chapter 11.

11.5 The Jeans Mass

We see from (11.34) that the sign of the quantity

$$\kappa_J \equiv \left(\frac{k^2 v_s^2}{a^2(t)} - 4\pi G \rho_0 \right) \tag{11.41}$$

will determine whether the solutions are growing or oscillating. The sudden drop in the velocity of sound during the epoch of decoupling has many interesting consequences. The term proportional to v_s^2 is the pressure force, which is felt by charged particles, i.e. the baryons. Before decoupling v_s is very large, which means that the baryons will only undergo damped oscillations.

As usual with Fourier transforms, to a given Fourier wave vector $\mathbf{k}$ corresponds a comoving wavelength $\lambda_{com} = 2\pi/|\mathbf{k}|$: that is, a physical wavelength

$$\lambda_{phys} = \frac{2\pi a(t)}{|\mathbf{k}|} \tag{11.42}$$

It is seen from (11.34) that only physical scales larger than the *Jeans length* λ_J can grow (for smaller values of λ_{phys} the solutions are damped oscillations), where

$$\lambda_J = v_s \sqrt{\frac{\pi}{G \rho_0}} \tag{11.43}$$

The mass within a sphere of radius $\lambda_J/2$ is called the Jeans mass and is given by

$$M_J = \frac{4\pi}{3} \left(\frac{\lambda_J}{2} \right)^3 \rho_0 = \frac{\pi^{5/2} v_s^3}{6 G^{3/2} \sqrt{\rho_0}} \tag{11.44}$$

11.6 Structure Growth in the Linear Regime

For small k (lengths much larger than the Jeans length), or for cold dark matter which does not feel the electromagnetic pressure, the term proportional to k^2 in (11.34) can be neglected. For a flat, matter dominated FLRW model, $a(t) \propto t^{2/3}$, $\dot{a}/a = 2/(3t)$ and $\rho_0 = 1/(6\pi G t^2)$, so that

$$\ddot{\delta}(t, \mathbf{k} \sim 0) + \frac{4}{3t} \dot{\delta}(t, \mathbf{k} \sim 0) - \frac{2}{3t^2} \delta(t, \mathbf{k} \sim 0) = 0 \tag{11.45}$$

which produces the solution

$$\delta(t, \mathbf{k} \sim 0) = c_+ t^{2/3} + c_- t^{-1} \tag{11.46}$$

Thus, one mode is growing with time as $t^{2/3}$, and one is decaying as $1/t$. Obviously the growing mode is the interesting one for structure formation: it makes a small initial density perturbation grow under the action of gravity. Since the growing mode has the same time dependence proportional to $t^{2/3}$ as the scale factor $a(t)$, the density contrast grows linearly with the scale factor. Perturbations in cold dark matter can grow during matter domination, whereas the baryons oscillate until decoupling. This means that cold dark matter plays a crucial role for structure formation - they form potential wells in which the baryons can fall as soon as they escape the pressure after decoupling.

An initially overdense region in a FLRW Universe with $\Omega_m = 1$ will behave as a 'mini-Universe' with $\Omega_m > 1$: that is, it will expand to a maximal radius, and then re-collapse. This re-collapsed region can be identified with the halo of a galaxy or of a galaxy cluster. The detailed dynamics after the density contrast has become greater than unity (that is, in the nonlinear regime) is quite complicated, since it involves an interplay of dark and ordinary matter. The only way so far to obtain a description of this phase is to carry out N-body simulations on large computers.

The density field $\delta(\mathbf{x})$ at a given time is generally assumed to be a Gaussian random field. In fact, this is what is predicted in models of inflation, where quantum fluctuations in the inflaton field are created by the exponential expansion as we will see in the next section, and could be the 'seeds' of cosmic structure. Defining the rms fluctuations by

$$\sigma^2 = \langle \delta(\mathbf{x})^2 \rangle \tag{11.47}$$

the autocorrelation function $\xi(|\mathbf{x}_2 - \mathbf{x}_1|) \equiv \xi(x)$ is defined as

$$\xi(x) = \frac{1}{\sigma^2} \langle \delta(\mathbf{x}_2)\delta(\mathbf{x}_1) \rangle \tag{11.48}$$

In the random phase approximation, all the Fourier modes $\delta(\mathbf{k})$ are uncorrelated, which means that

$$\langle \delta^*(\mathbf{k})\delta(\mathbf{k}') \rangle = (2\pi)^3 \delta^3(\mathbf{k} - \mathbf{k}')P(\mathbf{k}) \tag{11.49}$$

The function $P(\mathbf{k})$ is the *power spectrum* of the fluctuations. In most models $P(\mathbf{k}) = k^n$, with the spectral index n varying between 0.7 and 1.3. The value $n = 1$ corresponds to so-called scale-invariant fluctuations, since it can be shown that for such a spectrum the fluctuations δ are of the same amplitude for all length scales. As we will see, inflation predicts almost scale-invariant perturbations.

The fluctuations measured through the power spectrum $P(k)$ in the CMBR are usually described in terms of a 'primordial' part $P_i(k)$, which represents the curvature fluctuations generated (perhaps by inflation) at a very early epoch. To obtain the density field actually measured one needs to know how the fluctuations in the photon field have been influenced by various other effects near the time of decoupling and matter domination. This depends on the nature of the dark matter, but can be expressed in terms of a transfer function $T(k)$, defined such that

$$P(k) = T^2(k)P_i(k) \tag{11.50}$$

If the Universe were still radiation dominated, $T(k) = 1$ for all k, and the spectrum of fluctuations today would be the primordial spectrum. Before recombination, the speed of sound was of the order of the speed of light, and the Jeans mass was extremely large (see (11.44)). In fact, the Jeans length was larger than the horizon size, so all perturbations within the horizon grew very slowly (at most, logarithmically). At the time of equality of matter and radiation energy density ($z \sim 4 \cdot 10^3$), perturbations in the dark matter could start to grow, but the photons and baryons were still strongly coupled and affected by pressure until photon decoupling. At that point, the speed of sound dropped by a huge factor (from 10^8 m/s to 10^4 m/s) so that the Jeans length suddenly became smaller than the size of present galaxies. Thus, perturbations in the ordinary matter could start to grow on all scales.

As the Universe expands, the horizon continues to increase. This means that the comoving scale of any perturbation sooner or later becomes smaller than the horizon scale (it is said to enter the horizon). Cold dark matter fluctuations that entered the horizon after matter and radiation equality should not have been modified much from the primordial spectrum, so that $T(k) \sim 1$ for these modes. The comoving length scale corresponding to the horizon at equality is roughly

$$\lambda_{eq} \approx \frac{13}{\Omega_T h^2} \text{ Mpc} \tag{11.51}$$

For scales smaller than this, it can be shown that there is a suppression $T(k) \sim k^{-2}$. For hot dark matter, the suppression at small scales is much larger, since relativistic particles 'free-stream' out of small density enhancements. Recent data from the distribution of galaxies measured by the 2dF collaboration [13] agree excellently with cold dark matter. This has been used to put an upper limit on the relic density of neutrinos (see (9.10)) corresponding to around 2 eV for the sum of the masses of neutrinos.

11.7 Connection to Fluctuations in the CMBR

The density contrast field δ is defined over all space and time. However, when we study fluctuations in the CMBR temperature, we only sample a projection on the celestial sphere, at the time of last scattering of the photons. In this situation, it is convenient to use a spherical decomposition. As is well known, any function $f(\theta, \phi)$ (or equivalently $f(\hat{n})$ with $\hat{n}$ a unit vector) on the unit sphere can be expanded in spherical harmonics $Y_{lm}(\theta, \phi)$:

$$f(\hat{n}) = \sum_{l=0}^{\infty} \sum_{m=-l}^{m=l} a_{lm} Y_{lm}(\theta, \phi) \tag{11.52}$$

In particular, the temperature fluctuations $\Delta T / T(\hat{n})$ can be expanded

$$\frac{\Delta T(\hat{n})}{T} = \sum_{l=2}^{\infty} \sum_{m=-l}^{m=l} a_{lm} Y_{lm}(\theta, \phi). \tag{11.53}$$

(Here the dipole, $l = 1$, is usually excluded since it is indistinguishable from the effects of the peculiar motion of the Earth with respect to the cosmic rest frame.)

Observationally, from a measured set of a_{lm} we can form the angular average C_l:

$$C_l = \langle a_{lm} a_{lm}^* \rangle = \frac{1}{2l+1} \sum_{m=-l}^{l} a_{lm} a_{lm}^* \qquad (11.54)$$

The set of coefficients C_l for all l is the basic information-containing set of the microwave background. (There is also, a much smaller, effect on polarization which can be measured.) By comparing these measured C_l with the theoretical ones obtained by varying all cosmic parameters one may obtain very good accuracy on many parameters. The only way to compute these coefficients for all scales is to integrate the transport equations numerically. The result will depend on the initial (primordial) fluctuation spectrum, the nature of the dark matter, the fraction of baryonic matter, the Hubble constant and some other parameters. This means that if one has measurements of, say, several hundred of the C_l's, there would be enough redundancy to determine the cosmological parameters to high accuracy.

To a considerable extent this has now been done, both in balloon experiments and in the satellite experiment WMAP – something that has revolutionized cosmology. The Planck satellite is scheduled for launch in 2007 and will be equipped with an 1.3-meter mirror with an angular resolution of 10 arcminutes and a frequency range between 30 and 850 GHz. If all goes well, it should determine all the above-mentioned parameters to an accuracy of about 1 per cent.

The shape of the spectrum of the C_l contains several features – a plateau at low l, and then a series of peaks ('acoustic peaks') at higher l (that is, smaller angular scales). The physical reason for these peaks is that the cosmic fluid oscillates due to the competing actions of the gravitational attraction and the photon pressure. For this to happen, the initial conditions for the perturbations have to be set at an early time, so that gravity can act coherently. If perturbations were generated in the earliest Universe, at the epoch of inflation, so they were driven outside the horizon by the exponential expansion, we can understand the mechanism heuristically. As the Universe expands, successively larger scales of perturbations, dominated by dark matter, enter the horizon, at which point they start to grow. But baryons are coupled through gravity to the dark matter and through electromagnetic interactions to the photons and electrons. When the baryonic mass has been compressed enough, the photon pressure acts as a repulsive, harmonic force, causing the baryonic fluid to expand again. Then eventually gravity takes over again and the sequence is repeated. That the plasma of the early Universe should contain these 'acoustic' oscillations was predicted by Sakharov in the early 1960's, but the connection to observations of the microwave background was not made until the 1970's by several groups. After last scattering, the microwave background photons propagate essentially freely through the Universe, but are of course affected by the gravitational potential of the region they left (the Sachs-Wolfe effect), and regions they passed through on the way (the integrated Sachs-Wolfe effect), the Doppler motion of the plasma, and a small but important scattering by hot electrons in galaxy clusters they pass (the Sunyaev-Zeldovich effect). The CMBR is then like a snapshot picture of the oscillating photon field in the Universe at a redshift of around 1100. In this picture, or rather in the power spectrum of it, we would expect to see a series of peaks, corresponding to compression and rarefaction of the baryon fluid, with the first peak being a compression peak with a scale that is given by the size of the sound horizon at decoupling. It is a great triumph of cosmology that these predicted peaks have now definitively been observed.

A particularly simple prediction is that the location of the first peak is very sensitive to the total energy density of the Universe, $\Omega_T = \Omega_M + \Omega_\Lambda$. This is an effect of gravitational lensing of the horizon size at the epoch of the formation of the CMBR (the last scattering surface). For a low-density Universe, a geodesic bundle of photons from the last scattering surface of the CMBR would be focused less, causing the effects to be seen at a smaller angle, i.e. a higher value of l, than in a high-density Universe. Balloon and ground-based experiments found in 1998 - 2002 that the peak location corresponded to an excellent accuracy to a flat Universe, $\Omega_T = 1.0 \pm 0.1$, exactly as predicted by inflation. Taken together with the supernova data, which favour the existence of a cosmological constant, a 'concordance model', with $\Omega_M \sim 0.3$, $\Omega_\Lambda \sim 0.7$, has emerged during the last few years.

11.8 Primordial Density Fluctuations

The remaining question is what caused the primordial density fluctuations with $\delta\rho/\rho \approx 10^{-5}$, which could then be processed by causal physics once they entered the horizon. An intriguing possibility is that they were generated at the inflationary epoch, through quantum fluctuations of the inflaton field. This is one of the most exciting ideas of present cosmology: that the largest structures we see on the sky today were tiny quantum fluctuations that were stretched by inflation to gigantic scales. In Appendix E we give a brief account of how this mechanism works.[7]

According to Appendix E, the relation between the curvature perturbation generated by inflation $\mathcal{R}_\mathbf{k}$ and the Newtonian potential perturbation $\Phi_{1\mathbf{k}}$ (see (11.31)) is given by

$$\mathcal{R}_\mathbf{k} = -\frac{5 + 3\alpha}{3 + 3\alpha}\Phi_{1\mathbf{k}} \tag{11.55}$$

---

[7] This Appendix is somewhat more advanced than the rest of the book, and can be skipped by the casual reader.

where α is the equation of state parameter (4.22). The Newtonian potential perturbation is related to the density perturbation by (11.40) giving

$$\delta(\mathbf{k}) = \frac{2}{3} \left(\frac{k}{aH} \right)^2 \frac{3+3\alpha}{5+3\alpha} \mathcal{R}_{\mathbf{k}} \qquad (11.56)$$

At radiation domination after inflation, $\alpha = 1/3$, and

$$\delta(\mathbf{k}) = \frac{4}{9} \left(\frac{k}{aH} \right)^2 \mathcal{R}_{\mathbf{k}} \qquad (11.57)$$

On the very largest scales that enter the horizon after the Universe has become matter dominated, $\alpha = 0$ and

$$\delta(\mathbf{k}) = \frac{2}{5} \left(\frac{k}{aH} \right)^2 \mathcal{R}_{\mathbf{k}}. \qquad (11.58)$$

On these large scales, the COBE satellite measured the adiabatic fluctuation directly, and $\delta^2_{COBE} = 4/25\mathcal{P}_R$, with the power spectrum of the curvature perturbation given in (E.52). Using the COBE large-scale average rms value, $\delta_{COBE} \simeq 1.9 \cdot 10^{-5}$, gives the normalization of the slow-roll potential V as

$$\frac{1}{m_{Pl}^2} \sqrt{\frac{32V}{75\varepsilon}} = \delta_{COBE} = 1.9 \cdot 10^{-5}, \qquad (11.59)$$

where ε is the slow-roll parameter (10.25). Thus,

$$\frac{V^{\frac{1}{4}}}{\varepsilon^{\frac{1}{4}}} = 5.4 \cdot 10^{-3} m_{Pl} = 6.5 \cdot 10^{16} \text{ GeV}. \qquad (11.60)$$

In terms of the slow-roll parameters ε and η, the slope of the spectrum is (see (E.53))

$$n_s = 1 - 6\epsilon + 2\eta. \qquad (11.61)$$

This shows that the prediction of inflaton is generally a close to scale-invariant spectrum of density perturbations, which as explained in Appendix E are gaussian and adiabatic. This prediction has recently vindicated by observations, as we now will see.

11.9 Present Experimental Situation

The present situation (2003) regarding observations of the cosmic microwave background is that the regions of the first and the second peak have been quite accurately mapped out, with even the third peak starting to be seen. In February 2003, the WMAP satellite (Wilkinson Microwave Anisotropy Probe) released its first data sets, which cover the range from $l = 2$ (the quadrupole) over the first two acoustic peaks to $l \sim 800$. This was the first space project to map the cosmic microwave background since the COBE satellite. The WMAP data has shown a wealth of interesting data, and with some additional information from other experiments (such as demanding $h > 0.5$), many cosmological parameters have been determined to remarkable precision, most notably $\Omega_T = 1.02 \pm 0.02$.

In Table 11.1 are shown the values of some other cosmological parameters obtained by WMAP [5].

Table 11.1. The cosmological parameters as estimated by the first set of data from the WMAP satellite [5].

Parameter	Value
Ω_M	0.27 ± 0.04
Ω_Λ	0.73 ± 0.04
Ω_B	0.044 ± 0.004
Baryon to photon ratio, η	$(6.1 \pm 0.3) \cdot 10^{-10}$
$\Omega_\nu h^2$	< 0.0076 (95 % c.l.)
n_s (see (11.49) and (11.61))	0.93 ± 0.03
Age of Universe	13.7 ± 0.2 Gyr
Decoupling redshift, z_{dec}	1089 ± 1

Examples of predicted spectra (before any of the present data were known) in various models are shown in Fig. 11.3.

The WMAP data, and the best fit model using the parameters in Table 11.1 are shown in Fig. 11.4. The impressively detailed all-sky map of the CMBR presented by the WMAP team is shown as colour Plate 6.[8] In the last few years, cosmology has entered a new epoch, one of precision measurements. What is appearing from the measurements is a concordance model, which has less than 5 % baryons, almost 25 % cold dark matter and 70 % vacuum, or 'dark' energy. It is not unreasonable to claim that the golden age of cosmology has just started. While the observational content of the Universe is being pinned down, we are still waiting for cosmologists to find answers to the burning questions: what is the dark matter, and the dark energy?

11.10 Summary

- The observation of the cosmic microwave background (CMBR) constitutes a remarkable corroboration of the Big Bang theory
- The temperature of the relic radiation at present, $T_0 = T_e/(1+z)$, has been measured to be 2.725 ± 0.004 K (95 % CL).
- The observed anisotropy of the CMBR of about 10^{-5} at the angular scale of several degrees confirms the existence of small-density perturbations in the early Universe ($z \geq 1000$). These were the

---

[8] The colour plate section is positioned in the middle of the book.

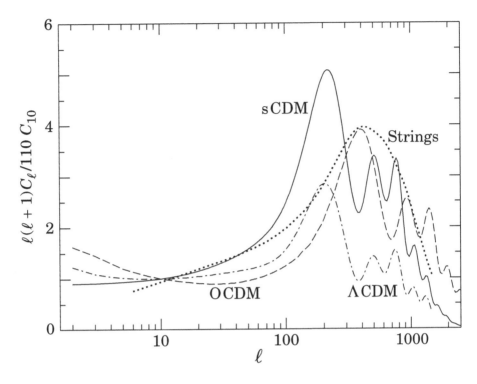

Fig. 11.3. Examples of theoretically predicted $\ell(\ell+1)C_\ell$ (normalized to $\ell = 10$) or CMBR anisotropy power spectra. sCDM is the standard cold dark matter model with $h = 0.5$ and $\Omega_B = 0.05$. ΛCDM is a model with $\Omega_T = \Omega_\Lambda + \Omega_M = 1$ where $\Omega_\Lambda = 0.3$ and $h = 0.8$. OCDM is an open model with $\Omega_M = 0.3$ and $h = 0.75$. 'Strings' is a model where cosmic strings are the primary source of large-scale structure. The plot indicates that precise measurements of the CMBR anisotropy power spectrum at $\ell > 100$, which will be possible with the Planck and WMAP satellites, could distinguish between current models. From G. Smoot, astro-ph/9705135 (1997). As shown in Fig. 11.4, only the model labeled ΛCDM is now viable, due to WMAP data.

seeds necessary to form the structure seen today in the Universe: for example, stars and galaxies.

- Ongoing and future measurements of the anisotropy of the CMBR at sub-arcminute scale are likely to improve significantly our knowledge of the cosmological parameters.
- Thanks to the small amplitude of the anisotropy of the CMBR, a linear analysis of the early growth of structure in the Universe can be used. The growing modes grow linearly with the scale factor $a(T)$.
- The size of the perturbations which can grow due to the action of gravity is given by the Jeans length,

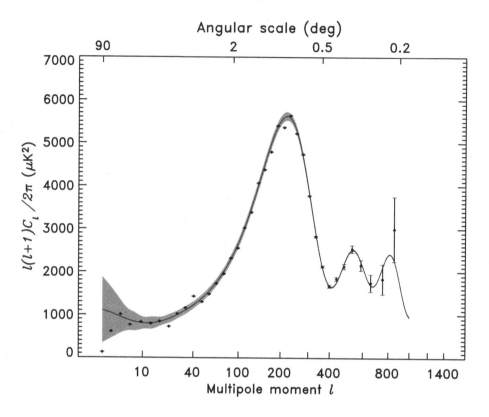

Fig. 11.4. The data on the CMBR measured by the WMAP satellite [5], together with the best fit ΛCDM model, with parameters given according to the list in Table 11.1. Courtesy of the NASA/WMAP Science Team.

$$\lambda_J = v_s \sqrt{\frac{\pi}{G\rho_0}}$$

which in turn depends on the speed of sound v_s, where

$$v_s^2 = \left(\frac{\partial p}{\partial \rho}\right)_{\text{adiabatic}}$$

After photon decoupling, the speed of sound, and therefore the Jeans length, drops suddenly.

- Cold dark matter gives power on all length scales after matter dominance, whereas hot dark matter (like neutrinos) free-stream out of small regions due to their relativistic velocities.
- The best way to observe the primordial density fluctuations is through the fluctuations of the temperature of the microwave background that they induce. New satellite experiments, of which WMAP is the first, may distinguish between dark matter models

and determine all the classical cosmological parameters to a per cent accuracy.

- Two important predictions of inflation, a total energy density close to unity, and a nearly scale-invariant spectrum of gaussian primordial fluctuations, have been verified by new measurements of the CMBR.

11.11 Problems

11.1 Integrate (11.6) over all frequencies and show that the total photon density is:

$$n_\gamma = 410(1+z)^3 \text{cm}^{-3}$$

11.2 Calculate the angular size of the region in causal contact at $z = 1100$ in a radiation dominated Universe.

11.3 Solve (11.29) for a constant $\rho = \rho_0$ using spherical coordinates.

11.4 Perform the intermediate steps leading to (11.34).

11.5 Compute, in SI units, the speed of sound after recombination.

11.6 Compute the Jeans mass before and after photon decoupling.

12 Cosmic Rays

12.1 Introduction

So far, we have mostly been dealing with information obtained about our Universe through the use of various forms of electromagnetic radiation. However, there exist other carriers of information, some of which we shall describe in the following chapters.

The existence of ionizing particles continually bombarding the Earth's atmosphere was first noticed during a balloon flight in 1912 by Victor Hess[1] and the term 'cosmic rays' was first used by Robert Millikan in 1925. Later on, it was discovered that the cosmic rays (CR) at the top of the atmosphere are mainly atomic nuclei (including hydrogen nuclei, protons, and helium nuclei, alpha particles). Only about 2 per cent of the particles are electrons (and positrons). At energies above a few TeV, where cosmic rays are undeflected by the magnetic fields within the solar system, their distribution is nearly isotropic (but see the later discussion about ultra-high energy CR).

Example 12.1.1 Show that protons above 5 TeV are essentially unaffected by a solar system magnetic field of 10 μG.

Answer: The largest deflection occurs if the proton's motion is perpendicular to the magnetic field. For a particle with charge Ze, the relation between the gyroradius r, the magnetic field B and the orthogonal particle momentum p becomes:

$$r = \frac{p}{ZeB} \tag{12.1}$$

Inserting numerical values for a proton (Z=1) one finds that the gyromagnetic radius becomes:

$$\left(\frac{r}{1 \text{ a.u.}}\right) = 2 \cdot 10^2 \left(\frac{p}{1 \text{ TeV}}\right) / \left(\frac{1 \ \mu G}{B}\right) \tag{12.2}$$

where 1 a.u. (astronomical unit) is the mean distance between the Earth and the Sun, $1.5 \cdot 10^{11}$ metres. For $p = 5$ TeV and $B = 10 \ \mu G$ the gyromagnetic

---

[1] Hess shared the 1936 Nobel prize with Carl D. Anderson, who discovered the positron in cosmic rays.

radius is $r \approx 100$ a.u., far greater than the distance from the Sun to Pluto (the planet furthest away from the Sun).

Cosmic rays have historically been very important for the development of particle physics. Before the emergence of man-made high-energy particle accelerators, they were the only means of studying energetic collision and decay processes. The muon, the pion, the positron and particles containing s quarks were first discovered in cosmic-ray induced reactions.

In spite of over 85 years of research, many important questions about cosmic rays remain unanswered. Where do they come from? What are the cosmic accelerators capable of emitting nuclei with energies exceeding 10^{19} eV? What is the nuclear composition at high energies?

12.2 The Abundance of Cosmic Rays

The energy spectrum of cosmic ray nuclei at the top of the atmosphere spans over many decades, and can be described by a segmented power-law formula above 10 GeV/nucleon, as shown in Fig. 12.1:

$$\frac{dN}{dE} \propto E^{-\alpha} \tag{12.3}$$

with the following values for the spectral index:

$$\alpha = \begin{cases} 2.7 & E < 10^{16} \text{ eV} \\ 3.0 \ 10^{16} < E < 10^{18} \text{ eV} \end{cases} \tag{12.4}$$

For the highest energies, above 10^{19} eV, the index appears to be somewhat smaller: that is, the distribution is flatter. The two breaks in the spectrum, around $10^{15} - 10^{16}$ eV (the 'knee') and at $10^{18} - 10^{19}$ eV (the 'ankle') may indicate the energy limits of different cosmic accelerators.

The chemical composition of cosmic rays is interesting as it may offer clues for understanding their origin. Data from mass spectrometers on board satellites and balloons exhibit clear differences when compared to solar system abundances (see Fig. 12.2):[2]

- The relative abundance of protons and helium nuclei in cosmic rays is smaller than in the solar system.
- Two groups of elements, Li, Be, B, and Sc, Ti, V, Cr, Mn are significantly more abundant in cosmic rays.

---

[2] The solar system abundances are derived from the spectral features in the photosphere of the Sun and from the studies of meteorites, which are believed to have the same chemical composition.

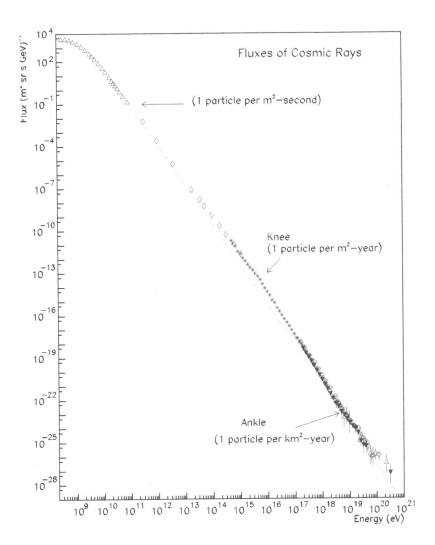

Fig. 12.1. Compilation of measurements of the differential energy spectrum of cosmic rays [47]. The dotted line shows an E^{-3} power-law for comparison. Approximate integral fluxes (per steradian) are also shown. Courtesy of Jim Matthews.

The relative underabundance of hydrogen and helium in the cosmic rays is not fully understood: it could either reflect the primordial composition of the cosmic ray sources or simply be due to the difference in propagation properties of the elements and the fact that the heavier elements are more easily ionized, thereby being more readily accessible for acceleration. The overabundance of Li, Be and B is known to be due to *spallation* of carbon and oxygen. As these common elements travel through the interstellar medium, they are fragmented in collisions with the hydrogen and helium gas in the interstellar medium into elements with somewhat smaller atomic number. Similarly, Sc, Ti, V, Cr and Mn result from the spallation of iron.

Among the interesting similarities between the solar system and cosmic ray abundances are:

1. There are peaks in the abundance for carbon, nitrogen, oxygen and iron.
2. An even odd mass number effect in the abundances is seen in both data sets. This is understood as being due to the relative stability of the nuclei according to their atomic numbers.

Although the results are inconclusive, it seems plausible that the primary composition of the source of cosmic rays at energies below the first break ($\sim 10^{15} - 10^{16}$ eV) is similar to the local abundance of elements.

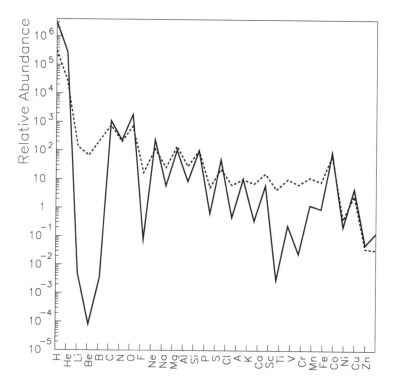

Fig. 12.2. The relative abundance of cosmic rays at the top of the atmosphere (dashed curve) compared with solar system and local interstellar abundance (solid curve), all arbitrarily normalized to silicon (=100). From [30] and references therein.

Example 12.2.1 Show that the number of cosmic rays above a certain energy, E_0, is also a power-law if $\alpha > 1$ in (12.3).

Answer: Integrating (12.3) yields:

$$N(E > E_0) = \int_{E_0}^{\infty} E^{-\alpha}dE \propto E_0^{-\alpha+1} \tag{12.5}$$

The small amounts of electrons and positrons in cosmic rays are thought to be of galactic origin. Energy depletion through Compton scattering with the cosmic microwave background radiation prevents their propagation over intergalactic distances (Problem 12.3). The positron fraction, $e^+/(e^+ + e^-)$, has only been measured at low energies (below 50 GeV) and found to be just a few per cent, indicating that electrons are accelerated by primary sources. If their origin had been secondary, i.e through hadronic decays ($\pi^{\pm} \rightarrow \mu^{\pm} \rightarrow e^{\pm}$) there should be comparable fractions of electrons and positrons. Positrons, on the other hand, are likely to be produced in secondary processes, like pair production $\gamma + \gamma \rightarrow e^+ + e^-$.

12.3 Ultra-High Energies

While balloon and satellite experiments are very well suited for the study of cosmic rays at low energies, the steep spectrum shown in Fig. 12.1 reveals that it is virtually impossible to gather sufficient statistics at energies above the knee of the spectrum with the relatively small detectors that may be accomodated in flown devices. The study of the highest energy events are, however, very interesting, as these particles are only weakly affected by galactic magnetic fields ($\sim \mu G$) and intergalactic ($\sim nG$) magnetic fields. For a source at a distance L, the trajectory of a charged particle in a uniform magnetic field is deflected by (Problem 12.2):

$$\theta \approx 3^{\circ} Z \left(\frac{L}{1 \text{ kpc}} \right) \times \left(\frac{B}{1 \text{ }\mu G} \right) \times \left(\frac{10^{19} \text{ eV}}{E} \right) \tag{12.6}$$

The mapping of the directionality of highest energy cosmic rays is of great interest for the understanding of cosmic accelerators.

12.3.1 Extensive Air-Showers

Although the Earth's atmosphere prevents primary cosmic rays and gamma-rays from reaching sea level or even mountain tops, there is a technique for studying high-energy cosmic rays from the ground. As the primary particles hit the atmosphere, their energy is transferred through subsequent collisions to a 'cascade' of particles, as shown schematically in Fig. 12.3. The picture can be understood as follows: a gamma-ray, for instance, will split into

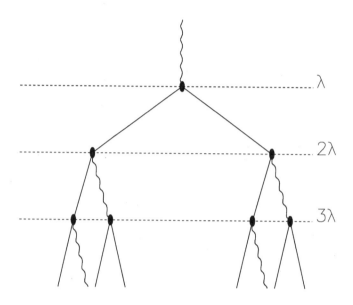

Fig. 12.3. Schematic view of the creation of a particle 'shower' in the atmosphere.

an electron positron pair within approximately one *interaction length*[3]. In
the next interaction length, the electron and positron will each radiate one
bremsstrahlung photon which, in turn, will interact in the next radiation
length, and so on. To simplify the situation, the energy is assumed to be split
equally at every vertex. The multiplication process stops when the energy of
the shower particles falls below some critical energy ϵ_c. For an electromagnetic shower – that is, one consisting only of photons, electrons and positrons
– the critical energy is reached when the cross-section for bremsstrahlung becomes smaller than for ionization. To summarize, the number of particles
along the shower profile coordinate x is

$$N(x) = 2^{\frac{x}{\lambda}} \tag{12.7}$$

where λ is the average distance between interactions. At 'shower maximum'
for a primary of energy E, the number of particles with the critical energy,
ϵ_c, is

$$N_{max} = \frac{E}{\epsilon_c} \tag{12.8}$$

The number of shower particles falls almost exponentially beyond maximum
as the particles successively stop. Combining (12.7) and (12.8) it can be seen

---

[3] For particles interacting electromagnetically, such as gamma-rays, the interaction
length is normally referred to as *radiation length*.

that the position of the shower maximum grows as the logarithm of the primary energy:

$$x_{max} = \frac{\lambda}{\ln 2} \cdot \ln \frac{E}{\epsilon_c} \tag{12.9}$$

The general features of the particle cascades are also approximately correct for hadronic showers, generated by nuclear primaries. In addition to the electromagnetic part, the cascades also includes hadrons and muons. Muons interact electromagnetically but, because of their higher mass, have a much greater range than electrons and positrons. Thus, at sea-level, the cosmic-ray fluxes are dominated by muons with a mean energy of ~ 2 GeV, but with a steep spectrum reaching up to nearly the same energies as the primary cosmic rays.

Extensive 'air-showers' can thus be used for studying primary cosmic rays above about one TeV. The main features of the particle cascades are:

(1) $N_{max} \propto E$
(2) $x_{max} \propto \ln E$
(3) The relative arrival times of the shower particles can be used to reconstruct the direction of the particles hitting the top of the atmosphere, as shown in Fig. 12.4.

While (3) indicates that extensive air-showers can be used to trace the origin of VHE cosmic rays, the shower properties (1) and (2) can be used to reconstruct the energy of the incident particles at the top of the atmosphere.

In practice, Monte Carlo simulation programs are used to model the development of particle cascades in the atmosphere and compare with experimental data. Two types of technique are used to study air-showers:

(I) Particle detectors (for example, scintillators) that count the muons, electrons and positrons at the surface. The AGASA experiment in Japan, with an area of 100 km², is the largest operating extended air-shower surface array.
(II) The charged particles in the shower excite nitrogen in the atmosphere causing fluorescence emission. Optical telescopes can be used to detect these photons allowing a full reconstruction of the particle cascade. An example of one such instrument in operation is the Fly's Eye experiment in Utah.

12.3.2 Interaction with CMBR

The Universe becomes opaque for cosmic-ray protons when the resonant reaction with CMBR photons

$$p + \gamma_{CMBR} \rightarrow \Delta^+ \rightarrow N + \pi$$

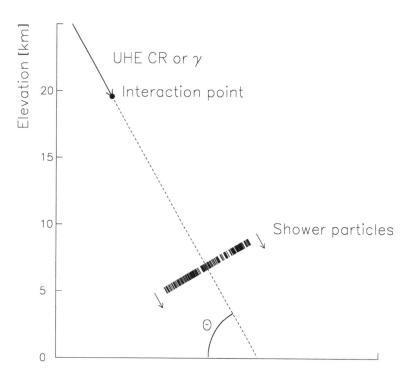

Fig. 12.4. Schematic view of an air-shower. The arrival times for the wavefront of shower particles at the detector surface are used to reconstruct the angle of the primary particle. The shower particles arrive at the ground in narrow 'pancake'-like sheets.

becomes energetically allowed. The final states could be either $p + \pi^0$ or $n + \pi^+$. To estimate the cut-off energy we assign four-momenta to the incoming particles as measured by an observer at rest:

$$
\begin{cases}
\gamma : & (q, q, 0, 0) \\
\text{proton} : & (\sqrt{p^2 + m_p^2}, p\cos\theta, p\sin\theta, 0)
\end{cases}
\tag{12.10}
$$

where the proton with momentum $\mathbf{p}$ hits the photon moving along the x-axis with momentum $\mathbf{q}$ at an angle θ in the xy-plane.

The energy requirement is met if the centre-of-mass energy is at least as large as the sum of the pion and proton mass: that is, $s \geq (m_p + m_\pi)^2$ (see (2.59)). With some algebra this can be shown to give the inequality (Problem 12.1):

$$
m_p m_\pi + \frac{m_\pi^2}{2} \leq q \left(\sqrt{p^2 + m_p^2} - p\cos\theta \right)
\tag{12.11}
$$

As the pion mass is much smaller than the proton (and neutron) mass the expression above can be simplified to

$$E_p - p\cos\theta \geq \frac{m_p m_\pi}{q} \tag{12.12}$$

Since, for a thermal gas of relativistic bosons $\langle q \rangle \sim 2.7\, T$ (see (8.22)) and $T_{CMBR} \simeq 2.74$ K, corresponding to only $2.36 \cdot 10^{-4}$ eV, the cut-off energy for nucleons becomes very large:

$$E_p \sim 10^{20} \text{ eV} \tag{12.13}$$

This is known as the *Greisen Zatsepin Kuzmin limit* or *GZK limit*.

Next we want to estimate the mean free path for nucleons above the GZK limit. The general expression for the mean free path for a particle in a scattering region is derived as follows: Consider a particle incident on a region of area A containing n scatterers per unit volume. The mean free path λ is defined as the average distance travelled by the incident particle before hitting a scatterer. If each such scatterer has a cross-section σ, the probability of a particular one being hit is $p_1 = \sigma/A$. Approximately, unit probability that at least one will be hit is reached when $N \cdot p_1 = 1$, where the total number of scatterers in the volume $A\lambda$ is $N = A\lambda n$. This happens when

$$\lambda = \frac{1}{n\sigma} \tag{12.14}$$

The cross-section for the $p\gamma_{CMBR}$ reaction is $\sigma_{p\gamma} \approx 10^{-28}$ cm^{-2} and the density of microwave photons was found to be $n = 410 \cdot (1+z)^3$ cm^{-3} (problem 11.1). Thus, the mean-free-path of protons with $E_p \sim 10^{20}$ eV is

$$\lambda_{GZK} \approx 8 \text{ Mpc} \tag{12.15}$$

about half the distance to the Virgo cluster, the nearest large cluster of galaxies. If the protons lose about 20 per cent of their energy in each encounter, almost all the energy is lost after about 100 Mpc.

Although still inconclusive, the observation of a handful of events above the GZK limit has been reported. One such case is shown in Fig. 12.5. It is intriguing that no candidate sources are found in the direction of these events. New larger experiments with close to two orders of magnitudes greater collection area are being planned to further investigate these results. The AUGER experiment, for example, a hybrid detector employing both surface arrays and atmospheric fluorescence telescopes to cover an area of 6000 km^2, is to be built in Argentina for studying the VHE CR southern sky.

12.4 Particle Acceleration

A major puzzle ever since the discovery of cosmic rays has been their exact origin. The fact that particles with energies exceeding 10^{20} eV have been

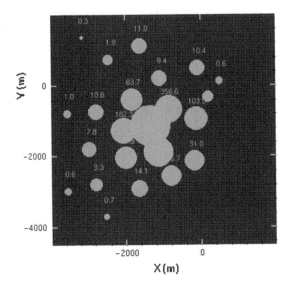

Fig. 12.5. Extensive air-shower event registered at the AGASA array [18]. The reconstructed energy of the air-shower spread over 4×4 km^2 is $2 \cdot 10^{20}$ eV. The radii of the circles are proportional to the logarithm of the particle density per m^2 at the detector location.

detected shows that there have to exist very powerful sites of acceleration in the Universe. In fact, it is plausible that particles with energies above the knee (10^{16} eV) originate from outside the Milky Way (since the magnetic fields are not strong enough to confine them), whereas galactic sources most likely have to be found for the lower-energy part.

A galactic event with great energy release is of course a supernova explosion. The total power in the form of cosmic rays leaving the galactic disk can be estimated to be of the order of 10^{34} W [8]. Since the average energy output of a supernova is around 10^{46} J, and the frequency of supernova explosions in the Galaxy is of the order of one per 50 to 100 years, only about 1 per cent of the energy released needs be used for cosmic-ray particle acceleration. The most promising mechanism is acceleration near the shock which is formed by the expanding envelope when it sweeps through the interstellar medium surrounding the exploding star. Recent X-ray data from the remnant of the supernova which was observed by Chinese astronomers to explode in the year 1006 confirm that acceleration of electrons still occurs, making it plausible that also protons and heavier nuclei are accelerated there (see Fig. 12.6). (Electrons are easier to detect since they radiate X-rays and γ-rays much more easily than do protons.)

It was Fermi who first realised that particles can be accelerated in a stochastic way to very high energies. Subsequently, several improvements to

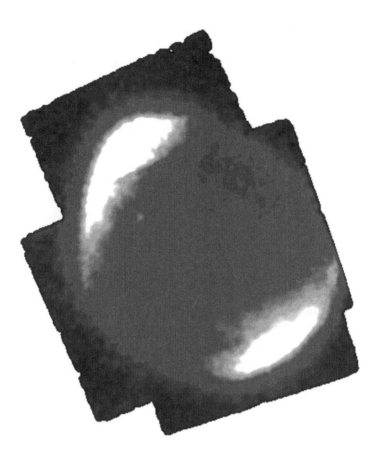

Fig. 12.6. X-ray emission from the supernova of 1006. The bright spots mark regions where intense emission occurs. The intensity and energy distribution is consistent with the hypothesis that electrons are accelerated to high energies at the shock front of the still-expanding envelope of the supernova. Credit: E. Gotthelf (GSFC), The ASCA Project, NASA.

his idea have been made, and there are now convincing models that describe acceleration in various environments. The basic idea is quite simple. Suppose that cosmic-ray particles travelling in interstellar space collide with much larger objects (for example, magnetized clouds) which move with random velocity and direction. Depending on the exact relative motion of the two types of object, the cosmic-ray particles can either lose or gain energy (to a good approximation, the energy of the large object is unchanged at each collision). A distribution of energies will then be obtained. Since there is a floor $E_{kin} = 0$ but no ceiling for the possible final kinetic energy, the average energy of the cosmic rays will tend to increase. The shape of the energy distribution will depend on the details, but Fermi showed that in many cases a power-law is expected.

A problem with the initial Fermi idea was that acceleration was quite slow and inefficient, with only $\mathcal{O}(v^2)$ gains of energy at each collision (so-called second-order Fermi acceleration). Due to losses at each stage (in particular ionization losses, which are very high for slow particles) it was difficult to obtain a working model. Much more efficient are first-order (that is, linear in v) processes, which may occur near shock fronts. Again, the basic idea is simple. Suppose that a strong shock is propagating through the interstellar medium surrounding, for example, a supernova. The shock represents a thin region where a pressure and density gradient exists, and it propagates much faster than the speed of sound in the medium. The medium in front of the shock and that behind the shock differ in density by a factor which depends on the equation of state (for a fully ionized gas, this pressure ratio is equal to 4). Also, kinetic energy is transferred from the shock to the gas, which means a bulk motion of the gas after the shock. However, on each side of the shock, particles diffuse (perform a random walk) with a diffusion constant that depends, among other things, on the energy of the particle and average value of magnetic fields present. A cosmic-ray particle downstream of the shock may pass the shock front, gain kinetic energy, and then scatter back downstream. This may then be repeated during many cycles. If a fraction ξ of energy is gained at each cycle, and the cycle time T is proportional to energy (and inversely proportional to the magnetic field strength), which is reasonable in a diffusion model, the equation

$$\frac{dE_p}{dt} = \frac{\xi E_p}{T} \tag{12.16}$$

implies a constant acceleration rate. Numerically, one finds [48]

$$\frac{dE_p}{dt} \sim 10^{12} \left(\frac{B}{1 \text{ Gauss}}\right) v_1^2 \text{ eV s}^{-1} \tag{12.17}$$

where v_1 is the flow velocity of the shocked gas.

This acceleration continues until energy losses (which depend on the ambient medium) balance the acceleration rate. The energy spectrum becomes a power law with spectral index around -2 in this type of acceleration. Taking into account energy-dependent containment time in the galactic disk (higher-energy particles escape more easily), this primary index would be modified, perhaps to the required -2.7 which is observed.

The shock-acceleration mechanism is expected to be active in many different types of shock: the termination shock of the solar wind and the galactic wind, accretion shock near a supermassive black hole (as is believed to exist at the centre of an AGN), intergalactic shock waves etc.

12.5 Summary

- The Earth's atmosphere is bombarded by a non-thermal power-law distributed flux of atomic nuclei reaching up to 10^{20} eV, about eight orders of magnitude higher than the most powerful man-made accelerators. However, the flux follows a steep power-law, $dN/dE \sim E^{-2.7}$ up to around 10^{16} eV. At the highest energies observed, $\sim 10^{21}$ eV, an excess of events has been detected, but so far with very low statistics.
- Ultra-high energy cosmic rays are of great interest, as they provide directional information about their origin.
- First-order Fermi acceleration near shock-fronts caused by super-nova explosions is a plausible mechanism which may explain the existence of galactic cosmic rays up to an energy of 10^{16} eV.

12.6 Problems

12.1 Derive the inequality (12.11).

12.2 Derive the expression for the deflection from a straight line for a charged track in a uniform magnetic field (12.6) assuming small deflections: that is, $\sin \theta \approx \theta$.

12.3 Show that the mean free path of electrons is smaller than typical galaxy sizes. The main source of interaction in the intergalactic medium is assumed to be through Compton scattering on the 2.7 K radiation, $\sigma \approx 10^{-24}$ cm^2.

13 Cosmic Gamma-Rays

13.1 The Sky of High-Energy Photons

The search for the origin of cosmic high-energy particles has naturally led to the investigation of the sky at the high energy end of the spectrum of electro-magnetic radiation, using *gamma-rays* as information carriers. Absorption by the Earth's atmosphere prevents the study from the ground of photons with far-ultraviolet or shorter wavelengths. The progress of the field has therefore followed the launch of dedicated satellite experiments. A major breakthrough was achieved when the COMPTON GAMMA-RAY OBSERVATORY (CGRO) was launched in April 1991.

The CGRO was a satellite mission carrying four instruments with large angular acceptance which together are sensitive in the energy range 30 keV $< E_\gamma < 30$ GeV, as shown in Table 13.1.

Table 13.1. Parameters for instruments on board the CGRO satellite.

Instrument	Energy range (MeV)	Field of view	Source location
BATSE	$0.03 - 1.2$	4π sr	$\sim 2°$
OSSE	$0.06 - 10$	$4° \times 11°$	-
COMPTEL	$1 - 30$	1 sr	~ 10 arcmin
EGRET	$20 - 30000$	0.6 sr	~ 5 arcmin

These detectors have detected gamma-rays from very distant corners of the Universe. The strongest feature of the all-sky map from the EGRET instrument, shown in Plate 7 (the colour plate section is positioned in the middle of the book), is the emission from the galactic plane, mainly due to interactions between cosmic rays and interstellar gas and photons (through processes such as $p + p \rightarrow p + p + \pi^0 + ...$, followed by $\pi^0 \rightarrow \gamma + \gamma$ decays). Other resolved objects in the Milky Way are the known Crab, Geminga and Vela pulsars, also observed in other wavelengths. Gamma-rays within the energy range of EGRET propagate freely through intergalactic space with very little absorption.

The detection of 100 MeV gamma-rays from several active galactic nuclei (AGN) with redshifts ranging from $z = 0.03$ to $z = 2.2$ has generated a great deal of interest. Most of these AGN are *blazars*. AGN are believed to be very massive black holes ($\sim 10^8$ solar masses). The violent processes near the black hole sometimes cause two relativistic, diametrically opposite jets to emanate from the vicinity of the black hole. Blazars are AGN which happen to have one of the jet axes pointing towards the Earth. One of them, 3C279, at a redshift $z = 0.538$, is particularly bright and variable. The peak flux is about 2 gamma-rays per m^2 per minute.

The emission has a characteristic time variability on scales as short as a day, in some cases even shorter than an hour, indicating that the region of emission is very small: that is, $R \leq c\Delta t/(1 + z)$, is very small compared with galaxy sizes. It is believed that the radiation originates either from small hot regions in the jets or from the accelerated matter in the accretion disk surrounding the very massive black hole.

The CGRO will be followed by a new satellite, GLAST, which is presently under construction, for a planned launch date at the end of 2006. GLAST will have an order of magnitude larger gamma-ray collection area than CGRO, and many other parameters such as the energy range, the energy resolution and the angular resolution will be far superior to those of CGRO.

13.2 Gamma-Ray Bursts

One of the most exciting fields in gamma-ray astronomy and arguably in all of contemporary high-energy astrophysics is the 'puzzle' of the origin of *Gamma-Ray Bursts* (GRB). The existence of rapid flares of gamma-rays from all directions, without association to any known astronomical object, has been known since the 1960s. The American *Vela* satellite, the purpose of which was to monitor possible clandestine tests of nuclear weapons, did in fact discover the first GRB. The information was, however, not disclosed to the astronomy community until 1973. The nature of GRB, whether they were galactic or of cosmological origin, remained unknown until the CGRO came into operation. The BATSE detector (see Table 13.1) showed very soon that, unlike the EGRET sources in colour Plate 7, GRB populated all the directions evenly, as shown in Fig. 13.1. The BATSE observations suggested very strongly that GRB sources are not within the Milky Way. Moreover, the strong isotropy of the signal indicated that the sources must come from truly large distances, comparable to the size of the Universe.

Definite proof that GRB originate at cosmological distances could only be achieved if the source could be resolved. This was not initially possible due to the poor angular resolution at gamma-ray wavelengths (several degrees of arc). This shortcoming was finally remedied when BeppoSAX, an Italian Dutch satellite, came into operation in 1997. BeppoSAX, equipped with wide-field X-ray cameras, could both detect X-ray counterparts of GRB and pinpoint the location of the source within a few arcminutes. This error box is small enough for it to be quickly scanned with CCD cameras on optical

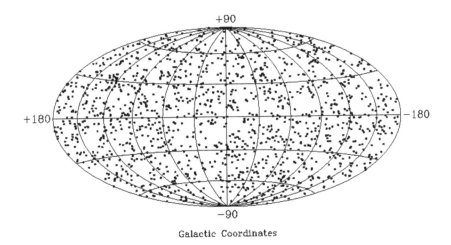

+90

+180

−180

−90

Galactic Coordinates

Fig. 13.1. More than 1600 bursts have been detected by the BATSE. The uniformity of the distribution of sources indicates that they have an extragalactic origin. Figure from http://cossc.gsfc.nasa.gov/cossc/burst.html.

telescopes on the ground. The first historical detection of a GRB with BeppoSAX took place on 28 February, 1997. The detection image and a similar exposure of the same spot just a few days later are shown in colour Plate 8.[1]

Less than 24 hours after the burst, a transient was seen in optical images. The signal faded very quickly both in X-rays and at optical wavelengths. Nevertheless, the source could be studied with the Hubble Space Telescope, and with the 10-metre Keck telescopes on Hawaii, which showed a faint nebula underlying the fading source. Spectroscopic studies revealed clear magnesium and iron absorption features at a redshift $z = 0.835$. Thus the cosmological nature of GRB was established, ending a scientific debate that had lasted for almost 25 years.

Since then, more observations have been made including the detection of one gamma-ray optical counterpart at $z = 3.4$ following a BeppoSAX observation.

13.2.1 What Are GRB?

Gamma-ray bursts are among the most energetic known objects, matched only by supernova explosions. Core collapse supernovae, however, release most of their energy as neutrinos, as will be discussed in Chapter 14. While the electromagnetic radiation (light) from supernovae is also large, often outshining the host galaxy, it is emitted over several weeks. Gamma-ray bursts often last for just a few seconds, as shown in Fig. 13.2.

---

[1] The colour plate section is positioned in the middle of the book.

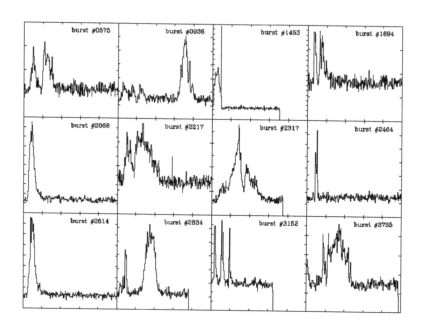

Fig. 13.2. GRB brightness (vertical axis) *versus* time (horizontal axis, 2 seconds) for twelve relatively short bursts. Bursts last a few tens of milliseconds to hundreds of seconds, with no two bursts showing exactly the same development in time. From http://cossc.gsfc.nasa.gov/cossc/burst.htm.

One possible explanation for the different time scales is that while a heavy star mantle acts as a 'shock absorber' in a supernova explosion, thereby stretching out the energy release over a long period of time, GRB could be associated with a much thinner matter envelope.

Assuming that a large amount of energy is transferred to a thin star envelope, the latter will expand at highly relativistic speeds, $v \approx c$. For example, for an envelope containing just 0.01 per cent of the star's mass, the energy of material in the expanding shell could be 1000 times larger than its rest mass.

The short time durations of GRB can be understood with the model shown in Fig. 13.3. A shell of matter is thrown with highly relativistic speed $\gamma \sim 10^3$ at the same time as a flash of light escapes the source. The next flash is emitted when the expanding envelope encounters material that decelerates its motion, typically a month later in the rest frame of the centre of the expansion. For an observer at the centre of the shell with radius R, it takes

$\Delta t = R/c$ for the original flash to reach the location of the second emission. For $\gamma \gg 1$, the speed of the expanding shell becomes (see (2.41)):

$$v \approx c(1 - \frac{1}{2\gamma^2}) \tag{13.1}$$

Thus, the time difference between the two flashes becomes:

$$\Delta\tau \approx \frac{R}{2\gamma^2 c} = \frac{\Delta t}{2\gamma^2} \tag{13.2}$$

which corresponds to a few seconds for $\Delta t \sim 1$ month and $\gamma \sim 10^3$. Finally, because of the cosmological time dilation factor $(1 + z)$, the duration of GRB will increase (statistically) with redshift.

Although the exact nature of the gamma-rays is not well understood, it is likely that the emitted gamma-rays originate from the shock-wave generated by the expanding shell. The 'fireball' model, first proposed by Martin Rees and Peter Meszaros, implies that as the shock-wave moves, it compresses and heats up the surrounding gas, and synchrotron radiation is emitted as electrons spiral in the magnetic fields thought to be present in these environments. As the shell expands it is slowed down by the interactions with the surroundings, and the radiation becomes less energetic. This phase is called the 'afterglow' and is first dominated by X-ray emission for a few minutes, followed by UV and visible emission for a period of some days or weeks. Infrared and radio waves can be emitted for several months after the GRB.

In at least one recently discovered GRB (030329) at redshift z=0.17, spectral studies of the optical afterglow about 1 week after the gamma ray emission show broad peaks in flux, characteristic of supernovae (SN2003dh). The spectra lack hydrogen, silicon and helium features and may therefore be associated with Type Ic supernovae.

Example 13.2.1 Show that if a star's rest energy can be converted to kinetic energy with about 10 % efficiency, a star envelope containing only 0.01% of the star's mass can be boosted with a Lorentz factor $\gamma = 10^3$.

Answer: The converted energy is $E_{kin} = (1 - 10^{-4}) \cdot Mc^2 \cdot \epsilon$, where ϵ is the efficiency of the system. If all of the kinetic energy is fed to the mantle with rest energy $E_{rest} = 10^{-4} \cdot Mc^2$ the Lorentz factor becomes:

$$\gamma = \frac{(1 - 10^{-4})\epsilon}{10^{-4}} = 10^3 \tag{13.3}$$

for $\epsilon = 0.1$.

It has been suggested that GRB might be the result of colliding neutron stars. The emission of gravitational radiation forces the orbits of binary neutron stars to shrink, as will be discussed in Chapter 15. It is therefore expected that at some point the two stars should merge. The expectation for

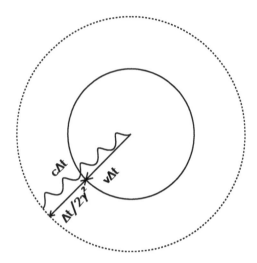

Fig. 13.3. Model to explain the short time duration of a gamma-ray burst. A shell of matter (solid circle) expands with a velocity v close to the speed of light: that is, $\gamma \gg 1$. At the same time a flash of radiation is emitted. After some time Δt, typically a month or so, the expanding shell will be slowed down by surrounding gas, causing a new emission of radiation. The original flash (dotted circle) has by then propagated a distance $c\Delta t$. As both flashes reach the Earth their time separation is just $\Delta t/(2\gamma^2)$, which is typically a few seconds for $\gamma \sim 10^3$.

such mergers is that they should take place once every 10^6 years in an average galaxy, about four orders of magnitude less often than supernova explosions. The rate of GRB detections are a good match with that rate.

13.3 Very High-Energy Gamma-Rays

Just as for hadronic cosmic rays, satellite-borne experiments can be used for the study of gamma-rays up to $\sim$100 GeV, while the higher energies are studied from the ground with detectors having larger collection area.

The most successful method for studying gamma-initiated atmospheric air-showers is through the detection of Cherenkov radiation, mainly UV and blue light, emitted from the very energetic electrons and positrons in the particle cascade described in Section 12.3.1. The origin of Cherenkov radiation and its application in building telescopes for high-energy particles will be discussed in detail in Section 14.10 (see also Problem 14.13).

Telescopes for detecting VHE gamma-rays consist of arrays of optical telescopes operated at night, such as the 10-metre Whipple telescope shown in Fig. 13.4. The particles in the electromagnetic shower have strong forward momentum, and the Cherenkov radiation is emitted within about one degree of these. Thus the collected light closely follows the direction of the incident photon.

An experimental difficulty for air Cherenkov telescopes is the identification of the incident particles. Hadronic air-showers are about 1000 times more frequent than those originating from gamma-rays of the same energy. On the other hand, gamma-rays do carry directional information while hadronic cosmic rays are nearly isotropic, as discussed in Chapter 12. Sources of gamma-rays are therefore identified through the accumulation of events in a fixed direction in the sky. The particle discrimination and energy resolution can be enhanced when several telescopes are used to view the shower simultaneously, in particular if the reflector information is gathered into an imaging camera which captures the shape of the particle cascade. Gamma-ray 'images' are typically smaller and more elongated than background hadronic showers.

Fig. 13.4. The 10-metre aperture Whipple telescope in Southern Arizona detects gamma-rays with energy above around 300 GeV.

13.3.1 Resolved Sources

TeV gamma-rays from the Crab nebula have been detected by several atmospheric Cherenkov detectors, similar to the Whipple telescope shown in Fig. 13.4. This was predicted in the mid-sixties, as it became clear that at the centre of the supernova remnant[2] was a pulsar emitting polarized synchrotron radiation at all wavelengths: radio, optical and X-rays. High-energy gamma-rays are produced as electrons transfer momentum to synchrotron produced photons through inverse Compton scattering. Thus, the flux of gamma-rays emitted depends on the magnetic field of the nebula, estimated to be of the order of $\sim 10^{-3}$ G.

More recently, high-energy gamma-rays from two extragalactic sources have been identified as being low redshift, $z \approx 0.03$, AGN blazars in the Markarian catalogue: Mkn421 and Mkn501. As opposed to the Crab nebula, the observed AGN showed large time variability in the gamma-ray flux.

The Whipple telescope, however, failed to detect TeV gamma-rays from most of the AGN observed at GeV energies by EGRET. This is despite the fact that some of them, for example 3C279, were much brighter at lower energies and had a similar energy dependence in the spectrum measured at low energies. Extrapolation indicated that they should have been well above threshold for detection at Whipple.

What makes Mkn421 and Mkn501 special is not that they are particularly bright but rather that they are at relatively low redshifts: that is, at small distances. The non-detection of 3C279 and other bright AGN at higher redshifts indicates that the TeV gamma flux is attenuated along the way.

Recently, a few more TeV blazars have been discovered, and the evolution of the field is rapid, with larger telescopes being built, such as the 17 m MAGIC telescope at the Canary Islands, which recently started to operate, and also groups of powerful telescopes such as HESS in Namibia, CANGA-ROO in Australia and VERITAS in the United States. These new telescopes and telescope arrays have a larger area and lower energy threshold which means that they will complement in a nice way the all-sky satellite gamma-ray telescope GLAST, which has a scheduled launch date of 2006.

13.3.2 Interaction with IR Photons

Since gamma-rays interact with electromagnetic coupling strength (in contrast, for example, to neutrinos which only interact weakly), they cannot travel arbitrarily long distances without being scattered or absorbed. In Section 6.11.3, we saw that for a given available energy in the centre of mass, Compton scattering and $\gamma\gamma \to e^+e^-$ have similar cross-sections (in particular, dropping quite rapidly at high energies). Of course, this is not true near the threshold for e^+e^- production, where the latter process is suppressed by the small available phase space. Therefore, absorption of high-energy gamma-rays will be determined by the Compton cross-section at energies below the pair production threshold. Above the threshold, the fact that photons are so much more numerous than electrons in the Universe (this is in particular the

---

[2] The explosion was observed by Chinese astronomers in 1054.

case for the cosmic microwave background) means that pair production will be the limiting factor determining the absorption length.

We saw in Fig. 6.9 that the pair production cross-section is maximal for $y \simeq 0.8$, where $y = (\sqrt{s} - 2m_e)/(2m_e)$.

For a head-on collision of a gamma-ray of energy E_γ on an ambient photon, this maximal cross-section occurs for a target photon energy of

$$E_{\text{target}} \sim \frac{(1.8m_e)^2}{E_\gamma} \sim 0.8 \left(\frac{1 \text{ TeV}}{E_\gamma} \right) \text{ eV} \tag{13.4}$$

This means that above 1 TeV, the optical and infrared (IR) photons will be the ones which determine the absorption length, while the CMBR will dominate completely above around 100 TeV. Using (6.35) for the cross-section, it is not difficult to compute the absorption length for a gamma-ray of a given energy passing through a target photon 'gas' of, say, IR photons of density $n(E_{IR})$, where the energy distribution is a power-law

$$n(E_{IR}) = \frac{n_0}{E_{IR}^0} \left(\frac{E_{IR}}{E_{IR}^0} \right)^{-\nu} \tag{13.5}$$

where E_{IR} is the energy of the IR photons.

The probability of absorption (the inverse absorption length) of high-energy photons with energy E_γ is given by [6]

$$\frac{dW}{dl} = \frac{4}{E_\gamma^2} \int_{m_e}^\infty d\omega_c \sigma(\omega_c) \omega_c^3 \int_{\omega_c^2/E_\gamma}^\infty d\omega \frac{n(\omega)}{\omega^2} \tag{13.6}$$

where ω_c is the photon energy in the centre of momentum system of two colliding photons, and $\sigma(\omega_c)$ is the cross-section (6.35) for $\gamma\gamma \to e^+e^-$.

After some calculations one obtains (see, for example, [46])

$$l_{abs}^{-1} = \frac{2\Phi_\nu}{1+\nu} n_0 \frac{\pi\alpha^2}{m_e^2} \left(\frac{E_\gamma \omega_0}{m_e^2} \right)^{\nu-1} \tag{13.7}$$

where

$$\Phi_\nu = \int_0^1 v dv (1 - v^2)^{\nu-1} \left[(3 - v^4) \ln \left(\frac{1+v}{1-v} \right) + 2v(v^2 - 2) \right] \tag{13.8}$$

For integer values of ν this integral can be solved, for example,

$$\Phi_1 = 14/9, \quad \Phi_2 = 22/45, \quad \Phi_3 = 56/225 \tag{13.9}$$

There are two ways in which these results can be used. For a known intensity and spectral index of the intergalactic IR background, (13.7) can be used to predict the furthest distance at which TeV gamma-rays can be expected to be detected. However, the IR distribution is presently not very well known, and one may use the fact that multi-TeV gamma-rays have been observed from several sources, such as the active galaxies Markarian 421 and 501, to limit the density of the IR photon field. Putting in reasonable numbers for the expected IR background flux, it seems difficult to detect

10 TeV gamma-rays from sources further than $z \sim 0.1$. A complication in this and many similar applications is that it is difficult to separate eventual absorption in the source of gamma-rays from absorption along the way (see problem 13.2).

A similar calculation to the one above shows that a gamma-ray of 80 TeV or above cannot propagate further than from $z \sim 0.01$ without being absorbed by the intense CMBR.

13.4 Summary

- Gamma-rays are photons of MeV energies and higher, and thus propagate on geodesics, making them suitable for detecting energetic processes in supernova remnants, AGNs and other astrophysical sources.
- Gamma-ray bursts were long an enigmatic phenomenon. With new observations their cosmological origin has been established. The exact mechanism for their generation is still unknown, however.
- Gamma-rays with energies exceeding several TeV have been detected from the Crab nebula (a supernova remnant) and a few active galactic nuclei of the blazar type.

13.5 Problems

13.1 Calculate the cut-off energy at which gamma-rays are absorbed by the CMBR ($\gamma\gamma_{CMBR} \to e^+ e^-$).

13.2 A high-energy photon, a gamma-ray of energy $E_\gamma \sim 100$ GeV, coming from a very distant galaxy, can interact with optical (starlight) photons in the Universe through reactions of the type $\gamma + \gamma \to e^+ + e^-$, with a cross-section roughly equal to the Thomson cross-section α^2/m_e^2. It can be assumed for simplicity that the number density n_0 of starlight photons has not changed, apart from the cosmic expansion.
(a) Use relativistic kinematics to deduce the threshold energy of the starlight photons needed to produce an $e^+ e^-$ pair at redshift z, if the high-energy and starlight photons collide at an angle θ.
(b) (More difficult) Give an order of magnitude estimate of how large the number density n_0 of photons with energy higher than the threshold energy can be, if sources at $z \sim 1$ are observed.

14 The Role of Neutrinos

14.1 Introduction

We have seen in Section 9.3 that neutrinos could provide an important contribution to the total energy density of the Universe if they have a mass in the 10 eV range. Neutrinos are also very important information carriers from violent astrophysical processes, which is why we now devote some time to the study of neutrino properties. For a complete review of weak interactions in astrophysical environments, see [37, 4].

As we saw in Section 6.10, neutrinos are the neutral counterparts of the charged leptons: e, μ and τ. There are therefore three types of neutrino in nature: ν_e, ν_μ and ν_τ. Neutrinos are *fermions* as are the rest of the leptons and quarks: that is, spin-$\frac{1}{2}$ particles. Apart from their possible gravitational interactions, νs interact with matter only through the exchange of the mediators of the weak force, the W and Z bosons. They are fundamental particles (without constituents, as far as is known), have extremely small masses (if any at all) and lack electric charge. Paradoxically, because neutrinos only interact weakly with matter, they are very important in astrophysics. Where other particles become trapped or can only propagate through very slow diffusive processes, neutrinos are able to escape. Neutrinos can thus connect regions of matter that would otherwise be isolated from each other. Because they are massless (or almost massless), they move at the speed of light, which makes the energy transfer very efficient.

For example, neutrinos produced near the centre of the Sun can be detected at the Earth after a time of flight of around 8 minutes, and permit the study of the nuclear reactions that take place in the core of 'our star'. The photons generated by the energy-producing nuclear reactions at the centre, however, diffuse slowly to the surface with an average diffusion time of around a million years!

The lack of electric charge allows neutrinos (like photons) to move in straight lines (or geodesics) even in the presence of strong magnetic fields. According to the Standard Model of particle physics, the magnetic moment of a neutrino is extremely small. They therefore point back to the sites of production, and offer a unique potential for obtaining information about where particle production and acceleration takes place in the Universe.

14.2 The History of Neutrinos

Neutrinos were postulated in 1930 by Wolfgang Pauli to explain a supposed anomaly in the energy spectrum of β-decay. While at the time only two emerging particles from the nuclear decay could be detected:

$$^A_Z X \rightarrow\ ^A_{Z+1}Y + e^-$$

the energy of the produced electron was not monochromatic, as it should be if the original nucleus decays at rest into only two bodies. To save the principle of energy conservation, Pauli suggested that a third particle was always produced in β-decay, but that it was not detectable. The third particle had to be neutral (or else the conservation of charge would have been violated!) and have very small mass, since the total energy of the detected particles accounted for almost all the mass of the parent nuclei. For these reasons, Fermi called the 'invisible' particle the *neutrino*.[1]

The proper description of β^--decay is thus

$$^A_Z X \rightarrow\ ^A_{Z+1}Y + e^- + \bar{\nu}_e \tag{14.1}$$

If energetically allowed, a proton within a nucleus can transform into a neutron under the emission of a positron:

$$^A_Z X \rightarrow\ ^A_{Z-1}Y + e^+ + \nu_e \tag{14.2}$$

The $\bar{\nu}_e$ in (14.1) is the antineutrino. The production of an antiparticle in β^- decay is necessary for the conservation of lepton number, as discussed in Section 6.3. The reverse happens in β^+ decay, where a positron (anti-electron) is emitted, thus a neutrino is needed to reset the lepton number to what it was prior to the decay.

The existence of neutrinos was finally demonstrated by the observation of the reaction:

$$p + \bar{\nu}_e \rightarrow n + e^+ \tag{14.3}$$

The original experiment was performed by Clyde Cowan and Fred Reines in 1955. For this fundamental discovery Reines was awarded the 1995 Nobel prize in physics.[2] The experiment was performed in an underground laboratory, 11 metres below one of the nuclear reactors at Savannah River. The signal of an antineutrino capture was the simultaneous detection of the positron, as it annihilated with an electron in the target,[3] and a recoiling neutron.

---

[1] Pauli was quite uneasy about predicting a particle whose existence nobody, including himself, thought could ever be proven. Only some 25 years later, when the first powerful nuclear reactors were put into operation, was the time ripe for its experimental discovery.

[2] The prize was shared between Reines (Cowan had died many years earlier) and Martin Perl for his discovery of the τ lepton.

[3] The result of the annihilation is two 511 keV photons in opposite directions.

Muon neutrinos were first observed in 1962 in an experiment led by Leon Lederman, Melvin Schwartz and Jack Steinberger. They received the Nobel prize in physics in 1988.

14.3 Neutrino Interactions with Matter

Neutrino interactions with matter are divided into two kinds. *Neutral current* (NC) interactions are mediated by the neutral Z bosons. *Charge current* (CC) interactions, on the other hand, involve the exchange of W$^+$ and W$^-$ bosons.

NC interactions are responsible for annihilation reactions involving neutrinos,

$$e^+ + e^- \rightarrow \nu_\mu + \bar{\nu}_\mu$$

for example, and interactions such as

$$\nu_\mu + e^- \rightarrow \nu_\mu + e^-$$

where neutrinos scatter with matter and thereby gain or lose energy from the collision partner without any additional matter being created or destroyed. Such collisions are called *elastic scatterings*. Both types of interaction are shown in Fig. 14.1.

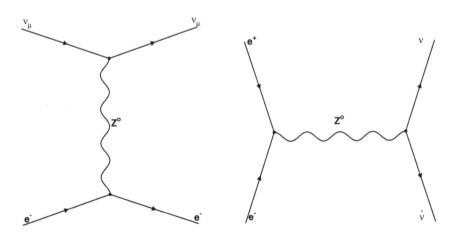

Fig. 14.1. Feynman graphs for NC interactions: elastic $e-\nu$ scattering and electron positron annihilation into neutrinos.

In CC interactions there is an exchange of lepton partners. For example, an antineutrino can be absorbed by a proton, producing a neutron and a positron in the final state as shown in Fig. 14.2.

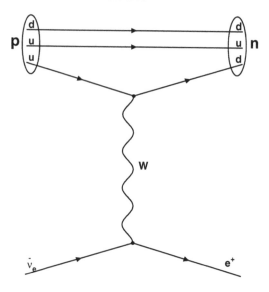

Fig. 14.2. Feynman diagram for antineutrino absorption. One of the quarks inside the proton changes flavour thereby transforming the nucleon into a neutron.

14.3.1 The Cross-Sections

Neutrino interactions involve the production of very high mass virtual particles, and are therefore heavily suppressed at low energies. The matrix element of the reaction has a propagator term which includes the mass squared of the mediator particle (see (6.23)):

$$\frac{1}{q^2 - M^2}$$

where q is the momentum transfer of the reaction.

The Z^0 mass is about 91 GeV: that is, almost 100 times heavier than the proton. W-bosons are somewhat lighter, around 80 GeV.

A rough estimate of the cross-section for weak processes, such as

$$\bar{\nu}_\mu \mu^- \rightarrow \bar{\nu}_e e^- \tag{14.4}$$

for energies well below the masses of the Z and W particles, can be derived from purely dimensional grounds. We found in Section 6.10.1 that the effective coupling constant in weak interactions is $g^2_{weak}/m^2_W \sim 10^{-5}$. Before the discovery of the W and Z bosons, the weak interactions had been parameterized by the so-called Fermi constant G_F, which governs the muon decay $\mu^- \rightarrow e^- + \nu_\mu + \bar{\nu}_e$ and is just of this order (see also Appendix D.3):

$$G_F = 1.1664 \cdot 10^{-5} \text{ GeV}^{-2}.$$

The approximate expression for high-energy neutrino interactions with matter can thus be written (see (6.26)):

$$\sigma \sim \frac{\alpha^2 s}{m_W^4} \sim G_F^2 E_{cm}^2 \tag{14.5}$$

where E_{cm} is the centre of momentum energy of the incoming neutrino.

Example 14.3.1 Give a numerical expression for (14.5)

Answer: We use the standard recipe to convert back from natural units. To obtain an answer in units of length squared we have to multiply (14.5) by $(\hbar c)^2$. A useful relationship to remember is:

$$\hbar c \approx 2 \cdot 10^{-11} \text{ MeV cm} \tag{14.6}$$

We therefore arrive at the numerical expression:

$$\sigma_{weak} = [2 \cdot 10^{-11} \times 1.1664 \cdot 10^{-11} \times E_{cm}]^2$$
$$\sim 5 \cdot 10^{-44} \left(\frac{E_{cm}}{1 \text{ MeV}} \right)^2 \text{ cm}^2 \tag{14.7}$$

Next we calculate the cross-section for neutrino interactions expressed in terms of the *laboratory* energy. In a target experiment designed to detect neutrino interactions through processes of the type[4] $\nu_X e^- \rightarrow \nu_X e^-$, only the beam particles (ν_X) have finite momentum. In a water tank, for example, the electrons in the target atoms have extremely small net velocities. The centre of momentum energy squared, s, is the sum of the four-momenta of the colliding particles squared:

$$s = (P_e + P_\nu)^2 = [(E_e, \bar{\mathbf{p}}_e) + (E_\nu, \bar{\mathbf{p}}_\nu)]^2$$
$$= [(m_e, \mathbf{0}) + (E_\nu, E_\nu \hat{p})]^2$$

where $\hat{p}$ indicates the direction of the neutrino momentum, and we have recognized that for a massless particle $|\bar{p}| = E$ in natural units ($c = 1$).

Thus, we find that in the *laboratory* frame:

$$s = 2m_e E_\nu + m_e^2 \tag{14.8}$$

For very high neutrino energies, $E_\nu \gg m_e$, the last term can be neglected, leaving us with the expression:

$$\sigma(\nu e) \approx 2G_F^2 m_e E_\nu \tag{14.9}$$

In other words: *the cross-section for neutrino interactions rises linearly with neutrino energy.*

---

[4] The experimental signature is given by the recoiling electron

A detailed calculation for neutrino energies above 5 MeV shows that the total cross-section for the reaction $\nu_X e^- \rightarrow \nu_X e^-$ is well approximated by [37]:

$$\sigma_{\nu e} = C_X \cdot 9.5 \cdot 10^{-45} \cdot \left(\frac{E_\nu}{1 \text{ MeV}} \right) \text{ cm}^2 \tag{14.10}$$

where the flavour-dependent constants C_X are

$$C_e = 1 \tag{14.11}$$

and

$$C_\mu = C_\tau = \frac{1}{6.2} \tag{14.12}$$

The cross-section is larger for electron neutrinos, as they can, unlike the other neutrino species, couple to the electrons in the target through both NC and CC interactions, as shown in Fig. 14.3.

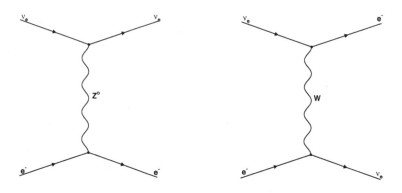

Fig. 14.3. Feynman diagrams for $\nu_e e$ interactions. The NC-interaction graph is shown in the left diagram. The right diagram shows the CC-interaction with the exchange of a charged W boson.

14.4 Neutrino Masses

Laboratory experiments have, so far, not succeeded in measuring the mass of any neutrino. Instead, the negative results have been expressed in the form of upper limits, due to the finite resolution of the experiments. The best (lowest) upper limits on the mass of the electron neutrino come from the studies of the electron energy spectrum in tritium decay:

$$^3\mathrm{H} \rightarrow ^3\mathrm{He} + e^- + \bar{\nu}_e$$

As the minimum amount of energy taken by the ν_e is its mass, the end-point energy of the emitted electron is a measurement of m_{ν_e}. According to these experiments the mass of the electron neutrino is lower than 3 eV at the 95 per cent confidence level.

Limits on the mass of the muon neutrino, ν_μ, are extracted from accurate measurements of the muon momentum p_μ in the decay of charged pions:

$$\pi^+ \rightarrow \mu^+ + \nu_\mu$$

For pions decaying at rest one obtains (Problem 14.1):

$$m_{\nu_\mu}^2 = m_\pi^2 + m_\mu^2 - 2m_\pi \sqrt{m_\mu^2 + p_\mu^2} \tag{14.13}$$

Inserting the measured masses for pions and muons, an upper limit on m_{ν_μ} is found to be 170 keV at the 90 per cent confidence level.

The existence of ν_τ is only established indirectly from the decay of τ leptons, as the ν_τ has not yet been observed directly. Mass limits on the ν_τ are set through studies of the missing energy and momentum in the charged τ lepton decays such as

$$\tau \rightarrow 5\pi^\pm + \nu_\tau$$

and

$$\tau \rightarrow 5\pi^\pm + \pi^0 + \nu_\tau$$

The current best laboratory mass limit is $m_{\nu_\tau} < 18$ MeV.

As shown in Table (11.1), a combination of the data sets from WMAP and the large scale structure survey 2dFGRS allows to set an upper bound to the total energy density contribution from neutrinos, $\Omega_\nu h^2 < 0.0076$. Equation (9.10) can be inverted to extract an upper limit on the mass of all neutrino species added:

$$\sum_i m_{\nu_i} < 94 \cdot \Omega_\nu h^2 (\mathrm{eV}) \sim 1 \text{ eV}. \tag{14.14}$$

14.5 Stellar Neutrinos

Neutrinos are very efficient in the process of the cooling of stars, in spite of their low probability of interaction with matter. Next we compute the mean free path of neutrinos in stellar environments.

As seen from (14.10), neutrinos in the MeV range have a cross-section of $\sigma_\nu \approx 10^{-44}$ cm^2 for matter interactions. The mean free path for neutrinos is related to the density of matter and the cross-section for interaction as (see (12.14)):

$$\lambda_\nu = \frac{1}{n\sigma_\nu} \tag{14.15}$$

where n is the number of nucleons per cubic centimetre.

For a hydrogen star with mass density ρ, the number density is given by

$$n = \frac{\rho}{m_H} = \left(\frac{\rho}{1\,\mathrm{g\,cm}^3}\right) \cdot 6 \cdot 10^{23}\ \mathrm{cm}^{-3} \tag{14.16}$$

Inserting (14.16) in (14.15) we find

$$\lambda_\nu \approx \frac{2 \cdot 10^{20}}{\rho/(1\,\mathrm{g\,cm}^3)}\ \mathrm{cm} \tag{14.17}$$

For normal stellar matter, $\rho \approx 1$ g/cm^3, the mean free path becomes 100 pc, and even at $\rho \approx 10^6$ g/cm^3 λ_ν is still 3000 solar radii! In other words, neutrinos with MeV-scale energies escape from the stellar interior without scattering even for extremely high densities. This is what makes neutrinos so special in astrophysical environments.

Example 14.5.1 In very hot stellar environments, photons are produced with enough energy to create large numbers of e^+e^- pairs.[5]

Estimate the fraction of events where an e^+e^- annihilation results in a production of neutrinos instead of the dominant two-photon production.

Answer: The relevant energy scale in a stellar environment where electron positron pairs are thermally produced is $E_{cm} \approx 2m_e = 1$ MeV. From Fig. 6.9 we see that the two-photon cross-section at these energies is roughly 10^{-25} cm^2. We can therefore obtain a rough estimate of the neutrino production ratio by taking the ratio of the cross-sections:

$$P = \frac{\sigma_{weak}(1\ \mathrm{MeV})}{\sigma_{em}(1\ \mathrm{MeV})}$$
$$\approx \frac{10^{-44}}{10^{-25}} \sim 10^{-19}$$

While the rate of electromagnetic processes is overwhelmingly larger, weak interactions are important in the energy balance of a star, as neutrinos, once created, escape from the star, thereby reducing its available energy.

---

[5] Eventually, an equilibrium density of positrons is reached as most of the electron positron pairs annihilate, mostly creating two photons.

14.5.1 Solar Neutrinos

The low cross-sections of weak interactions allow stars to burn their fuel slowly instead of exploding soon after formation. Our Sun, for example, is believed to be 4.5 billion years old, and is predicted to continue in its present luminous state for at least as long. The main source of energy in hydrogen-burning stars (as the Sun) is through the pp-fusion reaction:

$$4p + 2e^- \rightarrow {}^4He + 2\nu_e + 26.731 \text{ MeV}$$

where, on average, only about 2 per cent (0.6 MeV) of the energy is carried by the neutrinos according to the 'standard solar model', SSM. The details of the nuclear reaction – what is called the *pp* or *proton proton* chain – and the spectrum of the produced neutrinos are shown in Table 14.1 and Fig. 14.4.[6]

Table 14.1. Nuclear reaction for solar neutrinos [4]. The (hep) reaction, ${}^3He + p \rightarrow {}^4He + e^+ + \nu_e$, is several orders of magnitude less common than the others. Yet, it is important as it provides the high-energy tail of the solar neutrino spectrum.

	Reaction	Neutrino energy
1	$p + p \rightarrow {}^2H + e^+ + \nu_e$	≤ 0.42 MeV
	or	
	$p + e^- + p \rightarrow {}^2H + \nu_e$	1.442 MeV
2	${}^2H + p \rightarrow {}^3He + \gamma$	
	${}^3He + {}^3He \rightarrow {}^4He + p + p$	
3	or	
	${}^3He + {}^4He \rightarrow {}^7Be + \gamma$	
	or	
	${}^3He + p \rightarrow {}^4He + e^+ + \nu_e$	≤ 18.8 MeV
	${}^7Be + e^- \rightarrow {}^7Li + \nu_e$	0.86 MeV
	${}^7Li + p \rightarrow {}^4He + {}^4He$	
4	or	
	${}^7Be + p \rightarrow {}^8B + \gamma$	
	${}^8B \rightarrow {}^8Be^* + e^+ + \nu_e$	< 15 MeV
	${}^8Be^* \rightarrow {}^4He + {}^4He$	

Next, we calculate the expected flux of neutrinos from the Sun. The total luminosity of the Sun is:

---

[6] Here we may notice that the processes in stellar interiors are very similar to the ones that took place in the early Universe when the light elements ${}^4He$, ${}^3He$, deuterium and ${}^7Li$ were synthesized (Section 9.3). A difference is that the higher density and temperature in the interior of a star make the processes stay longer in thermal equilibrium, so that heavier elements (up to iron) can also be synthesized.

$$L_\odot = 3.92 \cdot 10^{26} \text{ Watts} = 2.4 \cdot 10^{39} \text{ MeV/sec}$$

Thus, according to the SSM, the Sun should emit around $2 \cdot 10^{38}$ electron neutrinos (ν_e) per second. At the Earth, $1.5 \cdot 10^8$ km away, the neutrino flux becomes:

$$\phi_{\nu_e} = 6.5 \cdot 10^{14} \text{ m}^{-2} \text{ s}^{-1}$$

Although this is a huge flux, the extremely low interaction cross-section makes its detection very difficult. Several detection techniques have been developed to observe solar neutrinos. The experiments can be divided into two classes: absorption and scattering experiments.

Neutrino Absorption Experiments

The field of observational neutrino astronomy started with the pioneering experiment of Ray Davis and collaborators in the early 1960s. A tank filled with 615 tons of a cleaning fluid, C_2Cl_4, was installed at the Homestake Mine in South Dakota, USA, about 1.5 km below ground. The nuclear reaction that makes neutrino detection possible is:

$$\nu_e + {}^{37}Cl \rightarrow {}^{37}Ar + e^-$$

The high threshold for the reaction, 0.814 MeV, permits the observation of only a small fraction of the solar spectrum, as shown in Fig. 14.4. From the accumulation of argon in the tank, the flux of electron neutrinos can be calculated. The argon is chemically extracted, and *single atoms* are counted through their subsequent decay.[7]

In about $4 \cdot 10^5$ litres of target, containing 130 tons of ${}^{37}Cl$, less than 15 argon atoms are produced every month! After subtracting the known background rates, mainly due to cosmic ray muons interacting with the liquid target and producing ${}^{37}Ar$, the flux of incoming ν_e can be measured. As the Sun is the dominant source of neutrinos, it is assumed that the measured flux of neutrinos originates from the Sun. There is no sensitivity of directionality in the absorption experiments. For every six counted argon atoms, one is estimated to be due to background.

The experimental and theoretical rates for absorption of solar neutrinos *per target atom* are expressed in *Solar Neutrino Units*, SNU. For convenience:

$$1 \text{ SNU} = 10^{-36} \text{ s}^{-1}$$

The estimate involves the integral over the solar neutrino flux and the absorption cross-section: $\int \frac{d\phi(E_\nu)}{dE_\nu} \sigma(E_\nu) dE_\nu$. In particular, the theoretical estimate requires detailed calculations of the cross-sections for neutrino absorption by

---

[7] The argon counting relies on detecting the 2.82 keV so-called Auger electrons from the electron capture decay of ${}^{37}Ar$.

the target atoms that take into account transitions to different states of the final nuclei.

While the predicted rate for neutrino capture at the chlorine experiment is 7.9±2.6 SNU, the observed rate averaged over more than 20 years is only 2.6± 0.2 SNU, as shown in Fig. 14.5. This long-standing disagreement has been called the 'solar neutrino problem'. Several other experimental techniques have been pursued to verify the nature of the disagreement.

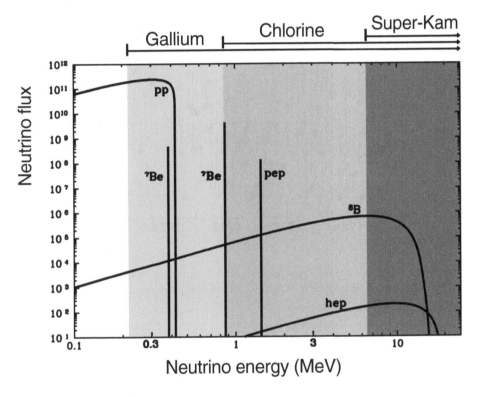

Fig. 14.4. Energy spectrum of neutrinos produced in the Sun by the various nuclear reactions, as predicted by the 'standard solar model' of J. Bahcall and collaborators [4].

The most efficient solar neutrino absorption experiments to date have been performed with the gallium detectors, SAGE and GALLEX. The nuclear reaction that takes place in these detectors is:

$$\nu_e + {}^{71}\mathrm{Ga} \rightarrow {}^{71}\mathrm{Ge} + e^-$$

The energy threshold for the reaction is only 0.233 MeV – significantly lower than for the ^{37}Cl experiment, as shown in Fig. 14.4. The gallium experiments are therefore sensitive to a much larger fraction of the total spectrum

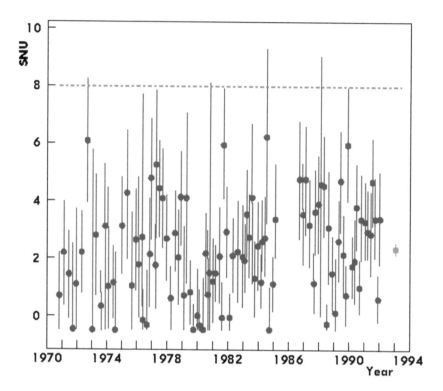

Fig. 14.5. Observational results from the Homestake neutrino experiment as a function of time. The line at 7.9 SNU shows the prediction of the 'standard solar model'.

of solar neutrinos. In particular, they probe the main reaction in the chain: $p + p \rightarrow {}^2\mathrm{H} + e^+ + \nu_e$, which produces neutrinos with $E \leq 0.420$ MeV.

The GALLEX experiment is located in the underground Gran Sasso Laboratory near Rome, Italy, and uses 30 tons of gallium in an aqueous solution of gallium chloride and hydrochloric acid. The SAGE experiment uses 55 tons of metallic gallium and is situated in the Baksan Neutrino Observatory, in the Caucasus Mountains, in Russia. SAGE has been taking data since 1990, and GALLEX (now discontinued) since 1991.

While the theoretical prediction for the gallium experiment is about 130 SNU for variants of solar models, the observed rates, combining the two experiments, is 70.3±7 SNU, at least 7.5 standard deviations from the standard solar model [4].

The deficit of neutrino captures in the gallium experiments, although smaller than in the chlorine detector, is significant, and has kept the 'solar neutrino problem' alive.

Neutrino Scattering Experiments

In the Japanese mine at Kamioka a large water tank was fitted with thousands of photomultipliers (PMs). This enabled the detection of solar neutrinos through the elastic scattering process:

$$\nu + e^- \rightarrow \nu + e^-$$

The recoiling electrons emit Cherenkov light, detectable with the PMs, and can thus be counted. (In Section 14.10 the Cherenkov process will be explained in detail.) The method is very attractive because of its relative simplicity (water is cheap) and because the scattered electrons, on average, follow the direction of the incoming neutrinos. That allows the experimenters to verify that the signal really comes from the Sun, as demonstrated in Fig. 14.6. One disadvantage is that there is an experimental threshold for the detectability of the recoiling electron corresponding to $E_\nu > 7 - 9$ MeV.[8] Therefore, only the highest-energy neutrinos emitted from the Sun can be measured, thereby reducing the detectable flux by about 10^4. Another difficulty is the indistinguishable detector signal from gamma-rays generated by radioactive impurities in the water or induced by cosmic rays, implying a high rate of background events, as shown in Fig. 14.6.

On the other hand, one clearly superior aspect of the scattering experiments is that they are sensitive to *all* neutrino species, although with different cross-sections (see equation (14.10)).

Example 14.5.2 Calculate the minimum necessary mass of water that a neutrino telescope must have in order to detect about one solar neutrino per day assuming that the detection threshold $E_\nu \sim 10$ MeV reduces the observable flux by 10^4 and that the cross-section $\sigma_{\nu e}$ where the electron recoil energy is at least 5 MeV is about 10 per cent of the total cross-section, given in (14.10).

Answer: The number of target electrons in M kg of water is:

$$N_{target} = \frac{M \cdot N_A \cdot 10}{A} \tag{14.18}$$

where $N_A = 6.022 \cdot 10^{26}$ kmol^{-1} is Avogadro's number, A is the atomic weight and the factor 10 appears because there are ten electrons for each H_2O molecule in the target. Given the neutrino flux ϕ_ν and the cross-section $\sigma_{\nu e}$, the number of interactions within a volume N_{target} electrons in a time interval Δt is:

$$N_{\nu e} = \phi_\nu \Delta t \sigma_{\nu e} N_{target}$$
$$= \phi_\nu \Delta t \sigma_{\nu e} \frac{M \cdot N_A \cdot 10}{A} \tag{14.19}$$

---

[8] For the new version of the experiment, Super-Kamiokande, which began in April 1996, the threshold has been lowered to 6.5 MeV.

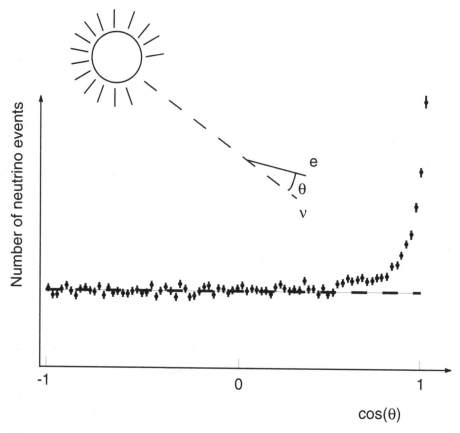

Fig. 14.6. Angular distribution of neutrino events in the Super-Kamiokande detector with respect to the angle to the Sun. The data points show the results of 504 days of exposure, with an electron energy threshold of 6.5 MeV. For an electron energy of 10 MeV the experimental resolution of the measurement of the electron direction is about 28 degrees. For that energy, the intrinsic spread of the angle between the original neutrino direction and the electron path is about 18 degrees. Adapted from Y. Suzuki, presentation on behalf of the Super-Kamiokande collaboration at the conference Neutrino98, Takayama, Japan, 1998.

The cross-section for electron neutrino interactions with electrons for the relevant energies is about 10^{-44} cm^2 (from (14.10)) and the reduced flux is $\phi_\nu = 6.5 \cdot 10^6$ cm^{-2}s^{-1}. Inserting A=18 kg/kmol and $\Delta t = 86400$ seconds (in one day) one finds

$$M \approx 0.5 \cdot 10^6 \ kg$$

In summary, about a kiloton of water is required in order to detect solar neutrinos at a reasonable rate. This points to yet another disadvantage for the water experiments: the scattering cross-section is much smaller than the neutrino absorption cross-section for suitable targets (only 30 tons of gallium are sufficient).

The Kamiokande II experiment contained a total 2.1 ktons of water, although only 680 tons were used to detect solar neutrinos. Super-Kamiokande, operating since April 1996, is about 10 times larger.

Just as for the absorption experiments, the water experiment at Kamioka has found a deficit of solar neutrinos. The observed flux is about half of that predicted by the SSM. This low flux has been verified by the Super-Kamiokande experiment. In fact, another important solar neutrino experiment has just started operating. This is the Sudbury Neutrino Observatory (SNO), which is using heavy water as the detector medium. This has the advantage of larger neutrino cross-sections, especially for neutral current interactions. By combining information on the CC and NC reactions, it has been conclusively determined that the electron neutrinos that seem to be missing from the Sun have been transformed into another neutrino species through quantum mechanical mixing (see Section 14.6).

The 'solar neutrino problem' constituted a fascinating puzzle, which has now been solved when a whole set of new experiments with increased sensitivity have acquired data.

14.5.2 Supernova Neutrinos

Young and middle-aged stars like the Sun produce energy mainly through the fusion of hydrogen nuclei into helium, thereby releasing nuclear binding energy. At high enough temperatures and densities, three helium nuclei may fuse to produce carbon. Through successive contractions and reheating, very massive stars ($M > 8 \, M_\odot$) also ignite carbon and, in turn, the remains of its combustion: Ne, Mg, O, Si. The sequence ends when iron is produced. Then, an 'onion-like' structure exists with the iron in the core followed by successive layers of lighter elements. Unlike the lighter elements, fusion of iron nuclei does not release binding energy. Radiation pressure can thus no longer balance the pressure from the outer layers. Instead, equilibrium is maintained by the pressure generated by the motion of degenerate electrons.

The iron core, steadily being 'fed' more and more mass, eventually reaches the Chandrasekhar mass (around 1.4 times the solar mass): as the velocity of the degenerate electrons approaches the speed of light, no further increase in electron pressure can be obtained. Instead, as the gravitational pressure increases, electrons and protons fuse through the weak interactions to produce neutrons and neutrinos. As the latter escape, a collapse of the core occurs until it reaches nuclear density: a neutron star with $15 - 20$ km radius is formed. The neutron pressure then prevents the star from becoming a black hole. The in-going wave bounces, generating the explosion that sweeps away the star's mantle: that is, the onset of a so-called type II (*core-collapse*) supernova.

One can make a rough estimate of the energy release and the number of neutrinos produced in the process of neutron star formation in connection with a type II supernova. The gravitational binding energy released as the star's radius shrinks is (Problem 14.4):

$$E_b \approx \frac{3}{5} \frac{G_N M_{ns}^2}{R_{ns}} = 2 \cdot 10^{46} \text{ J} \tag{14.20}$$

for $R_{ns}=15$ km and $M_{ns}=1.4$ M$_\odot$, the Chandrasekhar mass. In spite of the spectacular optical images of supernovae, sometimes outshining an entire galaxy, only a small fraction of the energy, about ~ 1 per cent, is transferred to the ejected star mantle, and a hundred times less to power the visible light curves. The 'easiest' way for the star to cool is through the emission of neutrinos. It is therefore a good approximation to assume that the total energy carried by neutrinos is approximately E_b. In the core, the density and temperature are high enough to make the weak interaction rates so fast as to keep all neutrino types roughly in thermal equilibrium for several seconds. If the energy is distributed evenly between ν_e, ν_μ, ν_τ, $\bar{\nu}_e$, $\bar{\nu}_\mu$ and $\bar{\nu}_\tau$, each kind of particle will carry a total of $\sim 0.3 \cdot 10^{46}$ J.

We estimate the thermal energies within the newborn neutron star by invoking the *virial theorem* which relates the thermal and potential energy of a self-gravitating system. For instance, for a nucleon on the surface of a neutron star, the average kinetic energy must be one half of its gravitational potential energy:

$$< E_k > = \frac{1}{2} \frac{G_N M_{ns} m_N}{R_{ns}} \approx 25 \text{ MeV} \tag{14.21}$$

Neutrinos in thermal equilibrium with their environment $(T = \frac{2}{3} < E_k >)$ will have similar energies. Thus, from (14.20) and (14.21) it follows that about 10^{58} neutrinos produced are produced in a supernova explosion!

Example 14.5.3 Estimate the total number of neutrinos that reach the Earth from a supernova explosion in the Large Magellanic Cloud, roughly 50 kpc distant.

Answer: At a distance of $5 \cdot 10^4$ parsecs: that is, $1.55 \cdot 10^{21}$ metres, the integrated flux of neutrinos becomes

$$F_\nu = \frac{10^{58}}{4\pi(1.55 \cdot 10^{21})^2} = 2 \cdot 10^{14} \text{ m}^{-2} \tag{14.22}$$

Example 14.5.4 Calculate the mean free path of 25 MeV neutrinos during the collapse that precedes a type II supernova for a density of $\rho = 10^{14}$ g·cm^{-3}.

Answer: With $\sigma_{weak}(25~MeV) \approx 10^{-41}$ cm^2 inserted in (14.15) together with m=m$_N$, we find $\lambda_\nu(25MeV) \approx 4$ metres.

Clearly, with a mean free path (for the energetic neutrinos) of only a few metres, the escape from the neutron star is slowed down to the level of a diffusive process. The number of scatterings is given by ratio between the star's radius and the mean free path, squared.

$$n = \left(\frac{R_c}{\lambda_\nu}\right)^2 \tag{14.23}$$

and the time scale for the neutrinos to diffuse out of the star is

$$\tau \approx \frac{n\lambda_\nu}{c} \sim 1 \text{ second} \tag{14.24}$$

Adding to this the actual collapse time, also about a second (not much longer than the free-fall collapse time), we can deduce that the burst of neutrinos from a new supernova should have a signal width of a few seconds.

As the density of nucleons increases, eventually reaching about $3{\cdot}10^{14}$ g/cm^3, the mean free path of neutrinos shrinks. Because of the energy dependence of neutrino cross-section, neutrinos of higher energy will scatter more times on average, thereby delaying their escape time from the high-density region.

The burst of neutrinos which first escape from the collapsing core will therefore have an energy corresponding to a neutrino mean free path $\lambda \simeq R_{ns}$:

$$\lambda = R_{ns} = \frac{m_n}{\rho_{ns}\sigma_{weak}} \tag{14.25}$$

Beyond $\rho_n \approx 10^{12}$ g/cm^3, the star becomes opaque for neutrinos. Inserting the cross-section formula from (14.7) and the nucleon mass, $m_n = 1.7 \cdot 10^{-24}$ g, one then finds that the typical energy for a burst of supernova neutrinos is

$$E_\nu \approx \sqrt{\frac{m_n}{\rho_{ns} \cdot R_{ns} \cdot 5 \cdot 10^{-44}}} = 10 \text{ MeV} \tag{14.26}$$

In other words, one expects the observable neutrino spectrum from a supernova explosion to be centered around a value just above 10 MeV: that is, just above the detection threshold for water Cherenkov detectors.

A spectacular confirmation of these expectations took place in February 1987. The closest supernova since the one sighted by Kepler in 1604 was spotted optically in the Large Magellanic Cloud, which together with the Small Magellanic Cloud is our nearest neighbour galaxy, about 50 kpc distant.

Core collapse neutrinos from this famous supernova (SN1987A) were detected in the underground water detectors at Kamioka and at the Irvine Michigan Brookhaven (IMB) 6.8 kton water Cherenkov detector.

Both detectors were originally designed to look for proton decays (predicted in Grand Unification Theories), such as $p \rightarrow e^+ + \pi^0$, which still have not been observed. Instead, they could confirm the model of supernova explosions by detecting the positrons produced in the absorption of antineutrinos by the protons in the water target:

$$\bar{\nu}_e + p \rightarrow n + e^+$$

The cross-section for antineutrino absorption at a few tens of MeV is between 20 and 100 times larger than the one for elastic scattering on electrons, and is therefore the dominant process for detection of supernova neutrinos. As opposed to the elastic scattering collisions, the $\bar{\nu}_e p$ interactions produce isotropic positrons. The primary neutrino direction can thus not be measured through this process.

A total of 11 detected positrons at Kamiokande and eight at IMB[9], with energies up to 40 MeV, are believed to be due to the capture of antineutrinos from SN1987A. The arrival times and measured energies of the recoiling positrons are shown in Fig. 14.7.

14.6 Neutrino Oscillations

The long-standing 'solar neutrino problem' has forced particle physicists and astronomers to review their calculations and look for possible scenarios that explain the deficit of detected ν_es in the solar neutrino experiments. The approaches have been twofold. On one hand, the 'astrophysics' part of the calculations have been challenged: that is, the model predictions of the rate of nuclear reactions in the interior of the Sun. However, new results on the different acoustic oscillation modes observed at the solar surface ('helioseismology') indicate that the standard solar model works very well. Particle physicists, on the other hand, have suggested an intriguing alternative explanation: ν_es are produced in the interior of the Sun just as all the astrophysics models predict, but they change identity somewhere between the Sun and the detector. For example, a fraction of ν_es convert into ν_μs, ν_τs (or something else!) thereby decreasing the flux of electron neutrinos at the Earth. Current solar neutrino experiments are mostly sensitive to ν_es. This weakness has been resolved as the next generation of solar neutrino experiments with different target materials, SNO (deuterium) (and soon BOREXINO (^{11}B)), has measured the *total* neutrino flux, integrated over all neutrino species. Their finding is that the total flux of neutrinos is in good agreement with

---

[9] The IMB detector had a higher detection threshold.

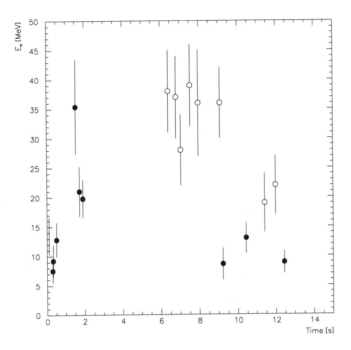

Fig. 14.7. Neutrino signals detected at Kamiokande (filled circles) and IMB (empty circles) from SN1987A. The figure show the positron energies. There is possibly an offset between the clocks of the two experiments [9, 21].

calculations based on the standard solar model, thus supporting the idea of oscillations, which we will now discuss.

If neutrinos are massless they are by definition stable: that is, they cannot decay into any other *lighter* particle. There is, however, no compelling reason why they cannot have finite masses. For ν masses less than an MeV or so, the radiative decay of a heavier ν_a to a lighter ν_b through $\nu_a \to \nu_b + \gamma$ is kinematically possible, but the estimated lifetime in the Standard Model is much longer than the age of the Universe.

In such a scenario, mixing of neutrino species may occur if the *weak-interaction eigenstates*, ν_e, ν_μ and ν_τ, are not *mass eigenstates*: that is, the states in which neutrinos propagate in vacuum.

In general, any flavour or weak-interaction neutrino eigenstate, ν_f, can be expressed as a linear superposition of orthogonal mass eigenstates, ν_m:

$$|\nu_f> = \sum_m c_{fm} |\nu_m>$$

For example, let us consider the situation where there are two neutrino mass eigenstates associated with two flavour eigenstates.

The unitary transformation matrix connecting the mass eigenstates with the flavour eigenstates can be described with one parameter, the mixing angle θ.

$$\begin{pmatrix} \nu_\mu \\ \nu_e \end{pmatrix} = \begin{pmatrix} \cos\theta & \sin\theta \\ -\sin\theta & \cos\theta \end{pmatrix} \begin{pmatrix} \nu_1 \\ \nu_2 \end{pmatrix} \tag{14.27}$$

Although the states $|\nu_e>$ and $|\nu_\mu>$ (and their antiparticles) are produced in weak interactions, such as in the decay $\mu^- \to e^- \bar{\nu}_e \nu_\mu$, the *physical* states: that is, the eigenstates of the Hamiltonian with definite masses, are ν_1 and ν_2.

Therefore, the time evolution of a muon neutrino wave function with momentum p is

$$|\nu_e(t)> = -\sin\theta e^{-iE_1 t}|\nu_1> + \cos\theta e^{-iE_2 t}|\nu_2> \tag{14.28}$$

where E_1 and E_2 are the energies of the two mass eigenstates. Two energy levels arise if ν_1 and ν_2 have different masses, as they must have the same momentum, p. Then, for very small neutrino masses: that is, $m_i \ll E_i$,

$$E_i = p + \frac{m_i^2}{2p} \tag{14.29}$$

The probability $P(\nu_e \to \nu_e) = |<\nu_e|\nu_e>|^2$, that an electron neutrino *remains* a ν_e after travelling a time t is (Problem 14.7):

$$P(\nu_e \to \nu_e) = 1 - \sin^2(2\theta)\sin^2\left[\frac{1}{2}(E_2 - E_1)t\right] \tag{14.30}$$

For very small neutrino masses, inserting (14.29) we get

$$P(\nu_e \to \nu_e) = 1 - \sin^2(2\theta)\sin^2\left[\left(\frac{m_2^2 - m_1^2}{4E}\right)t\right] \tag{14.31}$$

where E is the energy of ν_e.

It thus follows that the probability that the electron neutrino becomes a muon neutrino at a time t is

$$P(\nu_e \to \nu_\mu) = \sin^2(2\theta)\sin^2\left[\frac{\Delta m^2}{4E}t\right] \tag{14.32}$$

where $\Delta m^2 = |m_2^2 - m_1^2|$.

From (14.32) and Fig. 14.8 it is seen that the probability function for flavour change *oscillates*, with an amplitude given by $\sin^2(2\theta)$ and oscillation frequency $\sim \Delta m^2/E$. Therefore, for suitable neutrino masses and mixing angles, the presumed deficit of solar electron neutrinos can be explained by the oscillation phenomenon.

To summarize, the amplitude and oscillation length of the flavour oscillation are (reinserting factors of $\hbar$ and c):

$$\begin{cases} A = \sin^2(2\theta) \\ L_\nu = \frac{4\pi E \hbar}{\Delta m^2 c^3} \end{cases} \qquad (14.33)$$

Numerically, the oscillation length becomes

$$L_\nu = 1.27 \left(\frac{E}{1\,\text{MeV}} \right) \left(\frac{1\,\text{eV}^2}{\Delta m^2} \right) \quad \text{metres.} \qquad (14.34)$$

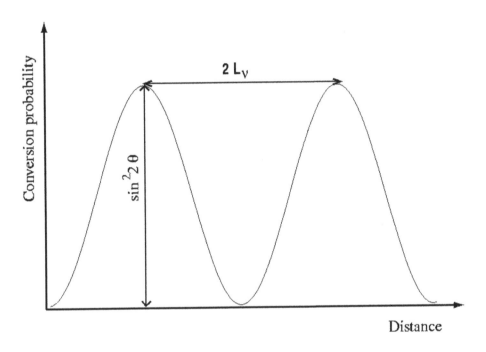

Fig. 14.8. Probability distribution for $\nu_e \to \nu_\mu$.

A number of terrestrial experiments have been designed to look for neutrino oscillations. They use a known source of neutrinos, either from a nuclear reactor or from an accelerator beam.

In the reactor experiments, the $\bar{\nu}_e$ energies are so small (typically a few MeV) that the charged current reaction $\bar{\nu}_\mu \to \mu$ is kinematically impossible even if a $\bar{\nu}_e \to \bar{\nu}_\mu$ conversion has taken place. The best oscillation signal in reactor experiments is therefore the *disappearance* of ν_es as a function of distance from the source.

In accelerator experiments ν_μ beams are created by a secondary beam of decaying pions, $\pi \to \mu\nu_\mu$. One then looks for the *appearance* of ν_e and $\bar{\nu}_e$. Currently, no convincing signal of oscillation has been seen, with the possible exception of an experiment at Los Alamos, which has not yet been independently confirmed by other experiments.

It is believed to be very likely that the answer to the solar neutrino problem involves neutrino oscillations. This issue has recently been emphasized when SNO has probed, with large statistics, all the neutrino flavours.

14.6.1 Neutrinos Propagating Through Matter

For the physics of solar neutrinos, there is another effect that is interesting, and according to present models preferred, namely the *Mikheev-Smirnov-Wolfenstein* (MSW) effect. It is a resonant conversion, in the solar matter, of the electron neutrino into the two other neutrinos depending on the mass differences and mixing angles. The reason for this conversion is the existence of two diagrams (the ones shown in Fig. 14.3) in the scattering process for an electron neutrino on an electron, one with Z^0 the other with $W^\pm$ exhange. If the electron neutrino has oscillated into a muon or tau neutrino, the graph corresponding to the $W^\pm$ exchange does not contribute. After being produced near the centre of the Sun, the average energy of the neutrinos will be affected by this. The 'charged current', $W^\pm$ exchange, gives a contribution $H_{CC} = \sqrt{2}G_F N_e$, where N_e is the electron density. (The Z^0 exchange is the same for all neutrinos and gives no contribution to neutrino oscillations.) The net effect of this oscillation in matter is that the vacuum mixing angle θ is replaced by a matter mixing θ_m:

$$\sin^2(2\theta_m) = \frac{\sin^2(2\theta)}{(\cos(2\theta) - a)^2 + \sin^2(2\theta)} \tag{14.35}$$

where

$$a = \frac{2EH_{CC}}{\Delta m^2}. \tag{14.36}$$

This formula shows that even if the vacuum mixing angle is very small, there will be maximum mixing ('resonance') if the electron density is such that $a = \cos(2\theta)$. If, as is the case in the Sun, the matter density varies, there is plausably a resonance, and the width of the resonance is roughly corresponding to matter densities such that $|a - \cos(2\theta)| < |\sin(2\theta)|$.

According to the present experimental situation, the MSW effect is taking place for solar neutrinos, with a 'large mixing angle' $\theta \approx 33$ degrees, and $\Delta m^2 \sim 7 \cdot 10^{-5}$ eV^2 preferred.

The recent data which has pinned down these properties for solar neutrinos are, besides the Davis, Super-Kamiokande, GALLEX and SAGE data, new results from SNO and KamLAND. KamLAND is a liquid scintillator neutrino detector near Super-Kamiokande, which is sensitive to the fluxes of antineutrinos produced in a number of Japanese nuclear power reactors. In a disappearance experiment, they confirmed the oscillation solution of the solar neutrino problem, namely, they observed the decrease in neutrino flux generated by oscillations $\bar{\nu}_e \rightarrow \bar{\nu}_{\mu,\tau}$ and obtained consistency with the values given by the MSW large mixing angle solution to the solar neutrino problem.

After decades of hard experimental work, the neutrino sector is now getting established, showing interesting patterns of mixing. To exactly understand the cause and magnitude of these mixing remains a very active field of research.

14.7 Atmospheric Neutrinos

Neutrinos are produced by hadronic and muonic decays following the interaction of cosmic ray nuclei with the Earth's atmosphere:

$$
\begin{cases}
p/n + N \rightarrow \pi^+/K^+ + \ldots \\
\qquad \pi^+/K^+ \rightarrow \mu^+ + \nu_\mu \\
\qquad\qquad \mu^+ \rightarrow e^+ + \bar{\nu}_\mu + \nu_e,
\end{cases}
$$

$$
\begin{cases}
p/n + N \rightarrow \pi^-/K^- + \ldots \\
\qquad \pi^-/K^- \rightarrow \mu^- + \bar{\nu}_\mu \\
\qquad\qquad \mu^- \rightarrow e^- + \nu_\mu + \bar{\nu}_e
\end{cases} \tag{14.37}
$$

Here $K^\pm$ are mesons containing the strange quark (K^+, for example, is composed of a u quark and an s-antiquark). From these cascade reactions one expects that there are about twice as many muon neutrinos than electron neutrinos produced in the atmosphere:

$$
\frac{\varphi_{\nu_\mu} + \varphi_{\bar{\nu}_\mu}}{\varphi_{\nu_e} + \varphi_{\bar{\nu}_e}} = 2 \tag{14.38}
$$

This expectation holds at low energies. At higher energies, additional effects have to be taken into account: for example, the competition between scattering and decay of the produced pions and the possibility that muons hit the ground before decaying, due to time dilation (see Problem 14.9).

As the energy spectrum of the primary nuclei extends out to $\sim 10^{20}$ eV (Chapter 12), one expects neutrinos to be produced to comparable energies. The spectrum must fall faster, though, as it seldom happens that the full energy of the primary is transferred to one of the particles in the cascade.

Due to the complicated chain of reactions in the cascade, a Monte Carlo simulation is needed in order to calculate the differential spectrum of atmospheric neutrinos. The general features are: a broad peak around 0.1 GeV (~ 1 cm^{-2} s^{-1}) and, at very high energies, $E_\nu \gg 1$ TeV, the flux falls as $E^{-3.7}$.

The cross-section for a neutrino nucleon interaction in a target can be calculated by inserting the nucleon mass (instead of m_e) in equation (14.9). In the region of the maximum flux of atmospheric neutrinos the cross-section is $\sigma_{\nu N} \sim 10^{-39}$ cm^2. We can thus estimate the minimum target size required to detect νs produced in atmospheric cascades. As for solar neutrinos, a 'kiloton size' detector is required (see Problem 14.6).

Atmospheric neutrinos are also useful to test the oscillation hypothesis. The relationship in (14.38) can be compared with the observed ratio. It is often possible to determine the direction of the neutrinos in a detector. The neutrinos that move 'upwards'– that is, those that have crossed the entire Earth diameter L≈ 10^4 km – can probe mass differences as small 10^{-5} eV^2, as shown in the example below.

Example 14.7.1 Show that 'upward-moving' atmospheric neutrinos of about 100 MeV are useful for looking for neutrino oscillations in the region $\Delta m^2 \geq 10^{-5}$ eV^2.

Answer: In order to be able to detect a significant deviation in the ratio between neutrino flavours (14.38), the path length that the neutrinos travel must be comparable or larger than the one-half oscillation length.

$$L \geq \frac{1}{2}L_\nu \tag{14.39}$$

Thus, combining (14.39) and (14.34) one finds that the 'minimal' detectable mass difference, $\Delta m^2{}_{min}$ is

$$\Delta m^2{}_{min} = 5 \cdot 10^{-5}\text{eV}^{-2} \tag{14.40}$$

In fact, there are at present indications that atmospheric neutrinos indeed do oscillate. Results from, among others, the Super-Kamiokande experiment seem to be best interpreted as a deficit of muon neutrinos, with one of the possibilities being oscillation $\nu_\mu \rightarrow \nu_\tau$ with a large mixing angle ($\sin^2(2\theta) \sim 0.8 - 1.0$) and $\Delta m^2 \sim 5 \cdot 10^{-3}$ eV^2. There are several other experiments which have confirmed a lack of muon neutrinos relative to electron neutrinos, which could be explained if oscillations $\nu_\mu \rightarrow \nu_\tau$ take place. The reason for this is that due to the high $\tau^\pm$ lepton mass (1.8 GeV) the ν_τs generated by mixing will not have enough energy on average to make the charged current interaction $\nu_\tau + N \rightarrow \tau + X$ kinematically possible. (Their contribution to the neutral current events is also too small to be easily detected.) Of course, there is a possibility that a 'sterile' neutrino exists which does not couple to the $W^\pm$ and Z^0 bosons. Maybe the muon neutrinos mix with such a neutrino, but this is a less economical model since it involves a hypothetical, not yet detected, particle species. When more detailed data on neutral current interactions become available this hypothesis may be tested experimentally.

The Super-Kamiokande collaboration has recently released data on the zenith-angle distribution of the detected muons. We see from (14.34) that for given values of the mixing angle and mass difference, the probability for finding a muon neutrino will depend on the ratio E/L of neutrino energy over distance travelled. The average energy E can be estimated from the energy of the detected muon, and the average value of L can be calculated from the direction of the detected muon. This is a simple consequence of the fact that if the neutrinos are produced at height d in the atmosphere (realistically, $d \sim 20$ km) then if they arrive at zenith angle θ_z, the length of travel is given by simple geometry:

$$L = \sqrt{2Rd + (R - d)^2 \cos^2 \theta_z} - (R - d) \cos \theta_z \qquad (14.41)$$

where R is the radius of the Earth (see Fig. 14.9). Since neutrinos are produced with a large spread of energies and angles, one has to integrate over these distributions to extract the predicted signal. The main effect for the parameters given above for $\nu_\mu \to \nu_\tau$ oscillations is a depletion of the muon neutrino flux by a factor of around 2 for upward-going muons compared to downward-going, with a smooth zenith-angle distribution for intermediate angles (see Fig. 14.10).

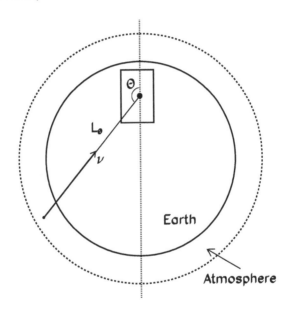

Fig. 14.9. Schematic picture of atmospheric neutrino oscillation experiments. While all neutrinos are produced in the Earth's atmosphere (thickness about 30 km), the distance to the detector varies with zenith angle, up to a maximum length of the Earth's diameter (13,000 km) for directly upward-moving neutrinos. The probability for flavour oscillation is therefore larger for upward-moving events.

14.8 Neutrinos as Tracers of Particle Acceleration

We have seen in the previous sections that a kiloton-size detector is necessary to observe neutrinos from sources as close as the Earth's atmosphere,

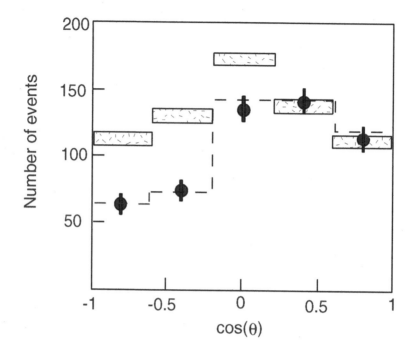

Fig. 14.10. The rate of muons detected in the Super-Kamiokande experiment as a function of zenith angle. If there were no neutrino oscillations the data points would have been expected to lie in the boxes. As can be seen, the experimental results agree better with the hypothesis of neutrino oscillations with mixing angle $\sin^2(2\theta) \sim 1$ and $\Delta m^2 \sim 2.2 \cdot 10^{-3}$ eV^2 (indicated by the dashed lines). Adapted from T. Kajita, talk presented on behalf of the Super-Kamiokande collaboration, Neutrino98, Takayama, 1998.

the Sun or a type II supernova as far as the nearest galactic neighbours to our own Galaxy. In order to attempt to study a broader class of astrophysical objects, the required detector volume becomes practically unattainable. Fortunately, there is a way to overcome this difficulty. The detector volume estimates above were carried out for *contained* events: that is, for neutrinos that interact *inside* the detector volume. Consider instead a process where a neutrino interaction results in the production of a particle that interacts electromagnetically and that, once created, traverses a very large distance. The detection of the newly created particle is an indirect neutrino detection, even at a great distance from the interaction point. In fact, this is the way very high-energy (VHE) neutrino detectors operate: by looking for neutrino-induced muons. The process is one of a charged current interaction

$$\nu_\mu + \mathrm{N} \to \mu + ...,$$

where N is a nucleon in the material surrounding the detector. The muon range rises with energy, and around 1 TeV (10^{12} eV) is more than one kilometre. The detection area is therefore greatly enhanced at high energies. In water, a good approximation of the muon range as a function of energy is given by

$$R_\mu \approx 2.5 \ln \left(2 \cdot \frac{E_\mu}{1 \text{ TeV}} + 1 \right) \text{ km} \tag{14.42}$$

Moreover, the muon produced conserves, on average, the direction of the incoming neutrino. The root-mean-squared of the ν_μ-μ angle is approximately:

$$\sqrt{<\theta^2>} \approx 2 \left(\frac{1 \text{ TeV}}{E_\nu} \right)^{\frac{1}{2}} \text{ deg.} \tag{14.43}$$

We have also seen that the cross-section for neutrino interaction with a target at rest rises linearly with energy. VHE neutrino telescopes become efficient at a few GeV, where the product of the neutrino-matter cross-section and the muon range rises approximately as E_ν^2. Above 1 GeV, the induced flux of muons from atmospheric neutrinos, for example, is about 1 m^{-2} year^{-1}.

This detection scheme does not work as well for other types of neutrinos. Electrons (from $\nu_e + \text{N} \rightarrow e + \ldots$), because of their much smaller mass, have a very short range since they lose energy through the emission of bremsstrahlung photons. τ leptons, the heaviest known charged leptons ($m_\tau = 1.78$ GeV), are produced in charged current interactions of ν_τs but are very short lived (the lifetime is $t_\tau \sim 3 \cdot 10^{-13}$ seconds) and therefore not suitable for detection (except for the fraction of times where the τ decays into $\mu \bar{\nu}_\mu \nu_\tau$, which happens in 18 per cent of the cases). However, if large neutrino-detectors are built, there may be a possibility of detecting contained ultra-high-energy electron and τ neutrino events by the intense cascade of light that is produced by secondary electrons, positrons and photons. In the case of τ neutrinos, special relativity may help to produce a good signature. If sources of PeV (10^{15} eV) τ neutrinos exist, the produced charged τ lepton would have a relativistic γ factor of (see (2.36))

$$\gamma \sim \frac{E_\nu}{m_\tau} \sim 10^6 \tag{14.44}$$

which means, thanks to time dilation, that in the detector reference frame the τ lepton will travel a distance

$$\ell \sim \gamma c t_\tau \sim 100 \text{ m} \tag{14.45}$$

The 'double bang' created by the charged current interaction (which breaks up the hit nucleon and gives rise to a hadronic cascade) and the subsequent decay of the τ lepton, separated by 100 m, would be the signature of PeV τ neutrinos [27].

If neutrinos oscillate, very energetic τ neutrinos could be produced by mixing with muon neutrinos created in high-energy pion decays in cosmic accelerators.

In present detectors, only neutrino-induced muons moving *upwards* in the detectors (or downwards but near the horizon) are safe tracers of neutrino interactions. Most muons moving downwards have their origin in cosmic-ray nuclei interacting with the Earth's atmosphere.

At the surface of the Earth, the flux of *downward-moving* muons produced in the atmosphere is about 10^6 times larger than the flux of neutrino-induced *upward-moving* muons.

By going underground, the material (rock, water, ice, etc.) above the detector attenuates the flux of atmospheric muons. In addition, if it is experimentally possible to select events where a muon is moving upwards,[10] the Earth itself acts as a perfect filter: only neutrino-induced muons can be produced close enough to the detector. Atmospheric muons produced in the opposite hemisphere, a whole Earth diameter away, have no chance of reaching the detector.

14.9 Indirect Detection of CDM Particles

Neutrinos may give clues to the dark matter problem in another way than just being a part of the dark matter (if neutrinos have a mass). If the dark matter has a component that is massive and weakly coupled (electrically neutral) it will be non-relativistic at freeze-out: cold dark matter (CDM).[11] The prime example of such a dark matter candidate is the lightest supersymmetric particle – the neutralino χ (see Section 6.9.1).

Neutralinos (or other WIMPs) have interactions with ordinary matter which are equally as small as those of neutrinos. However, since they move with non-relativistic velocity there is a chance that they become gravitationally trapped inside, for example, the Sun or the Earth. A neutralino which scatters on the ordinary particles that make up the celestial body in question will lose energy and fall further inside the body. In the end, neutralinos will assemble near the centre of the Earth or the Sun. Since they are their own antiparticles, they can annihilate with each other, resulting in ordinary particles (quarks, leptons, gauge particles). Most of these annihilation products create no measurable effect; they are just stopped and contribute somewhat to the energy generation.[12] However, neutrinos have the unique property that they can penetrate the whole Sun (and/or Earth) without hardly being absorbed.

---

[10] This is by no means an easy task, especially since a bundle of downward-going muons can, in some cases, mimic the signals from a single upward-moving muon.

[11] As we have seen, sometimes the acronym WIMP – Weakly Interacting Massive Particle – is given to this kind of hypothetical dark matter particle.

[12] When WIMPs were first discussed, in the late 1970s, it was thought that they could be numerous enough that the temperature of the interior of the Sun would

An annihilating neutralino pair of mass m_χ would thus give rise to *high-energy* neutrinos of energy around $m_\chi/3$ or so (the reason that $E_\nu \neq m_\chi$ is that other particles created in the annihilation process share the energy). The signal of high-energy neutrinos (tens to hundreds of GeV – to be compared with the 'ordinary' MeV solar neutrinos) from the centre of the Sun or Earth would be an unmistakable signature of WIMP annihilation (see Fig. 1.3).

Calculations show, however, that the detection of such neutrinos requires a neutrino detector ('neutrino telescope') with an area of more than 10^5 m^2.

14.10 Neutrino Telescopes: the *Cherenkov* Effect

The detection of the neutrino burst from supernova 1987A marked a new era in the field of observational high-energy astronomy. Several new neutrino telescopes were born in the wake of this event. Most experiments are set up to detect the 'blue flashes' radiated by charged particles produced in neutrino interactions: for example, muons. The coherent emission of UV and optical photons from a charged track is known as the *Cherenkov* effect.

In the middle of the 1930s Pavel Cherenkov discovered the emission of blue light from radioactive sources in water. The interpretation of the 'Cherenkov effect' was provided by two of his colleagues in Moscow, Ilja Frank and Igor Tamm.[13] Charged particles moving faster than the speed of light in the medium, c/n, where n is the index of refraction of the medium, generate an electromagnetic shock-wave along their path: that is, a coherent wavefront of radiation similar to the more familiar effect of a sonic boom from supersonic aircraft.

The coherent emission follows a characteristic angle given by the Mach relation:

$$\cos\theta = \frac{1}{\beta n} \tag{14.46}$$

where β is the speed of the particle traversing the medium in units of the speed of light ($v = \beta \cdot c$). Thus, the condition for the Cherenkov effect to take place is that

$$\beta > \frac{1}{n} \tag{14.47}$$

The relation in (14.46) can be easily visualized through the Huygens construction in Fig. 14.11. The numbers in the figure indicate the order in which

---

rise, thereby perhaps explaining the solar neutrino problem. However, the experimental limits on their coupling strength makes this solution of the solar neutrino problem impossible.

[13] Frank, Tamm and Cherenkov were awarded the 1958 Nobel prize in physics for the discovery and explanation of the 'Cherenkov effect'.

the radiation is emitted, which in turn corresponds to the direction of the moving charged particle.

The direction of the track can be deduced from the Cherenkov wave front cone, making this effect very useful for building telescopes. The intensity of photons, on the other hand, is low. The number of photons per unit wavelength and unit distance travelled by the charged particle is[14]

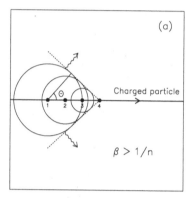
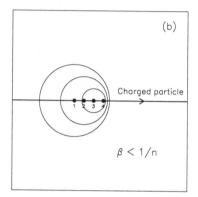

Fig. 14.11. (a) Cherenkov emission for a charged particle moving above threshold, $\beta \geq \frac{1}{n}$. The circles (spheres) show the isotropic emission of light along the charged track. After time t the charged particle has moved a distance βct. In that time the light sphere has grown by $\frac{c}{n}t$. The angle of propagation of the Cherenkov wavefront is thus $\cos\theta = \frac{1}{\beta n}$, as stated in equation (14.46). (b) Below threshold, $\beta < \frac{1}{n}$, the light spheres do not support coherent emission.

$$\frac{dN}{d\lambda dx} = \frac{2\pi\alpha Z^2}{\lambda^2}\left(1 - \frac{1}{n^2\beta^2}\right) \qquad (14.48)$$

where Z is the charge of the moving particle and α is the fine structure constant.

Example 14.10.1 Most photomultipliers are sensitive to photons in the wavelength range $300 - 600$ nm. Calculate the number of photons per unit length over that wavelength interval emitted along the path of a muon with $\beta \approx 1$, assuming that the index of refraction is constant over that wavelength range.

Answer:

Integrating the expression in (14.48) over wavelength yields

---

[14] A complete, classical derivation can be found in [23].

$$\frac{dN}{dx} = 2\pi\alpha \sin^2\theta \left(\frac{1}{\lambda_1} - \frac{1}{\lambda_2} \right) \tag{14.49}$$

which for $300 - 600$ nm corresponds to:

$$\frac{dN}{dx} = 764 \cdot \left(1 - \frac{1}{\beta^2 n^2} \right) \text{ photons/cm} \tag{14.50}$$

Cherenkov radiation constitutes a very small fraction of the total energy loss of a charged particle as it crosses a medium. The superluminal condition is fulfilled only between the UV and near-infrared region of the electromagnetic spectrum. In water or ice, for example, where the index of refraction for UV and optical wavelengths averages around 1.3, the Cherenkov radiation cut-off in the UV region is around 70 nm. For shorter wavelengths the index of refraction is smaller than 1, indicating that the *phase velocity* of the radiation is larger than c.[15] The differential energy loss into Cherenkov photons in water or ice is just a few per cent of the total differential energy loss of a charged particle moving with a speed very close to c.

14.10.1 Water and Ice Cherenkov Telescopes

Neutrinos can be detected indirectly by the Cherenkov radiation from scattered, fast, charged leptons and hadrons produced in neutrino interactions with matter. Water and ice are convenient detector materials because of the low cost, suitable index of refraction and low absorption for UV and optical photons. The extremely large detector volumes needed to detect neutrinos from distances beyond our Sun makes the use of any other material practically impossible.

A detector typically consists of an array of photomultipliers with good time resolution ($\sim$ 1 ns) distributed in the medium. The pattern of the hit PMs, as well as the relative times of arrival, are used to fit the direction of the particle that generated the Cherenkov cone, as shown in Fig. 14.12.

Because of the correlation between the original direction of the neutrino and the produced charged lepton ((14.43)) it is possible to reconstruct the direction of the neutrino source.

---

[15] But not the signal velocity!

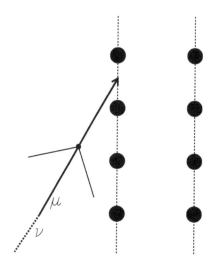

Fig. 14.12. Schematic picture of a detector for a high-energy neutrinos. The difference in arrival time of the wave front at the photomultipliers is used to reconstruct the muon track and, indirectly, the neutrino direction.

Fig. 14.13. Typical model for a source of high-energy neutrinos: a compact star with a companion. Protons are accelerated towards the massive companion, and neutrinos as well as gamma-rays are produced in the interactions. While gamma-rays can penetrate only a small amount of material, neutrinos can cross an entire star.

14.11 Potential Sources of High-energy Neutrinos

Where could high-energy neutrinos ($E_\nu \gg 1$ GeV) be produced? Whereas photons are produced by both hadronic and electromagnetic interactions, VHE neutrinos can only be produced in hadronic processes such as those in the Earth's atmosphere: that is, through cascades following the interaction of fast nucleons on some target. One can therefore expect neutrinos to be produced in the neighbourhood of astrophysical sources of acceleration of nuclei, such as binary stars, supernova remnants and accreting black holes. The detection of neutrinos can thus be used to understand the nature of cosmic ray particle acceleration, as shown schematically in Fig. 14.13.

Since gamma-rays can be absorbed inside or around the acceleration source, neutrinos offer a new observational window with the potential to discover unknown point sources.

There are also extragalactic sources, such as gamma-ray bursts and active galactic nuclei (AGN), which could produce high-energy neutrino radiation. AGN are generally believed to be massive ($\sim 10^8$ solar mass) black holes which accrete matter from the galaxy in which they reside. In this process strong shocks are formed, in which particles may be accelerated to enormous energies. Protons that are accelerated may interact with photons or other nucleons to produce very energetic pions. The pions decay in flight, producing photons, charged leptons and neutrinos.

14.12 Status of High-energy Neutrino Telescopes

The 'small' (area of order 1000 m^2) neutrino telescopes that are presently acquiring data, the US Japanese Super-Kamiokande experiment being the prime example, have an excellent energy and angular resolution, and detect atmospheric neutrinos at the rate of one or a few per day. These are, however, mainly of low energy – at energies above a few hundred GeV the flux of atmospheric neutrinos is simply too small for detection. For the study of more energetic atmospheric neutrinos, and also the search for neutrinos from AGN and other sources, the effective area has to be much larger.

Three approaches are currently being followed for the observation of high-energy neutrinos. Because of the enormous volumes that are required to explore the most likely sources, around 1 km^3, only naturally existing targets are under consideration. That limits the possibilities to deep lakes, ocean water or glacier ice.

The first possible detection of upward-moving neutrino induced muons in a natural water detector has been reported by the BAIKAL collaboration operating in Lake Baikal. Strings of photomultipliers are deployed at about 1.4 km depth.

Two collaborations are pursuing the use of ocean water: NESTOR and ANTARES, both in the Mediterranean. The DUMAND collaboration, which was the first group to attempt to build a VHE neutrino telescope (outside the island of Hawaii), had to be discontinued due to various difficulties with deployment in the demanding open-ocean environment.

The AMANDA experiment (in which groups from the USA, Sweden and Germany are involved) is situated in the geographical south pole in Antarctica. The disadvantages related to the remote location of the telescope are compensated by the virtues of the glacier ice, found to be the clearest natural solid on Earth. The Cherenkov photons emitted along the path of a muon can be seen, at some wavelengths, hundreds of metres away from the charged track. Some neutrinos have already been detected, proving the principle of operation (see Fig. 14.14).

The AMANDA experiment will be surrounded by an even larger detector, the IceCUBE, with 80 strings that will encompass a cubic kilometer of ice. Construction will start in 2004.

14.13 Summary

- Neutrinos play an important role in astrophysics because of their weak coupling with matter. This allows them to escape from dense regions, whereas photons are trapped.
- The cross-section for neutrino interactions with ordinary matter is approximately $5 \cdot 10^{-32} \left(\frac{E_{cm}}{1 \text{ MeV}} \right)^2 \text{ cm}^2$.
- MeV neutrinos of astronomical origin have been detected from the Sun and from supernova 1987A.
- Solar neutrino experiments, as well as the measurements of the fluxes of atmospheric neutrinos, are used to bound the neutrino masses and mixing between the flavours. There are some indications from both types of experiment that such oscillations indeed occur.
- If neutrinos are massive they may play an important role in cosmology and structure formation.
- High-energy neutrino astronomy, based on the Cherenkov technique, might provide fundamental information about acceleration sites in the Universe, as well as probe the particle physics solutions to the dark matter puzzle.

14.14 Problems

14.1 Derive (14.13)

14.2 Discuss the advantage of building a neutrino detector deep underground.

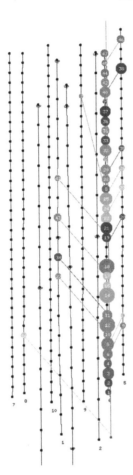

Fig. 14.14. One of the events recorded in 1997 by the AMANDA detector, interpreted as being due to an upward-moving atmospheric muon neutrino. The muon was created by a charged current weak interaction below the detector, and the Cherenkov light emitted by the muon was detected by the indicated optical modules of the 10-string detector. The time sequence of the hits is shown as well as the relative intensity of photons, indicated by the radius of the coloured spheres. (More strings with optical modules have since been added to the detector.)

14.3 Calculate the neutrino energy for which the interaction length is as large as the Earth's diameter. What does that mean for experiments looking for neutrinos coming 'from below'?

14.4 Derive (14.20). A Newtonian analysis is sufficient.

14.5 Calculate the expected number of positrons from $\bar{\nu}_e p \rightarrow n e^+$ in the Kamiokande II detector (2.1 kton) from type II supernovae 10 and 50 kpc away. The cross-section for the reaction is $\sigma = 10^{-41}$ cm^{-2}.

14.6 Estimate the volume of water necessary to detect 100 atmospheric neutrinos a year.

14.7 Derive (14.30) using (14.28)

14.8 a) What regions in Δm^2 can be studied with down-moving atmospheric neutrinos, $L \sim 10$ km ?
b) What regions in Δm^2 can be studied with solar neutrinos?

14.9 The average muon lifetime at rest is 2.2 μs. Estimate at what energy a vertically travelling muon, produced at 20 km height in the atmosphere, has a probability higher than 50 per cent of reaching the ground before decaying in flight.

14.10 a) Show that the muon range should have the analytical form stated in quation (14.42) if the muon energy loss per unit length is described by $\frac{dE}{dx} \approx -[\alpha + \beta E_\mu]$, where α and β are material constants.
b) Compare the muon range for a 1 TeV muon as given by the formula in (14.42) with the pathlength the muon would reach before decaying in vacuum.

14.11 Write a Monte Carlo programme to simulate the angular response to electrons produced by solar neutrinos in a water experiment. Assume that the angle between the incoming neutrino and the outgoing electron is normally distributed with $\sigma=18$ degrees and that the instrumental resolution is $\sigma=28$ degrees. Estimate then the signal-to-background ratio necessary to describe the angular distribution in Fig. 14.6. The background is assumed to be from radioactive impurities in the water and therefore isotropic.

14.12 a) Show that the differential energy loss in the form of Cherenkov radiation along a relativistic charged particle track in the wavelength interval $[\lambda_1, \lambda_2]$ is

$$\frac{dE}{dx} = 2\pi^2 r_e m_e c^2 \sin^2\theta \left(\frac{1}{\lambda_1^2} - \frac{1}{\lambda_2^2} \right) \tag{14.51}$$

where r_e $(=e^2/m_e c^2)$ is the classical electron radius.
b) Show that for water or ice, with a cut-off at 70 nm, the energy loss into Cherenkov photons is 25 keV/cm.

14.13 a) Estimate the relative yield of Cherenkov photons for an electron moving in water and air with $\beta \approx 1$. At sea-level $n_{air} - 1 = 2.7 \cdot 10^{-4}$.
b) What is the direction of the Cherenkov photons emitted in air?

15 Gravitational Waves

15.1 Introduction

In the preceding chapters we have discussed several types of radiation which, besides the extraordinarily useful electromagnetic quanta – photons – may convey information to us about processes in the Universe. We shall now discuss one more type of radiation which is deeply linked to the theory of general relativity on which modern cosmology rests: gravitational radiation.

As we have mentioned, there does not yet exist a full theory of quantum gravity. Therefore we cannot be fully sure about the existence and detailed properties of quantized mediator particles – gravitons (although it would be very surprising if they do not exist). However, this does not mean that we can say nothing about gravitational radiation. The situation is somewhat similar to that of electromagnetism when Maxwell proposed his equations but before quantum mechanics was developed. Of course, much could be deduced about electromagnetic radiation without knowing anything about photons. In particular, by analysing the solutions of his equations Maxwell could make the probable connection between electromagnetism and electromagnetic wave radiation such as light. We shall follow a similar approach here, and analyse Einstein's classical equations for the gravitational field in the search for, and finding of, wave solutions.

As we shall see, there exists already convincing indirect evidence for the existence of gravitational waves, and several large sophisticated detectors of gravitational waves are presently under construction which could give the first direct detection of such waves in the next few years.

15.2 Derivation of the Gravitational Wave Equation

Due to the nonlinearity of Einstein's equations, it is virtually impossible to find exact solutions to the metric tensor $g^{\mu\nu}(\mathbf{r}, t)$ corresponding to the dynamics, for example, of a massive star which collapses to a black hole near the strong gravitational field of the star (using supercomputers, numerical studies can, however, be made). Far from the source of the gravitational field, it is on the other hand reasonable to use a first-order approximation along the lines we indicated in connection with the Newtonian approximation (3.39) for the case of a static source. As we shall see, the gravitational deformation of space-time at the Earth due to conceivable astrophysical processes is indeed extremely small, which justifies such a perturbative approach.

We first recall the way we derived the existence of electromagnetic waves in Maxwell's theory in Section 2.6. There, we inserted the vector potential A^μ in the equations of motion (2.74) for a vanishing current j^μ (that is, in vacuum) to obtain

$$\Box A^\mu - \partial^\mu \left(\partial_\nu A^\nu \right) = 0 \tag{15.1}$$

Through the use of the gauge freedom $A^\mu \to A^\mu + \partial^\mu f$, we could choose A^μ to fulfil the axial condition $A^0 = 0$ and the Lorentz condition $\partial_\nu A^\nu = 0$. This immediately led to the simple wave equation

$$\Box A^\mu = 0 \tag{15.2}$$

which was found to have solutions of the form

$$A^\mu(\mathbf{r}, t) = \epsilon^\mu e^{\pm i(\omega t - \mathbf{k} \cdot \mathbf{r})} = \epsilon^\mu e^{\pm i k^\mu x_\mu} \tag{15.3}$$

where $k^\mu k_\mu = 0$ (light-like propagation) and the gauge conditions $A^0 = 0$ and $\partial_\nu A^\nu = \nabla \cdot \mathbf{A} = 0$ translate into $\epsilon^0 = 0$ and $\mathbf{k} \cdot \epsilon = 0$, showing that the two physical degrees of freedom are transverse to the direction of propagation. By superposition of, for example, a wave linearly polarized in the x-direction and one in the y-direction phase shifted by 90 degrees (obtained by multiplication of the amplitude by i), we obtained circularly polarized states, corresponding to definite helicity.

In the case of gravity waves, we make a first-order expansion of the dynamical degrees of freedom, the components of the metric tensor field $g_{\mu\nu}$, around the constant Minkowski metric $\eta_{\mu\nu}$:

$$g_{\mu\nu} = \eta_{\mu\nu} + h_{\mu\nu} \tag{15.4}$$

where we work only to first non-vanishing order in $h_{\mu\nu}$.

Inserting this expression into the Einstein field equations (3.63) appropriate for vacuum: that is, $T_{\mu\nu} = 0$, we find simply that

$$R_{\mu\nu} = 0 \tag{15.5}$$

Now we have to compute the Ricci tensor $R_{\mu\nu}$ in terms of the perturbations $h_{\mu\nu}$. With the help of the formulae in Appendix A, this can be shown to be (Problem 15.1)

$$2R_{\rho\nu} = \partial_\rho \partial_\mu h^\mu{}_\nu + \partial_\nu \partial_\mu h^\mu{}_\rho - \Box h_{\nu\rho} - \partial_\rho \partial_\nu h^\mu{}_\mu = 0 \tag{15.6}$$

In analogy with the electromagnetic case, we now try to make some of these terms vanish by choosing a particular gauge. It is not difficult to prove (see Problem 15.2) that under a local coordinate change $x_\mu \to x_\mu + \xi_\mu(x)$, the metric transforms as

$$h_{\mu\nu} \to h_{\mu\nu} - \partial_\mu \xi_\nu - \partial_\nu \xi_\mu \tag{15.7}$$

We now use this freedom to demand that the trace of $h^{\mu\nu}$ vanishes:

$$h^{\mu}{}_{\mu} = 0 \tag{15.8}$$

and transverse (similar to the Lorentz condition for A^{μ}):

$$\partial_{\mu} h^{\mu\nu} = \partial_{\mu} h^{\nu\mu} = 0 \tag{15.9}$$

Then (15.6) reduces to the simple wave-equation form

$$\Box h_{\mu\nu} = 0 \tag{15.10}$$

In fact we can also, as in the electromagnetic case, impose an axial-like gauge condition

$$h_{0\nu} = h_{\nu 0} = 0 \tag{15.11}$$

Exactly as for the electromagnetic waves we can now search for solutions of the type

$$h_{\mu\nu} = E_{\mu\nu} e^{\pm i k_{\rho} x^{\rho}} \tag{15.12}$$

where insertion in (15.10) shows that $k_{\mu} k^{\mu} = 0$: that is, the propagation vector is light-like. Gravitational waves thus propagate with the speed of light (the graviton, if it exists, is massless like the photon). As usual, we may of course combine terms with the two signs of the exponential into a real expression which will oscillate like, for example,

$$h_{\mu\nu} = E_{\mu\nu} \cos(\omega t - \mathbf{k} \cdot \mathbf{r}) \tag{15.13}$$

The constant polarization tensor $E_{\mu\nu}$ has to be traceless, $E^{\mu}{}_{\mu} = 0$, and transverse, $k^{\mu} E_{\mu\nu} = 0$, because of the gauge conditions. Also, $E_{0\nu} = 0$ (and E has to be symmetric because the metric and hence h are symmetric). It is not difficult to construct constant polarization tensors of this kind. If we again choose the z-direction as the direction of propagation, we find only two possible polarization basis states:

$$E^{+}_{\mu\nu} = \begin{pmatrix} 0 & 0 & 0 & 0 \\ 0 & 1 & 0 & 0 \\ 0 & 0 & -1 & 0 \\ 0 & 0 & 0 & 0 \end{pmatrix} \tag{15.14}$$

and

$$E^{\times}_{\mu\nu} = \begin{pmatrix} 0 & 0 & 0 & 0 \\ 0 & 0 & 1 & 0 \\ 0 & 1 & 0 & 0 \\ 0 & 0 & 0 & 0 \end{pmatrix} \tag{15.15}$$

We can now write an arbitrary wave amplitude at a fixed location, for example at a detector, as a time-dependent $E_{\mu\nu}(t)$ which is a linear combination of these two fundamental quadrupole modes:

$$E_{\mu\nu}(t) = h_{+}(t) E^{+}_{\mu\nu} + h_{\times}(t) E^{\times}_{\mu\nu} \tag{15.16}$$

15.3 Properties of Gravitational Waves

Having found plane-wave solutions characterized by the amplitudes h_+ and $h_\times$, we now investigate the physical meaning of these distortions of Minkowski space-time. We suppose that the distances between, for example, various parts of a gravity wave detector we consider are much smaller than the wavelength of the gravitational wave, so that we do not need to deal with retardation effects.

Let us see how the unit circle in the xy plane is distorted by a gravitational wave of the h_+ type. The distance between a diametrically opposed pair of points $(\cos\theta, \sin\theta)$ and $(-\cos\theta, -\sin\theta)$ is in the unperturbed case

$$d_0 = \sqrt{-\sum_{i,k=1,2} \eta_{ik}\Delta x^i \Delta x^k} = 2\sqrt{\left(\cos^2\theta + \sin^2\theta\right)} = 2 \tag{15.17}$$

regardless of the position on the unit circle. In the presence of h_+, the result is for $t = r = 0$,

$$d_+ = \sqrt{-\sum_{i,k=1,2} g_{ik}\Delta x^i \Delta x^k} = \sqrt{-\sum_{i,k=1,2} \left(\eta_{ik} + h_+ E_{ik}^+\right)\Delta x^i \Delta x^k} \tag{15.18}$$

Here we use (15.14) and the expansion for small δ, $\sqrt{4-\delta} \simeq 2 - \delta/4$, to obtain

$$d_+ \simeq 2 - h_+\left(\cos^2\theta - \sin^2\theta\right) = 2 - h_+\cos 2\theta \tag{15.19}$$

From this we see that the distance between two points on the x-axis is larger by the relative amount $h_+/2$, while the distance between two points on the y-axis is smaller by the same amount: see the middle diagram of Fig. 15.1 (a). Since the sign and size of the perturbation will oscillate with time according to (15.13), there will be an oscillating distortion of points on the unit circle into an ellipse, with its major and minor axes alternatingly being the x and y axes (but never any other axis).

For the $h_\times$ case, the behaviour is very similar. This is most easily seen by rotating the coordinate system by an angle $\pi/4$ around the z-axis so that $E_{\mu\nu}^\times$ becomes diagonal (exercise: perform this diagonalization). In that new frame, $E_{\mu\nu}^\times$ looks exactly like $E_{\mu\nu}^+$ did in the old frame. The pattern of deformation is thus the same, except that the main axes of the elliptical deformation make a 45-degree angle to the x and y axes (see Fig. 15.1 (b)). By superposing the two types of fundamental mode, phase-displaced by 90 degrees (just as in the case of circularly polarized light), one may construct circularly polarized gravity waves which represent deformations in the form of an ellipse which rotates either clockwise or anti-clockwise with angular frequency ω, maintaining its shape.

Sources which can excite these waves should ideally have the same type of quadrupole symmetry, such that the energy-momentum tensor (or rather

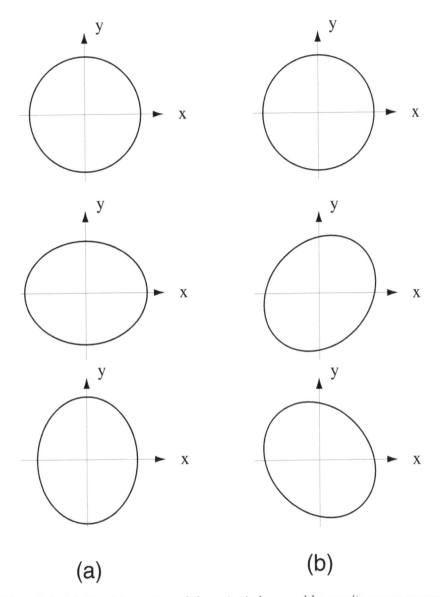

Fig. 15.1. (a) The deformation of the unit circle caused by gravity waves proportional to the polarization amplitude h_+. Shown are the unperturbed circle and the maximally stretched configurations along the two axes of symmetry, the x and y axes. (b) The corresponding pattern for the orthogonal polarization state described by the amplitude $h_\times$. Note that the axes along which stretching and compression occur form 45-degree angles to the x and y axes.

transverse and symmetric part of it) which is to be inserted on the right-hand side of (15.5) should represent a time-varying quadrupole moment. (In particular, a spherically symmetric source does not contribute.) An order-of-magnitude estimate (exercise: motivate this on dimensional grounds) produces

$$h \sim \frac{G\ddot{Q}}{d} \tag{15.20}$$

where Q is the quadrupole moment of the source and d is the distance to the source. A non-symmetric source of mass M and size l has a quadrupole moment $Q = Ml^2$, which means $\ddot{Q} \sim 2Mv^2$ with v the internal velocity. Thus, since the internal (non-spherically symmetric) kinetic energy is $Mv^2/2$, we can estimate

$$h \sim \frac{4GE_{kin}}{d} \tag{15.21}$$

However, there is in most cases a direct proportionality between the gravitational energy and the kinetic energy (through the virial theorem), as we shall show below. The most promising sites for the generation of gravitational radiation should therefore be very compact objects, where the gravitational fields are strong.

A prime example is that of a coalescing binary star system. In some cases, it is expected that the non-symmetric kinetic energy may be as large as the rest energy of the Sun. For $d \sim 3$ Gpc: that is, of the order of the Hubble radius, (15.21) gives $h \sim 10^{-22}$, for the Virgo galaxy cluster ($d \sim 15$ Mpc) $h \sim 10^{-20}$, and for the Milky Way ($d \sim 10$ kpc) $h \sim 10^{-17}$. Note the extremely tiny amplitudes even for Milky Way sources: a 100 m rod would stretch and compress with an amplitude around one nuclear diameter!

Not only are the conceivable sources very weak: they are also expected to be transient due to the rapid energy loss from gravitational radiation (and maybe the process stops through the formation of a black hole).

15.4 The Binary Pulsar

The fact that binary systems (where either or both companions can be a neutron star or a black hole) lose energy as a result of the emission of gravitational radiation has turned out to be a very useful tool for testing Einstein's theory of relativity, including its prediction of gravitational radiation.

If the typical separation between two neutron stars in a binary system is l, the mass of each of them is M, and they rotate around each other with angular frequency ω, the luminosity (energy lost in the form of gravitational radiation per unit time) can be shown to be (see [51])

$$L = \frac{16G\omega^6 M^2 l^4}{5} \tag{15.22}$$

It was with the pioneering discovery in 1974 of the binary pulsar PSR 1913+16 by R.A. Hulse and J.H. Taylor, using the 300-metre radio telescope at Arecibo, Puerto Rico, that the first useful test of the hypothesis of gravitational radiation was possible.[1]

The unique feature of pulsars is the very regular emission of radio waves, which makes an accurate determination of orbital and spin parameters possible. For the pair PSR 1913+16, a steady decrease in the orbital period time has been observed [49]:

$$\frac{dP}{dt} = (-2.4225 \pm 0.0056) \cdot 10^{-12} \tag{15.23}$$

If the pair loses energy due to gravitational radiation, such a decrease in orbital time can be expected. As we are dealing with a rather weak process, a Newtonian analysis of the energy loss should be sufficient. The total energy of the pair is

$$E_{tot} = Mv^2 - \frac{GM^2}{l} \tag{15.24}$$

and from the virial theorem (or from the results in Example 15.4.1 below) one can deduce that the first term, the kinetic energy, has a magnitude which is half that of the potential term. Thus,

$$E_{tot} = -\frac{GM^2}{2l} \tag{15.25}$$

So, if the total energy decreases due to gravitational radiation, we see from (15.25) that the distance l between the two stars will decrease, and therefore the angular frequency ω will increase.

Example 15.4.1 How does the angular frequency ω change when the distance l between the two neutron stars, each of mass M, decreases?

Answer: By demanding the balancing of the centrifugal and attractive gravitational forces on one of the masses in the pair one obtains

$$\frac{Mv^2}{l/2} = \frac{GM^2}{l^2}$$

or $v^2 = GM/2l$, which produces

$$\omega^2 = \frac{v^2}{(l/2)^2} = \frac{2GM}{l^3} \tag{15.26}$$

Thus,

$$\omega \sim l^{-3/2}$$

---

[1] Hulse and Taylor were awarded the 1993 Nobel prize in physics for their discovery.

Note that our result $v^2 = GM/2l$ can be inserted into (15.24) to derive the virial theorem result (15.25).

Since $E \sim l^{-1}$ and (Example 15.4.1) the orbital period $P \sim \omega^{-1} \sim l^{3/2}$, we see that $P \sim |E|^{-3/2}$, so that by taking the time derivative of the logarithm of this last relation,

$$\frac{1}{P}\frac{dP}{dt} = -\frac{3}{2}\frac{1}{|E|}\frac{d|E|}{dt} = -\frac{3}{2}\frac{L}{|E|} \tag{15.27}$$

where the luminosity L was given in (15.22). Thus,

$$\frac{1}{P}\frac{dP}{dt} = -\frac{24}{5}\omega^6 l^5 \tag{15.28}$$

With the measured period $P \simeq 7.75$ hours, and diameter of the orbit (measured from time delay) $l \sim 4$ lightseconds, this simple estimate gives a factor of about 10 smaller value than that measured according to (15.23). However, this discrepancy can be fully explained by the non-equality of the two masses and the non-circular shape of the orbit. By measuring the Doppler shift of the pulsation rate in various parts of the orbit, the eccentricity of the elliptical orbit has been measured to be around 0.62. A full calculation including the effects of this as well as so-called post-Newtonian corrections (i.e., beyond the linear approximation) [36] gives

$$\frac{dP_{GR}}{dt} = -2.40 \cdot 10^{-12} \tag{15.29}$$

in striking agreement with (15.23). In fact the agreement is so good that it can be used to constrain various proposed modifications to general relativity.

15.5 Gravitational Wave Detectors

We see from (15.22), (15.26) and (15.28) that the more energy that is radiated in the form of gravitational waves for a binary system, the closer the two stars approach one another, and the faster they orbit, increasing the luminosity. The situation is thus unstable, and will eventually have to end, perhaps in a dramatic way such as the collapse to a black hole. Of course, for a system such as PSR 1913+13, it will take a long time before this regime is entered. However, there may be other pairs that are about to coalesce in such a violent way. Those events could be likely sources of gravitational waves that would possibly be detected at Earth. However, we saw in Section 15.3 that the amplitude of such waves is in the range $h \sim 10^{-22} - 10^{-20}$ for the extragalactic distances which are necessary to have a reasonable probability of detecting at least a handful of events per year.

We have seen that for a detector with arm length a, the amplitude of a gravitational wave is proportional to

$$\frac{\Delta a(t)}{a} = c_+ h_+(t) + c_\times h_\times(t) \tag{15.30}$$

with c_+ and $c_\times$ constants of order unity which depend on the exact orientation of the detector arm with respect to the two polarization directions. Detectors therefore require a sensitivity reaching one part in 10^{22} to establish a signal. The most promising technique today for acquiring this phenomenal sensitivity is through the use of laser interferometry of the Michelson type (see Fig. 15.2). Light from a powerful laser is split into two long orthogonal paths and is reflected against mirrors attached to test weights at the end of the two arms. The two returning light beams are then made to interfere with each other, creating interference fringes which would be stationary if the test bodies were perfectly at rest, and the distance between them did not change. In the presence of gravitational radiation, the disturbances of the metric caused by the h_+ and $h_\times$ amplitudes will, as we have seen, typically stretch the length of one of the two perpendicular arms and squeeze the other, causing a shift in the pattern of interference fringes.

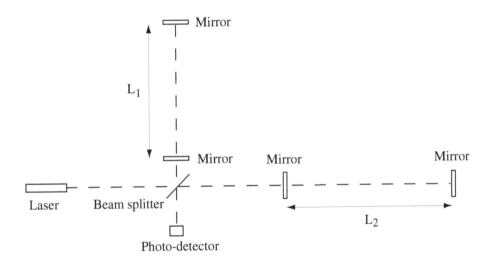

Fig. 15.2. A schematic view of the LIGO and VIRGO type of laser interferometer. The beam from the laser is split into two perpendicular long arms, of kilometre-scale length L_1 and L_2, each forming a resonant cavity. A small portion of the laser light is taken out from the beams, and the phases of the two beams are compared at the photo-detector. A passing gravitational wave will periodically change the length difference $L_1 - L_2$ (the sensitive frequency being between 10 Hz and 1 kHz) which causes an oscillating phase difference at the photo-detector.

To increase the sensitivity, the two cavities are of resonant (Fabry Perot) type. This means that light is allowed to travel many times back and forth between the laser and the mirrors, increasing the number of photons in the beams, which enables a higher resolution of the relative phase of the two beams (that is, the location of the interference fringes). Present-day technology allows measurements to an accuracy reaching 10^{-18} m, which means that with arms of $1-10$ km length the accuracy may be sufficient for the detection of coalescing neutron stars or black holes. There are several such detectors currently being constructed, with the most ambitious ones, the American LIGO and the French Italian VIRGO projects, recently starting to acquire data.

LIGO (see Fig. 15.2) will consist of two facilities with 4-km long arms, one in Hanford, Washington (in the north-western United States) and the other in Livingston, Louisiana (in the south-eastern United States). The VIRGO collaboration is building one facility near Pisa in Italy, with 3-km long arms.

The main background problem is that of various types of noise. The dominant contributions are noise from the environment (seismic noise) at low frequencies, thermal noise from the test weights and mirrors at intermediate frequencies, and shot noise from the laser system at high frequencies. The latter is related to the fact that the definition of the location of the laser interference fringes necessitates the detection of many photons. When the number of photons decreases, stochastic effects will smear the measurements. To keep all the possible sources of background noise below the tiny gravitational wave signals represents a formidable technological challenge. The increase in both the amplitude and frequency of the gravitational waves originating from the final stages of a coalescing binary system may be used as a 'template' to discriminate against this noise.

One way to avoid the seismic noise, and also to have longer interferometer arms, would obviously be to develop a satellite-borne system. There are plans (LISA – laser interferometer space antenna) to launch three satellites with laser emitters and receivers which would constitute a giant interferometer with the arms in an equilateral triangle of side length 5 million km. If approved, the launch could take place in 2008.

According to estimates (see [50]), LIGO's first interferometers should be able to detect waves from the inwards spiral of a binary pulsar out to a distance of 30 Mpc (90 million light years) and from the final collision and merger of two 25 solar mass black holes out to about 300 Mpc. LIGO and VIRGO together, operating as a coordinated international network, will be able to locate a source through time delays and the beam patterns of the interferometers to within a few degrees, the exact value depending on the source direction and on the amount of high-frequency structure in the waveforms. They will also be able to monitor both waveforms $h_+(t)$ and $h_\times(t)$ (except for frequency components above about 1 kHz and below about 10 Hz, where noise becomes the limiting factor).

If gravitational waves can be experimentally discovered, a whole new field of applications would open up, and many interesting new facts about the most violent processes in the Universe would be known.

Finally, we remark that there is also a possibility of detecting gravitational waves that are relics of dramatic processes in the early Universe, such as during the epoch of inflation or during the formation of cosmic strings, if such exist. In that case, the most promising method is through analysing the imprints they have made in the cosmic microwave background radiation (CMBR). Because gravitational waves carry a quadrupole moment it is possible to distinguish their effects through studies of CMBR polarization. With the planned Planck satellite there will be a possibility of searching for gravitational waves of very long wavelength generated through these hypothetical processes.

15.6 Summary

- The existence of gravitational radiation is a firm prediction from Einstein's theory of general relativity. In the transverse and traceless gauge, the equation of free propagation is simply

$$\Box h_{\mu\nu} = 0$$

where a first-order expansion of the metric $g_{\mu\nu} = \eta_{\mu\nu} + h_{\mu\nu}$ has been made.
- Gravitational waves are of quadrupole type with two independent polarization modes, with their amplitudes labeled h_+ and $h_\times$, respectively.
- Strong indirect evidence for the existence of gravitational radiation comes from the study of binary pulsars, where the energy loss due to gravity waves agrees with observations.
- Gravitational wave detectors are being built which may be sensitive enough to detect coalescing binary neutron stars or black holes at distances of several Mpc. The two most ambitious detector projects are LIGO in the United States and VIRGO in Europe. They will use laser interferometry over distances of several kilometres to detect quadrupole deformations of space due to gravity waves.

15.7 Problems

15.1 Show equation (15.6).

15.2 By using (A.17), show that to first order the metric transforms as (15.7) under a coordinate change $x_\mu \rightarrow x_\mu + \xi_\mu$.

15.3 Suppose that you have at your disposal a device for measuring lengths with the same accuracy as LIGO or VIRGO. How far away would an object have to be to produce an uncertainty of 1 mm in the distance determination?

A Some More General Relativity

A.1 Metric for Curved Space-Time

We know from the strong equivalence principle that even if space-time has a complicated structure due to curvature effects, we can always *locally* at a space-time point P find a reference frame with coordinates $x^\mu = (t, x^1, x^2, x^3) = (x^0, x^1, x^2, x^3)$ such that (see (3.10) and (3.13))

$$ds^2 = g_{\mu\nu} dx^\mu dx^\nu \tag{A.1}$$

with

$$g_{\mu\nu}(P) = \eta_{\mu\nu} \tag{A.2}$$

$$\frac{\partial g_{\mu\nu}}{\partial x^\rho}(P) = 0 \tag{A.3}$$

This is the free-fall frame at P. To transfer to such a frame from any arbitrary space-time coordinate system we need to perform a coordinate transformation of a more general type than given by the Lorentz transformations of special relativity. For large space-time regions, these transformations are usually highly non-linear and very difficult to perform. However, for small regions around a given space-time point P we can use much of the tensor machinery developed for special relativity. Thus, changing from coordinates x^μ to coordinates x'^μ, small coordinate distances dx'^μ will transform linearly:

$$dx'^\mu = \frac{\partial x'^\mu}{\partial x^\nu} dx^\nu \tag{A.4}$$

where as usual we use the summation convention (in this case over the index ν). We see that this is of the form (2.18)

$$dx'^\mu = \Lambda^\mu_{\ \nu} dx^\nu \tag{A.5}$$

if we define

$$\Lambda^\mu_{\ \nu} = \frac{\partial x'^\mu}{\partial x^\nu} \tag{A.6}$$

Using the chain rule

$$\frac{\partial x'^\mu}{\partial x^\nu} \frac{\partial x^\nu}{\partial x'^\rho} = \delta^\mu_{\ \rho} \tag{A.7}$$

where δ^μ_ρ is the usual Kronecker δ (which has the value 1 if $\mu = \rho$, the value 0 if $\mu \neq \rho$), we see that we can write the inverse transformation matrix

$$\Lambda_\mu{}^\nu = \frac{\partial x^\nu}{\partial x'^\mu} \tag{A.8}$$

that is

$$\Lambda^\mu{}_\sigma \Lambda_\nu{}^\sigma = \delta^\mu_\nu \tag{A.9}$$

Let us look at the motion of a test particle moving freely in a gravitational field. According to the equivalence principle, we can at each moment find a frame (the free-fall frame) where the motion is that of a free particle in special relativity: that is, according to (2.48) it moves in a straight line in those space-time coordinates. If we call the free-fall coordinates ξ^μ, the equations of motion are thus

$$\frac{d^2\xi^\mu}{d\tau^2} = 0 \tag{A.10}$$

whereas in (2.43)

$$d\tau^2 = \eta_{\mu\nu} d\xi^\mu d\xi^\nu \tag{A.11}$$

Now suppose that we have another set of coordinates x^μ that may be curvilinear. What are the local equations of motion for the particle in these coordinates? Again, we just use the transformation

$$d\xi^\mu = \frac{\partial \xi^\mu}{\partial x^\nu} dx^\nu \tag{A.12}$$

to obtain

$$\frac{d\xi^\mu}{d\tau} = \frac{\partial \xi^\mu}{\partial x^\nu} \frac{dx^\nu}{d\tau} \tag{A.13}$$

and

$$0 = \frac{d^2\xi^\mu}{d\tau^2} = \frac{d}{d\tau}\left(\frac{\partial \xi^\mu}{\partial x^\nu} \frac{dx^\nu}{d\tau}\right) = \frac{\partial \xi^\mu}{\partial x^\nu} \frac{d^2 x^\nu}{d\tau^2} + \frac{\partial^2 \xi^\mu}{\partial x^\nu \partial x^\rho} \frac{dx^\nu}{d\tau} \frac{dx^\rho}{d\tau} \tag{A.14}$$

We can also express the proper time τ in the new coordinates by using

$$d\tau^2 = \eta_{\mu\nu} d\xi^\mu d\xi^\nu = \eta_{\mu\nu} \frac{\partial \xi^\mu}{\partial x^\rho} \frac{\partial \xi^\nu}{\partial x^\sigma} dx^\rho dx^\sigma \tag{A.15}$$

or

$$d\tau^2 = g_{\rho\sigma} dx^\rho dx^\sigma \tag{A.16}$$

with the metric $g_{\rho\sigma}$ given by

$$g_{\rho\sigma} = \frac{\partial \xi^\mu}{\partial x^\rho} \frac{\partial \xi^\nu}{\partial x^\sigma} \eta_{\mu\nu} \tag{A.17}$$

The first term in (A.14) is almost the same as in (A.10). To make it exactly the same, we need to dispose of the factor $\partial \xi^\mu / \partial x^\nu$. But this we can do (see (A.7)) by multiplying the whole equation (A.14) with $\partial x^\sigma / \partial \xi^\mu$ (and summing over μ). Doing this, we obtain the *geodesic equation*

$$\frac{d^2 x^\sigma}{d\tau^2} + \Gamma^\sigma_{\mu\nu} \frac{dx^\mu}{d\tau} \frac{dx^\nu}{d\tau} = 0 \tag{A.18}$$

where the *metric connections* (sometimes called affine connections or Christoffel symbols) $\Gamma^\sigma_{\mu\nu}$ are given by

$$\Gamma^\sigma_{\mu\nu} = \frac{\partial x^\sigma}{\partial \xi^\rho} \frac{\partial^2 \xi^\rho}{\partial x^\mu \partial x^\nu} \tag{A.19}$$

Note that the metric connections are symmetric in μ and ν:

$$\Gamma^\sigma_{\mu\nu} = \Gamma^\sigma_{\nu\mu} \tag{A.20}$$

We see that the geodesic equation (A.18) can be interpreted as a kind of force equation

$$\frac{d^2 x^\sigma}{d\tau^2} = f^\sigma \tag{A.21}$$

with

$$f^\sigma = -\Gamma^\sigma_{\mu\nu} \frac{dx^\mu}{d\tau} \frac{dx^\nu}{d\tau} \tag{A.22}$$

The brilliant observation of Einstein was that since $\Gamma^\sigma_{\mu\nu}$ are purely geometrical objects (they depend only on the metric and its derivatives), it will be possible to view gravity as not really an ordinary force but the result of an influence of massive bodies on space-time, making it curved in a particular way. If the ratio between inertial and gravitational mass were not the same for all massive bodies this would not have been possible, because then we would need different free-fall coordinates for different bodies, preventing the elegant, unified description of general relativity.

To arrive at a geometrical view of gravity it remains to show that $\Gamma^\sigma_{\mu\nu}$ are indeed geometrical, that is, related to the metric $g_{\mu\nu}$, and also to find the physical law that gives $g_{\mu\nu}$ for a given distribution of mass. The latter problem was solved by Einstein who proposed a set of equations, the *Einstein equations*, which we studied in Section 3.7. The proof that $\Gamma^\sigma_{\mu\nu}$ can be expressed in terms of $g_{\mu\nu}$ is straightforward, although technically a little involved (see Problem A.2). It is found that

$$\Gamma^\sigma_{\mu\nu} = \frac{g^{\rho\sigma}}{2} \left(\frac{\partial g_{\nu\rho}}{\partial x^\mu} + \frac{\partial g_{\mu\rho}}{\partial x^\nu} - \frac{\partial g_{\nu\mu}}{\partial x^\rho} \right) \tag{A.23}$$

where $g^{\mu\nu}$ is the inverse of $g_{\mu\nu}$:

$$g_{\rho\mu} g^{\mu\nu} = \delta^\nu_\rho \tag{A.24}$$

From linear algebra, it is known that a solution to (A.24) is provided by

$$g^{\mu\nu} = C^{\mu\nu} / g \tag{A.25}$$

with g the determinant of $g_{\mu\nu}$, and $C^{\mu\nu}$ the cofactor of $g_{\mu\nu}$ in this determinant. In the particular case when $g_{\mu\nu}$ is diagonal, $g^{\mu\mu}$ (no summation over μ here) is simply given by $g^{\mu\mu} = 1/g_{\mu\mu}$ (simple exercise: prove this!).

A.2 The Newtonian Limit

Let us study our expression for the geodesic equation for a particle moving slowly (compared to the speed of light) in a weak, stationary (that is, time-independent) gravitational field such as that of the Earth. We know from Newton's result that we should obtain a force equation in this limit which is of the form

$$\frac{d^2 x^i}{dt^2} = -\frac{\partial \phi}{\partial x^i} \tag{A.26}$$

with ϕ the gravitational potential, which for a spherically symmetric body of mass M is given by

$$\phi = -\frac{GM}{r} \tag{A.27}$$

where G is Newton's gravitational constant.[1] Look at the geodesic equation

$$\frac{d^2 x^\sigma}{d\tau^2} + \Gamma^\sigma_{\mu\nu} \frac{dx^\mu}{d\tau} \frac{dx^\nu}{d\tau} = 0 \tag{A.28}$$

Since the particle is slow, $dx^i \ll dx^0 = cdt$, so the dominant components of the geodesic equation are

$$\frac{d^2 x^\sigma}{d\tau^2} + \Gamma^\sigma_{00} \left(\frac{dt}{d\tau}\right)^2 = 0 \tag{A.29}$$

Since the field is stationary, all time derivatives of $g_{\mu\nu}$ vanish. Also, since the field is weak we should be able to expand around the Minkowski metric $\eta_{\mu\nu}$:

$$g_{\mu\nu} = \eta_{\mu\nu} + h_{\mu\nu} \tag{A.30}$$

and we see from (A.23) that to first order in the small quantities $h_{\mu\nu}$ the metric connection is

$$\Gamma^\sigma_{00} = -\frac{\eta^{\sigma\rho}}{2} \frac{\partial h_{00}}{\partial x^\rho}$$

and the equations of motion become

$$\frac{d^2 x^i}{d\tau^2} = -\frac{1}{2} \left(\frac{dt}{d\tau}\right)^2 \frac{\partial h_{00}}{\partial x^i} \tag{A.31}$$

$$\frac{d^2 t}{d\tau^2} = 0 \tag{A.32}$$

---

[1] G has the value $6.67 \cdot 10^{-11}$ m^{3}kg^{-1}s^{-2} in SI units. In the set of units where $c = \hbar = 1$, $G = 1/m_{Pl}^2$, with $m_{Pl} = 1.221 \cdot 10^{19}$ GeV, the so-called Planck mass.

Thus, (A.32) tells us that $dt/d\tau$ is constant, so dividing (A.31) by $(dt/d\tau)^2$ we obtain

$$\frac{d^2 x^i}{dt^2} = -\frac{1}{2}\frac{\partial h_{00}}{\partial x^i} \tag{A.33}$$

This is exactly of the same form as the Newtonian gravitational force (A.26) if

$$h_{00} = 2\phi + const \tag{A.34}$$

Far from the gravitating body the coordinate system should become pseudo-Euclidean, which means that the constant has to be set to zero. Reinserting c, the metric is thus (see (A.30))

$$g_{00} = 1 + \frac{2\phi}{c^2}. \tag{A.35}$$

A.3 The Curvature Tensor

Let us look at another way for arriving at the geodesic equation (A.18). To begin with, let us consider ordinary flat, Euclidean space. When forming the derivative of a space-time dependent contravariant vector quantity V^μ along a curve parameterized by the path length s in a cartesian coordinate system, we just compute the rate of change of the components ΔV^μ and divide by Δs. However, if we instead use curvilinear coordinates (for example, spherical coordinates), there will be an additional contribution to the change of the components of V measured in these coordinates due to the fact that the coordinate system has changed when moving along the curve. This is particularly apparent if we consider a constant vector field in Euclidean space. In cartesian coordinates the derivative will be trivially zero since ΔV^μ vanishes (see Fig. A.1 (a)). In polar coordinates there is a change $\delta V^\mu \neq 0$ (Fig. A.1 (b)). The latter change is thus 'unphysical' in the sense that it depends on the coordinate system – different curvilinear coordinate systems will give different values of $\delta V^\mu/\Delta s$ for a given Δs.

The natural definition of the derivative in this situation (and which generalizes to curved spaces) is to subtract the frame-dependent quantity $\delta V^\mu/\Delta s$ from $\Delta V^\mu/\Delta s$ to obtain the *covariant derivative*

$$\frac{DV^\mu}{Ds} = \lim_{\Delta s \to 0} \frac{\Delta V^\mu - \delta V^\mu}{\Delta s} \tag{A.36}$$

To compute the quantities δV^μ we need to determine the rate of change of the components of a vector field that does not change its 'physical' direction as we move along the curve s. We say that we need to *parallel transport* it along the curve. This is what we considered in Section 3.5.1 when we discussed curvature of the sphere, but now we have to define the procedure of parallel transport in mathematical terms. In a space of the Riemann type, we can

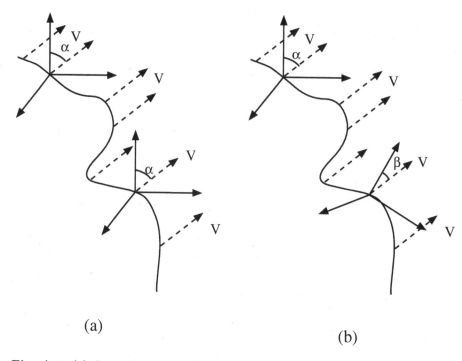

Fig. A.1. (a) A constant vector field in cartesian coordinates has the derivative $\Delta V^\mu / \Delta s = 0$ since $\Delta V^\mu = 0$ everywhere, in particular along the curve. (b) The same vector field has a non-vanishing derivative when computed in curvilinear coordinates, since the direction of the basis vectors changes from point to point along the curve.

always do this locally by choosing locally cartesian coordinates at a point A, and carrying the vector V^μ to a neighbouring point A' without changing its cartesian coordinates. Then the vector components are re-transformed to the relevant frame at A'. This defines parallel transport of a local vector in space-time.

Mathematically, it works as follows. If V^μ is a vector, the change δV^μ when V is parallel transported Δx^ν in the ν-direction should be bilinear in V and Δx:

$$\delta V^\mu = -\Gamma^\mu_{\rho\nu} V^\rho \Delta x^\nu \tag{A.37}$$

where it can be shown (see the derivation of (A.18)) that $\Gamma^\mu_{\rho\nu}$ are precisely the metric connections we encountered previously.

The covariant derivative along the curve can thus be written as

$$\frac{DV^\mu}{Ds} = \frac{dV^\mu}{ds} + \Gamma^\mu_{\rho\nu} V^\rho \frac{dx^\nu}{ds} \tag{A.38}$$

and it can be shown that it transforms as a contravariant vector under the change $x^\mu \to x'^\mu$ of coordinate system:

$$\frac{DV'^{\mu}}{Ds} = \frac{\partial x'^{\mu}}{\partial x^{\nu}} \frac{DV^{\nu}}{Ds} \tag{A.39}$$

Using the fact that the variation of the scalar product $\delta(V^{\mu}V_{\mu}) = 0$, one finds similarly (see Problem A.1) that

$$\frac{DV_{\mu}}{Ds} = \frac{dV_{\mu}}{ds} - \Gamma^{\nu}_{\rho\mu}V_{\nu}\frac{dx^{\rho}}{ds} \tag{A.40}$$

transforms like a covariant vector. For higher-rank tensors, the rules are obvious generalizations of these rules, for instance:

$$\frac{DT^{\mu}_{\ \nu}}{Ds} = \frac{dT^{\mu}_{\ \nu}}{ds} + \Gamma^{\mu}_{\ \rho\sigma}T^{\sigma}_{\ \nu}\frac{dx^{\rho}}{ds} - \Gamma^{\sigma}_{\ \rho\nu}T^{\mu}_{\ \sigma}\frac{dx^{\rho}}{ds} \tag{A.41}$$

Returning to the geodesic equation (A.18), we see that since the four-momentum $p^{\mu} = m\,dx^{\mu}/d\tau$, it can be interpreted as the condition that the particle's momentum p^{μ} is parallel transported along itself,

$$\frac{Dp^{\mu}}{D\tau} = 0 \tag{A.42}$$

As a particular case, we may consider the covariant derivative in one of the coordinate directions, usually written in the so-called semicolon convention:

$$V^{\mu}_{\ ;\nu} = \frac{\partial V^{\mu}}{\partial x^{\nu}} + \Gamma^{\mu}_{\ \nu\rho}V^{\rho} \tag{A.43}$$

By using the results of Problems A.6 and A.7 this can be rewritten in a form that is much easier to use in practice:

$$V^{\mu}_{\ ;\mu} = \frac{1}{\sqrt{-g}}\frac{\partial}{\partial x^{\mu}}\left(\sqrt{-g}V^{\mu}\right) \tag{A.44}$$

where $g = \det(g_{\mu\nu})$.

Unlike the first term on the right-hand side of (A.43), $V^{\mu}_{\ ;\nu}$ transforms as a tensor quantity in both the indices μ and ν. This is the basic key to setting up formulae that are general relativistic: that is, in accordance with the strong equivalence principle: According to this principle, the laws of nature in a free-fall frame are the usual tensor formulae of special relativity. To make them applicable to any frame, we just have to substitute the Minkowski metric $\eta_{\mu\nu}$ by $g_{\mu\nu}$, and the ordinary derivatives like (2.33) by covariant derivatives like (A.43).

Now we are prepared to investigate whether or not a given space-time is curved, using the infinitesimal version of the round-trip parallel transport discussed in Section 3.4. If we parallel transport a vector V^{μ} over a small rectangle $P_1P_2P_3P_4$ (Fig. A.2), the total change δV^{μ} will be

$$\delta V^{\mu} = -\Gamma^{\mu}_{\ \beta\nu}(x)V^{\nu}(x)\Delta a^{\beta} - \Gamma^{\mu}_{\ \beta\nu}(x+\Delta a)V^{\nu}(x+\Delta a)\Delta b^{\beta}$$
$$+\Gamma^{\mu}_{\ \beta\nu}(x+\Delta b)V^{\nu}(x+\Delta b)\Delta a^{\beta} + \Gamma^{\mu}_{\ \beta\nu}(x)V^{\nu}(x)\Delta b^{\beta} \tag{A.45}$$

This can be rewritten

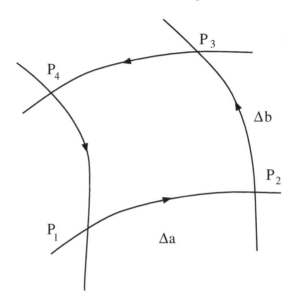

Fig. A.2. Closed path $P_1 P_2 P_3 P_4$ around which a vector V^μ is parallel transported to determine the local curvature.

$$\delta V^\mu = \frac{\partial(\Gamma^\mu_{\beta\nu} V^\nu)}{\partial x^\alpha} \Delta b^\alpha \Delta a^\beta - \frac{\partial(\Gamma^\mu_{\beta\nu} V^\nu)}{\partial x^\alpha} \Delta a^\alpha \Delta b^\beta \tag{A.46}$$

Expanding a partial derivative like $\partial(\Gamma^\mu_{\beta\nu} V^\nu)/\partial x^\alpha$ one finds, using the fact that for parallel transport $DV^\mu = 0$, so that $dV^\mu = -\Gamma^\mu_{\nu\rho} V^\rho dx^\nu$,

$$\delta V^\mu = \Delta a^\alpha \Delta b^\beta V^\sigma R^\mu_{\sigma\beta\alpha} \tag{A.47}$$

where *the Riemann curvature tensor* $R^\mu_{\sigma\beta\alpha}$ is given by

$$R^\mu_{\sigma\beta\alpha} = \frac{\partial \Gamma^\mu_{\sigma\alpha}}{\partial x^\beta} - \frac{\partial \Gamma^\mu_{\sigma\beta}}{\partial x^\alpha} + \Gamma^\mu_{\rho\beta} \Gamma^\rho_{\sigma\alpha} - \Gamma^\mu_{\rho\alpha} \Gamma^\rho_{\sigma\beta} \tag{A.48}$$

The Riemann tensor appears to be a complicated object, and in fact for a given metric $g_{\mu\nu}(x)$ it is wise to use any of the symbolic algebra computer programs available for its calculation. However, due to a large number of symmetries, the number of independent components which naively looks like $4^4 = 256$ is reduced to 20. These symmetries are most easily summarized for the associated tensor $R_{\alpha\beta\gamma\delta} = g_{\alpha\rho} R^\rho_{\beta\gamma\delta}$ formed by lowering the first index:

- Symmetry in the exchange of the first and second pairs of indices:

$$R_{\alpha\beta\gamma\delta} = R_{\gamma\delta\alpha\beta} \tag{A.49}$$

- Antisymmetry:

$$R_{\alpha\beta\gamma\delta} = -R_{\beta\alpha\gamma\delta} = -R_{\alpha\beta\delta\gamma} \tag{A.50}$$

- Cyclic property:

$$R_{\alpha\beta\gamma\delta} + R_{\alpha\delta\beta\gamma} + R_{\alpha\gamma\delta\beta} = 0 \tag{A.51}$$

Through contraction of the first and third index of the Riemann tensor one obtains the *Ricci tensor*

$$R_{\mu\nu} = g^{\alpha\gamma} R_{\alpha\mu\gamma\nu} \tag{A.52}$$

which is easily seen to be symmetric in its indices. By contracting the two indices of the Ricci tensor one obtains the *Ricci scalar*

$$R = g^{\mu\nu} R_{\mu\nu}. \tag{A.53}$$

From the Ricci tensor and the Ricci scalar one can form another symmetric tensor, which by construction has vanishing covariant divergence, the *Einstein tensor* $G_{\mu\nu}$:

$$G_{\mu\nu} = R_{\mu\nu} - \frac{1}{2} g_{\mu\nu} R \tag{A.54}$$

If we look at the three-dimensional curvature: that is, we disregard the time coordinate, we obtain, using (A.48) the three-dimensional version of the Riemann tensor for the Friedmann-Lemaître-Robertson-Walker metric

$$^3 R_{ijkl} = \frac{k}{a^2(t)} [g_{ik} g_{jl} - g_{il} g_{kj}] \tag{A.55}$$

for the Ricci tensor (using the spatial part of the metric g^{ij} to make the contraction of indices)

$$^3 R_{ij} = \frac{2k}{a^2(t)} g_{ij} \tag{A.56}$$

and for the spatial curvature three-scalar

$$^3 R = \frac{6k}{a^2(t)} \tag{A.57}$$

Using the Friedmann equation we can write this as

$$^3 R = 6 H^2 (\Omega - 1) \tag{A.58}$$

A.4 Summary

- The geodesic equation for a particle travelling freely along a trajectory with proper time τ reads

$$\frac{d^2 x^\sigma}{d\tau^2} + \Gamma^\sigma_{\mu\nu} \frac{dx^\mu}{d\tau} \frac{dx^\nu}{d\tau} = 0$$

Here the metric connections, or Christoffel symbols, are given by

$$\Gamma^\sigma_{\mu\nu} = \frac{\partial x^\sigma}{\partial \xi^\rho} \frac{\partial^2 \xi^\rho}{\partial x^\mu \partial x^\nu}$$

- The Christoffel symbols can be computed for a given metric through the formula

$$\Gamma^{\sigma}_{\mu\nu} = \frac{g^{\rho\sigma}}{2}\left(\frac{\partial g_{\nu\rho}}{\partial x^{\mu}} + \frac{\partial g_{\mu\rho}}{\partial x^{\nu}} - \frac{\partial g_{\nu\mu}}{\partial x^{\rho}}\right)$$

- Writing the metric $g_{\mu\nu} = \eta_{\mu\nu} + h_{\mu\nu}$, where $h_{\mu\nu}$ represent small perturbations, the effects of a Newtonian gravitational potential φ is to replace $g_{00} = 1$ by

$$g_{00} = 1 + 2\varphi$$

- The Riemann curvature tensor is given by

$$R^{\mu}_{\sigma\beta\alpha} = \frac{\partial \Gamma^{\mu}_{\sigma\alpha}}{\partial x^{\beta}} - \frac{\partial \Gamma^{\mu}_{\sigma\beta}}{\partial x^{\alpha}} + \Gamma^{\mu}_{\rho\beta}\Gamma^{\rho}_{\sigma\alpha} - \Gamma^{\mu}_{\rho\alpha}\Gamma^{\rho}_{\sigma\beta}$$

- The Ricci tensor is

$$R_{\mu\nu} = g^{\alpha\gamma}R_{\alpha\mu\gamma\nu}$$

- The Ricci, or curvature, scalar is

$$R = g^{\mu\nu}R_{\mu\nu}$$

- The Einstein tensor

$$G_{\mu\nu} = R_{\mu\nu} - \frac{1}{2}g_{\mu\nu}R$$

is symmetric and has vanishing covariant divergence.

A.5 Problems

A.1 Use the fact that the scalar quantity $V^{\mu}V_{\mu}$ has vanishing variation to derive (A.40).

A.2 Derive (A.23). Hint: start with

$$\frac{Dg_{\mu\nu}}{Dx^{\sigma}} = 0$$

which is easy to prove in a free-fall frame, and is valid in any frame since it is a tensor equation. Use the equation that follows from this, and the two related equations obtained by cyclic permutation. Finally, use the symmetry in the indices of Γ to obtain (A.23).

A.3 Show that the metric tensor $g_{\mu\nu}$ is covariantly constant, meaning that $g_{\mu\nu;\sigma} = 0$.

A.4 (a) Show that $R^{\mu}{}_{\nu\rho\sigma} = 0$ for the two-dimensional Euclidean plane, using any suitable coordinates.
(b) Show that $R^{\mu}{}_{\nu\rho\sigma} = 0$ for the Minkowski space-time parameterized by spherical coordinates.

A.5 Calculate $g^{\mu\nu}$ for the two-dimensional surface of a sphere of radius a, using the metric (3.18).

A.6 Let $g = \det(g_{\mu\nu})$. By writing the determinant in terms of co-factors, and taking into account that $g^{\mu\nu}$ is the inverse of $g_{\mu\nu}$, show that

$$\frac{\partial g}{\partial x^{\mu}} = g g^{\rho\sigma} \frac{\partial g_{\rho\sigma}}{\partial x^{\mu}}$$

A.7 Use the result of Problem A.6 to show that

$$\Gamma^{\mu}_{\nu\mu} = \frac{\partial}{\partial x^{\nu}} \log \sqrt{-g}$$

B Relativistic Dynamics

B.1 Classical Mechanics

In classical mechanics, we are usually interested in how some coordinates change with time. This may be the positions of a set of point particles, but also, for instance, the angle that a pendulum makes with respect to the vertical direction. Thus, we use some set of *generalized coordinates* to describe the motion as a function of time. For instance, if we consider N particles moving in three space dimensions, we describe their positions by generalized coordinates q_i $(i = 1, 2, \ldots N)$. For example, we can use

$$\{q_i\} = \{\mathbf{r}_i\} \tag{B.1}$$

or

$$\{q_i\} = \{r_i, \varphi_i, \theta_i\} \tag{B.2}$$

For each of the generalized coordinates, we define the generalized velocities by

$$\dot{q}_i \equiv \frac{dq_i}{dt} \tag{B.3}$$

We can thus formulate the fundamental dynamical problem in classical mechanics:

- Given $\{q_i, \dot{q}_i\}$ at a time $t = t_0$, what is the time development that follows?

It is shown in textbooks in classical mechanics (for example [17]) that the solution is given by *Hamilton's principle:*

There is a quantity L (called the Lagrangian) such that the integral (the *action*)

$$S = \int_{t_0}^{t_1} L(q_i, \dot{q}_i, t)dt \tag{B.4}$$

takes an extremum when the system moves from $\{q_i(t_0)\}$ to $\{q_i(t_1)\}$. The path followed by the system along this extremal solution is called the *classical path* $q_i^{cl}(t)$.

Let us see how Hamilton's principle will give us the equations of motion of the system. We shall perturb the classical path in the action integral by an amount $\delta q_i(t)$ (see Fig. B.1):

$$q_i(t) = q_i^{cl}(t) + \delta q_i(t) \tag{B.5}$$

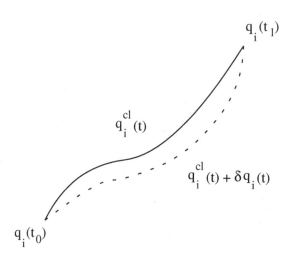

Fig. B.1. Variation of the path followed by a system, around the classical path. Note that the end-points are required to be fixed.

According to Hamilton's principle, the classical path is at an extremum, and therefore $\delta S = 0$,

$$\delta S = \int_{t_0}^{t_1} \left(\frac{\partial L}{\partial q_i} \delta q_i + \frac{\partial L}{\partial \dot{q}_i} \delta \dot{q}_i \right) dt = 0 \tag{B.6}$$

Noting that $\delta \dot{q}_i = d/dt \delta q_i$, integrating by parts, and using that $\delta q_i = 0$ at the end-points, this becomes

$$\delta S = \int_{t_0}^{t_1} \left(\frac{\partial L}{\partial q_i} - \frac{d}{dt} \frac{\partial L}{\partial \dot{q}_i} \right) \delta q_i dt = 0 \tag{B.7}$$

Since this has to be true for an arbitrary variation $\delta q_i(t)$ the only possibility is that the integrand in (B.7) vanishes: that is

$$\frac{\partial L}{\partial q_i} - \frac{d}{dt} \frac{\partial L}{\partial \dot{q}_i} = 0 \tag{B.8}$$

These are the equations of motion of the system, the *Euler Lagrange* equations. Since they are a system of second-order differential equations, the solutions are specified by giving both $\{q_i\}$ and $\{\dot{q}_i\}$ at a given time t_0.

The fundamental problem of classical mechanics is now reduced to the problem of finding the Lagrangian and solving the Euler Lagrange equations. As a simple example we look at a single particle of mass m moving in a potential $V(q)$. In this case the Lagrangian is given by [17]

$$L = \frac{1}{2}m\dot{q}^2 - V(q) \tag{B.9}$$

(according to the general rule $L = T - V$ where T is the kinetic and V the potential energy). The Euler Lagrange equation (B.8) then gives

$$m\ddot{q} = -\frac{\partial V(q)}{\partial q} \equiv F \tag{B.10}$$

which is nothing but Newton's law (F is the force).

We have treated coordinates and velocities in this Lagrangian formulation of classical mechanics. Sometimes we also require the *Hamiltonian* formulation, which involves coordinates and momenta. We therefore start by defining canonically conjugate momenta to $\{q_i\}$ by

$$p_i \equiv \frac{\partial L(q, \dot{q})}{\partial \dot{q}_i} \tag{B.11}$$

We assume that this relation is invertible so that the $\dot{q}_i$ can be expressed in terms of the q_i and p_i. We now form the *Hamiltonian* by

$$H \equiv \sum_i p_i \dot{q}_i - L(q, \dot{q}(p, q)) \tag{B.12}$$

We now compute how H is changed if we make a small change of the coordinates and momenta:

$$dH = \left[\dot{q}_i + \frac{\partial \dot{q}_j}{\partial p_i}\left(p_j - \frac{\partial L}{\partial \dot{q}_j}\right)\right]dp_i + \left[-\frac{\partial L}{\partial q_i} - \frac{\partial \dot{q}_j}{\partial q_i}\left(\frac{\partial L}{\partial \dot{q}_j} - p_j\right)\right]dq_i \tag{B.13}$$

Both the expressions in round brackets vanish because of (B.11), and by comparing (B.13) with the general expression

$$dH = \frac{\partial H}{\partial p_i}dp_i + \frac{\partial H}{\partial q_i}dq_i \tag{B.14}$$

we find

$$\left\{\dot{q}_i = \frac{\partial H}{\partial p_i}; \quad -\frac{\partial L}{\partial q_i} = \frac{\partial H}{\partial q_i}\right. \tag{B.15}$$

But according to (B.8) and (B.11), we can replace $\partial L/\partial q_i$ by $\dot{p}_i$ to arrive at the Euler Lagrange equations in the Hamiltonian formulation:

$$\left\{\dot{q}_i = \frac{\partial H}{\partial p_i}; \quad \dot{p}_i = -\frac{\partial H}{\partial q_i}\right. \tag{B.16}$$

Let us return to our simple example. We had

$$L = \frac{1}{2}m\dot{q}^2 - V(q) \tag{B.17}$$

which gives

$$p = \frac{\partial L}{\partial \dot{q}} = m\dot{q} \tag{B.18}$$

that is, $\dot{q} = p/m$. Then

$$H = p\dot{q} - L = \frac{p^2}{2m} + V(q) = T + V \tag{B.19}$$

The Euler Lagrange equations

$$\left\{ \dot{q} = \frac{\partial H}{\partial p} = \frac{p}{m}; \quad \dot{p} = -\frac{\partial H}{\partial q} = \frac{-\partial V}{\partial q} = F \right. \tag{B.20}$$

then give $m\ddot{q} = F$, as before.

B.2 Classical Fields

Let us consider an interesting application of our formalism. We look at a system of N particles, moving in one dimension, connected to each other by almost massless springs with force constant k (see Fig. B.2). The particles are separated from each other by a distance a in equilibrium. Let ϕ_i be the displacement from the equilibrium position of particle i. Then the Lagrangian is

$$L = \sum_i (T_i - V_i) = \frac{1}{2} \sum_i \left[m\dot{\phi}_i^2 - k(\phi_{i+1} - \phi_i)^2 \right] =$$

$$a \sum_i \frac{1}{2} \left[\frac{m}{a} \dot{\phi}_i^2 - ka \left(\frac{\phi_{i+1} - \phi_i}{a} \right)^2 \right] = \sum_i a\mathcal{L}_i \tag{B.21}$$

where $\mathcal{L}_i$ is the Lagrangian density (Lagrangian per unit length) contributed by particle i. We can now take the continuum limit, $i \to x$, $\sum_i \to \int dx$, $\phi_i \to \phi(x)$, $(\phi_{i+1} - \phi_i)/a \to \partial\phi/\partial x$, $m/a \to \rho$ (mass density) and $ka \to \kappa$ (Young's modulus). The continuum Lagrangian then becomes

$$L = \int dx \frac{1}{2} \left[\rho\dot{\phi}^2 - \kappa \left(\frac{\partial\phi}{\partial x} \right)^2 \right] \tag{B.22}$$

We can thus write the action

$$S = \int dt L = \int dt \int dx \mathcal{L} \left(\phi, \dot{\phi}, \frac{\partial\phi}{\partial x} \right) \tag{B.23}$$

with the Lagrangian density

$$\mathcal{L} \left(\phi, \dot{\phi}, \frac{\partial\phi}{\partial x} \right) = \frac{1}{2} \left[\rho\dot{\phi}^2 - \kappa \left(\frac{\partial\phi}{\partial x} \right)^2 \right] \tag{B.24}$$

Fig. B.2. A system of identical classical particles of mass m, connected by springs with force constant k. The distance between the particles in equilibrium is a.

In this example, $\phi(x, t)$ is the displacement *field*. A field is a function of space and time: that is, it contains an infinite number of degrees of freedom. (To completely specify a field, we have to give its value in every point in space, at every time.) The construction we made in this one-dimensional example is easy to generalize to three space dimensions. Then

$$S = \int dt L = \int dt \int d^3x \mathcal{L} \left(\phi, \dot{\phi}, \nabla \phi \right) \tag{B.25}$$

which gives rise to the Euler Lagrange equations

$$\sum_{k=1}^{3} \frac{\partial}{\partial x^k} \frac{\partial \mathcal{L}}{\partial (\partial \phi / \partial x^k)} + \frac{\partial}{\partial t} \frac{\partial \mathcal{L}}{\partial (\partial \phi / \partial t)} - \frac{\partial \mathcal{L}}{\partial \phi} = 0 \tag{B.26}$$

In fact, we notice that (B.26) can be written in the relativistically invariant form

$$\frac{\partial}{\partial x^\mu} \left[\frac{\partial \mathcal{L}}{\partial (\partial \phi / \partial x^\mu)} \right] - \frac{\partial \mathcal{L}}{\partial \phi} = 0 \tag{B.27}$$

This is Lorentz invariant if $\mathcal{L}$ is a scalar density (that is, if $\mathcal{L}'(x'^\mu) = \mathcal{L}(x^\mu)$).
As the first realistic example we consider a real scalar field $\varphi(x)$ (this can, for instance, be the so-called Higgs field of particle physics). The simplest non-trivial Lagrangian density is given by

$$\mathcal{L} = \frac{1}{2} \left[(\partial_\mu \varphi)(\partial^\mu \varphi) - \mu^2 \varphi^2 \right] \tag{B.28}$$

The Euler Lagrange equation (B.27) gives the equation of motion

$$\partial_\mu \partial^\mu \varphi + \mu^2 \varphi = 0 \tag{B.29}$$

or

$$(\Box + \mu^2)\varphi = 0 \tag{B.30}$$

which, as we remarked before, is a relativistic wave equation, the *Klein Gordon equation*.

There is also a Hamiltonian formulation of classical field theory. Corresponding to the generalized coordinate $\varphi(\mathbf{r}, t)$ at the space-time point $(t, \mathbf{r})$ there is a canonical momentum, $\pi(\mathbf{r}, t)$, defined by

$$\pi(\mathbf{r}, t) = \pi(\mathbf{r}, t) = \frac{\partial \mathcal{L}}{\partial \dot{\varphi}(\mathbf{r}, t)} \tag{B.31}$$

For the real scalar field with time-independent potential this gives

$$\pi(\mathbf{r}, t) = \frac{\partial \mathcal{L}}{\partial \dot{\varphi}(\mathbf{r}, t)} = \dot{\varphi}(\mathbf{r}, t) \tag{B.32}$$

The Hamiltonian density is now

$$\mathcal{H} = \pi \dot{\varphi} - \mathcal{L} = \frac{1}{2}[\pi^2 + (\nabla \varphi)^2 + \mu^2 \varphi^2] \tag{B.33}$$

By integrating this density over space, we obtain the full Hamiltonian, $H = \int_V d^3\mathbf{r}\mathcal{H}$, which produces the total energy of the field.

B.3 Relativistic Quantum Fields

An important application of scalar field theories is in condensed matter physics. It should be clear from the derivation of the classical scalar field theory in Section B.2 that it describes vibrations in a crystal: that is, sound waves. If we replace the speed of light with the speed of sound, we can use the scalar quantum field theory to be developed below to describe the quantized vibrations called phonons..

B.3.1 The Klein Gordon Field

When we quantize a point particle in elementary quantum mechanics, we treat the generalized coordinates q (in the simplest case, the three cartesian coordinates x_i) and momenta p_i as operators. They fulfil the quantum mechanical commutation relations

$$[x_i, p_j] = i\hbar \delta_{ij} \tag{B.34}$$

and the wave function Ψ is a representation of the state vector, on which these operators act.

The Klein Gordon equation, being a wave equation, should have plane-wave solutions. We can imagine that our system is enclosed in a large box of volume V, and we impose periodic boundary conditions (we will eventually take the limit $V \to \infty$, of course). We then expand the real scalar field φ at a given time t in terms of plane waves:

$$\varphi(\mathbf{r}, t) = \sum_{\mathbf{k}} \frac{1}{\sqrt{2V\omega}}[a_{\mathbf{k}}e^{-i(\omega t - \mathbf{k} \cdot \mathbf{r})} + a_{\mathbf{k}}^* e^{i(\omega t - \mathbf{k} \cdot \mathbf{r})}] \tag{B.35}$$

where we have introduced $1/\sqrt{2V\omega}$ as a convenient normalization, and where insertion into the Klein Gordon equation shows that $\omega = \omega(\mathbf{k}) = \sqrt{\mu^2 + \mathbf{k}^2}$, which describes the energy of a relativistic particle with mass μ and momentum $\mathbf{k}$. We can thus write the factors in the exponentials as $kx \equiv k_\mu x^\mu$ with $k_0 = \omega$.

Since the K-G field is written as a sum of independent components which all fulfil a wave equation – that is, have harmonic motion – it is natural to quantize each mode in the same way as we usually quantize the harmonic oscillator. By inserting (B.35) into the expression (B.33) for the Hamiltonian density and performing the integration over the whole volume V, we obtain

$$H = \sum_{\mathbf{k}} \hbar\omega(\mathbf{k}) a_{\mathbf{k}}^* a_{\mathbf{k}} \tag{B.36}$$

where we have temporarily reinserted a factor $\hbar$ to show the similarity with the expression of the energy of the quantum harmonical oscillator. Indeed, if we interpret $a_{\mathbf{k}}$ and $a_{\mathbf{k}}^* \rightarrow a_{\mathbf{k}}^\dagger$ as lowering and raising operators, fulfilling the commutation relations

$$\left[a_{\mathbf{k}}, a_{\mathbf{k}'}^\dagger\right] = \delta_{\mathbf{k},\mathbf{k}'} \tag{B.37}$$

then φ and π will fulfil the commutations relations (B.34). Re-computing the Hamiltonian using (B.37) one finds

$$H = \sum_{\mathbf{k}} \hbar\omega(\mathbf{k}) \left(a_{\mathbf{k}}^\dagger a_{\mathbf{k}} + \frac{1}{2}\right) \tag{B.38}$$

Usually, we will just ignore the constant contribution coming from summing the zero-modes (that is, the terms $\hbar\omega/2$). There are cases, however, when they should be kept, as we shall see later.

Now we can use all the machinery that we have learned when studying the harmonic oscillator in non-relativistic quantum mechanics. We define the ground state as the one annihilated by all $a_{\mathbf{k}}$: that is

$$a_{\mathbf{k}}|0\rangle = 0 \tag{B.39}$$

for all $\mathbf{k}$. A normalized state with $n_{\mathbf{k}}$ excitations in the mode $\mathbf{k}$ is given by

$$|n_{\mathbf{k}}\rangle = \frac{[a_{\mathbf{k}}]^{n_{\mathbf{k}}}}{\sqrt{n_{\mathbf{k}}!}} |0\rangle \tag{B.40}$$

Since H is a sum of non-interacting harmonic oscillators, the eigenstates are direct products:

$$|\ldots n_{\mathbf{k_i}}, \ldots n_{\mathbf{k_j}}, \ldots\rangle = \prod_{\mathbf{k_i}} |n_{\mathbf{k_i}}\rangle \tag{B.41}$$

The (enormous) Hilbert space spanned by all of these basis states is called a Fock space.

To produce a physical interpretation of the states (B.41), we note

$$H|\ldots n_{\mathbf{k_i}}, \ldots n_{\mathbf{k_j}}, \ldots\rangle = \sum_{\mathbf{k}} n_{\mathbf{k}} \epsilon(\mathbf{k})|\ldots n_{\mathbf{k_i}}, \ldots n_{\mathbf{k_j}} \ldots\rangle \qquad \text{(B.42)}$$

where $\epsilon(\mathbf{k}) = \hbar\omega(\mathbf{k})$. Thus we can interpret (B.41) as a state with many particles, each of mass m, where $n_{\mathbf{k_1}}$ have momentum $\mathbf{k_1}$, $n_{\mathbf{k_2}}$ have momentum $\mathbf{k_2}$, etc. This is also why the raising and lowering operators, $a_{\mathbf{k}}^{\dagger}$ and $a_{\mathbf{k}}$, are usually referred to as creation and annihilation operators: $a_{\mathbf{k}}^{\dagger}$ acting on the vacuum state creates one particle with wave vector k. With this formalism we will be able to treat processes where particles of given momenta are created or destroyed: for example, in collision processes.

B.3.2 Electromagnetic Field

The method used to quantize the real scalar field is very general. We can use it almost without change to quantize the electromagnetic field $A^{\mu}(\mathbf{r}, t)$. We saw in (2.82) how we could describe the classical four-vector potential field in the radiation gauge $k.A = 0$, $A^0 = 0$. It contains only two physical degrees of freedom for a given four-momentum k^{μ}. We introduced polarization vectors $\epsilon_{1,2}^{\mu}$ which are orthogonal to each other and to the direction of propagation $\mathbf{k}$. The Fourier expansion of A^{μ} can thus be written

$$A^{\mu}(\mathbf{r}, t) = \sum_{i=1,2} \sum_{\mathbf{k}} \frac{1}{\sqrt{2V\omega}} [\epsilon_i^{\mu} a_{i,\mathbf{k}} e^{-i(\omega t - \mathbf{k}\cdot\mathbf{r})} + \epsilon_i^{\mu*} a_{i,\mathbf{k}}^* e^{i(\omega t - \mathbf{k}\cdot\mathbf{r})}] \qquad \text{(B.43)}$$

We now let the Fourier coefficients become annihilation and creation operators, fulfilling

$$\left[a_{i\mathbf{k}}, a_{j\mathbf{k'}}^{\dagger}\right] = \delta_{\mathbf{k},\mathbf{k'}}\delta_{ij} \qquad \text{(B.44)}$$

The physical interpretation is thus that $a_{i\mathbf{k}}^{\dagger}$ creates a photon with polarization vector ϵ_1^{μ} and wave vector $\mathbf{k}$. Then Fock states can be built in exactly the same way as for the scalar field.

B.3.3 Charged Scalar Field

It turns out that we cannot describe electrically charged particles with a real scalar field. To do so, we must use a complex scalar field. The calculations are very similar, however. A suitable classical Lagrangian density is given by

$$\mathcal{L} = [(\partial_{\mu}\varphi^*)(\partial^{\mu}\varphi) - \mu^2|\varphi|^2] \qquad \text{(B.45)}$$

Treating φ and φ^* as independent fields, the Euler Lagrange equations become

$$\left(\Box + \mu^2\right)\varphi(x) = 0 \qquad \text{(B.46)}$$

(x now denotes x^{μ}) and

$$\left(\Box + \mu^2\right)\varphi^*(x) = 0 \qquad \text{(B.47)}$$

The Fourier expansion must now describe a non-hermitian field, and takes the form

$$\varphi(x) = \sum_{\mathbf{k}} \frac{1}{\sqrt{2V\omega}} [a_{\mathbf{k}} e^{-ikx} + b_{\mathbf{k}}^{\dagger} e^{ikx}] \tag{B.48}$$

and from the canonical commutation relations of the classical fields φ and φ^* one deduces

$$[a_{\mathbf{k}}, a_{\mathbf{k}'}^{\dagger}] = [b_{\mathbf{k}}, b_{\mathbf{k}'}^{\dagger}] = \delta_{\mathbf{k}\mathbf{k}'} \tag{B.49}$$

It thus appears that we have two types of 'particles', of the same mass μ, created by $a^{\dagger}$ and $b^{\dagger}$, respectively. When we couple electromagnetism to this complex scalar field, it can be seen that the a-particles and b-particles have opposite signs for the electric charge. This formalism can thus include (and in fact even predicts) the existence of antiparticles!

Coupling to electromagnetism is most easily performed by using the minimal coupling prescription $p^{\mu} \to p^{\mu} - eA^{\mu}$, where according to quantum mechanics $p^{\mu} \to i\partial^{\mu}$. Inserting this into (B.45) and using the expansions in terms of creation and annihilation operators both for φ and A^{μ}, we find interaction terms of many different types. When integrating over space-time, some of these terms disappear (due to energy and momentum conservation), but we obtain, for instance, a term which can destroy a photon, and create a positive and a negative scalar boson or vice versa. These terms carry a factor of e. From the squared terms in A^{μ} comes a contribution which corresponds to a coupling between two photons and two φ bosons at one point (a so-called seagull or contact term). This is proportional to e^2. These types of basic coupling form the basis of determining the Feynman rules of the theory. For scalar quantum electrodynamics (scalar QED) we thus have the Feynman diagrams shown in Fig. B.3.

In addition, the propagators which we encountered in Section 6.10 are given by the inverse (in Fourier space) of the quadratic forms in the fields. Thus, they are the Green's functions for the theory. For instance, the Klein Gordon equation

$$\left(\Box + m^2\right) \varphi(x) = i\delta^4(x) \tag{B.50}$$

can be inverted trivially, since the Fourier transform of the δ function is unity. Thus

$$P(k) = \frac{-i}{k^2 - m^2 + i\epsilon} \tag{B.51}$$

Here the imaginary part is added to define how to treat the propagator 'on-shell': that is, when the mass shell condition $k^2 = m^2$ is fulfilled. It turns out that by adding a small positive imaginary part we arrive at a state corresponding to a free particle with positive energy.

Since we have seen that the photon field in the radiation (or Feynman) gauge obeys the simple wave equation $\Box A^{\mu} = 0$ (2.81), it corresponds to the propagator

$$P^{\mu\nu}(k) = \frac{-ig^{\mu\nu}}{k^2 + i\epsilon} \qquad\qquad\qquad (B.52)$$

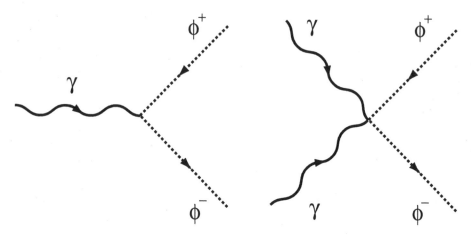

Fig. B.3. Feynman diagrams for the interaction between photons and a charged scalar field (scalar QED). The diagram on the left is proportional to e, and the one on the right is proportional to e^2.

B.4 Summary

- The Euler Lagrange equations for a field ϕ read

$$\frac{\partial}{\partial x^\mu}\left[\frac{\partial\mathcal{L}}{\partial(\partial\phi/\partial x^\mu)}\right] - \frac{\partial\mathcal{L}}{\partial\phi} = 0.$$

- For a scalar field, the simplest relativistically invariant Lagrangian density is

$$\mathcal{L} = \frac{1}{2}\left[(\partial_\mu\varphi)(\partial^\mu\varphi) - \mu^2\varphi^2\right].$$

  The equation of motion that follows from this Lagrangian is the Klein Gordon equation

$$(\Box + \mu^2)\varphi = 0$$

  a relativistic wave equation for a scalar (spinless) field of mass μ.
- Quantization is most easily performed by making a Fourier mode expansion, where each mode with wave vector $\mathbf{k}$ contributes independently to the Hamiltonian:

$$H = \sum_{\mathbf{k}} \hbar\omega(\mathbf{k})(a_{\mathbf{k}}^{\dagger}a_{\mathbf{k}} + \frac{1}{2})$$

- An electrically charged scalar field is best described by a complex-value ϕ with Lagrangian density

$$\mathcal{L} = \left[(\partial_{\mu}\varphi^{*})\,(\partial^{\mu}\varphi) - \mu^{2}|\varphi|^{2}\right]$$

The physical interpretation requires two types of particle, both of mass μ. The appearance of such antiparticles is a generic feature of relativistic quantum field theories.

C The Dirac Equation

C.1 Introduction

With the Friedmann-Lemaître-Robertson-Walker model of Chapter 4 we have a tool for studying the evolution in space and time of the Universe. Since $\dot{a} > 0$ today, and $\ddot{a} < 0$, it is clear that at an earlier time, conveniently chosen as $t = 0$, a was vanishingly small. This was the moment of the Big Bang, a singularity in the present models which we have no tools to handle (and which may not be possible to treat even using Einstein's theory of gravitation). Not only the energy density but also the temperature was singular in that limit, which means that the thermodynamics was undefined. By necessity, the description of the very first moments of time after the Big Bang involve elements of speculation. However, with accelerators, particle reactions up to TeV energies have been studied, which means that the particle reactions that took place in the early Universe after around 10^{-10} seconds after the Big Bang can be described by laboratory-tested physical laws.

During the first few seconds after the Big Bang the Universe went through many different phases, some of which have left traces that may still be studied today. We shall see that a thermodynamical treatment of early Universe physics shows that the temperature must have been extremely high: for example, around 10^{10} K one second a.b. (after the Bang), and 10^{15} K at the earliest time we can describe with tested laws of nature, 10^{-10} s a.b. At these high temperatures, the kinetic energies of particles were so high that particle production was possible, and the cosmic plasma contained many different types of particles in near equilibrium. In Chapter 8 we calculated how important quantities like number density, energy density and entropy depend on temperature, and how temperature has evolved with time in the Universe.

In the Universe today, only stable or very long-lived particles remain as relics of the Big Bang. However, at the earliest times all fundamental particles of nature were active players in cosmic evolution. We therefore need to familiarize ourselves with some basic facts about the modern theories of particles and fields, which is the topic of this appendix. Some particles, like neutrinos, may act as messengers from astrophysical processes and are well worth studying for that reason.

As the constituents of matter we know today all have spin 1/2 (in units of $\hbar$), in contrast to the force carriers (like the photon) which have spin 1 (and in the case of the hypothetical carrier of the gravitational force – the graviton – spin 2), it is worthwhile to first discuss some relevant features of the relativistic equation describing spin-1/2 particles. This is the Dirac equation.

The English physicist P.A.M. Dirac (1902-1984) was one of the founders of the modern quantum field theory. Many of his papers belong to the category of 'classical papers' today, and are still well worth reading. His most important contribution to physics is the relativistic equation for spin-1/2 particles (like the electron, the quarks, the neutrinos and so on) which bears his name. It is ironic that it was introduced partly for the wrong reasons, but the equation itself has stood the test of time. We will follow partly the historical path in which the equation was for quite some time regarded as a Schrödinger-like single-particle equation, before being seen as the equations of motion for a relativistic quantum field, similar to the scalar field $\varphi(x^\mu)$ that we treated in appendix B.3. It is amazing that for spin-1/2 particles (fermions) which obey the Pauli principle, one can in fact obtain a relativistic single-particle equation which makes sense, by postulating that all the negative-energy states, which caused the problems for the scalar field, are already occupied (forming the so-called 'Dirac sea'). The Pauli principle will then forbid positive-energy particles from making transitions to the negative-energy states in the Dirac sea. For many applications, there is therefore no need to use the full machinery of quantum field theory. We will thus treat Dirac particles similarly to how we treated the classical radiation field in elementary quantum mechanics (that is, before turning the classical vector potential field into a field operator). We will present a couple of simple examples with astrophysical applications using this approach. Towards the end of the appendix we shall, however, carry out the quantization of the Dirac field. Then you will have all the basic tools needed for a full course in quantum field theory.

C.2 Constructing the Dirac Equation

The Schrödinger equation for a free particle ($V(\mathbf{r}) = 0$) is based on the non-relativistic expression of kinetic energy

$$E_{kin} = \frac{\mathbf{p}^2}{2m} \tag{C.1}$$

which inserted into the energy equation $H = E_{kin}$ produces, with the prescription $H \to i\hbar\partial/\partial t$, $\mathbf{p} \to -i\hbar\nabla$,

$$\frac{-\hbar^2\nabla^2}{2m}\psi(\mathbf{r}, t) = i\hbar\frac{\partial\psi(\mathbf{r}, t)}{\partial t} \tag{C.2}$$

An important property of this equation, and the reason why we can interpret it as a single-particle equation, is that it implies a conserved current (the probability current). The continuity equation for this current can be derived by multiplying (C.2) from the left by ψ^*, its complex conjugate from the right by ψ, and subtracting:

$$\frac{\partial \rho}{\partial t} + \nabla \cdot \mathbf{j} = 0 \tag{C.3}$$

where the probability current is

$$\mathbf{j}(\mathbf{r}, t) = \frac{-i\hbar}{2m} \left(\psi^* \nabla \psi - \psi \nabla \psi^* \right) \tag{C.4}$$

and the probability density

$$\rho(\mathbf{r}, t) = |\psi(\mathbf{r}, t)|^2 \tag{C.5}$$

Thus, if we start with a particle described by a wave function normalized to unity, it will remain normalized during the time evolution.

We now check what will happen if we instead use the relativistic Klein Gordon equation (B.30)

$$\left(-\hbar^2 \nabla^2 + m^2 c^2 \right) \varphi = -\hbar^2 \frac{\partial^2 \varphi}{c^2 \partial t^2} \tag{C.6}$$

thus

$$\left(\Box + \mu^2 \right) \varphi = 0 \tag{C.7}$$

with $\mu = mc/\hbar$. We now set $\hbar = c = 1$, so that $\mu = m$. Multiplying (C.7) from the left by φ^*, the complex conjugate of (C.7) from the right by φ, and subtracting, we find

$$\frac{\partial \rho}{\partial t} + \nabla \cdot \mathbf{j} = 0 \tag{C.8}$$

with

$$\mathbf{j}(\mathbf{r}, t) = \frac{-i}{2m} \left(\varphi^* \nabla \varphi - \varphi \nabla \varphi^* \right) \tag{C.9}$$

and

$$\rho(\mathbf{r}, t) = i \left(\varphi^* \frac{\partial \varphi}{\partial t} - \varphi \frac{\partial \varphi^*}{\partial t} \right) \tag{C.10}$$

Although (C.8) looks like a continuity equation, it is not possible to interpret ρ in (C.10) as a probability density since it is not positive definite. For instance, a plane wave with time dependence e^{iEt} produces a negative value of ρ, whereas the time dependence e^{-iEt} does not have this problem. The solution is, as we saw in appendix B.3, to interpret φ as a quantum field whose excitations can be an arbitrary number of particles. Since the number of particles can change, there is no reason to have a single-particle equation. Also, the Hamiltonian of the quantum field (B.38) is in fact positive definite. This was not known to Dirac, however, and he tried another solution. The problem of the negative energies arises from the fact that the time derivative in (C.7) is of second order. In the Schrödinger equation, we have a first-order time derivative but a second order space derivative. Could we perhaps find a first-order equation? For a long time, this was thought to be impossible, because relativity theory demands that we treat space and time in a similar way, and a linear equation in space derivatives is not, for instance, invariant under space rotations, as is the Schrödinger equation.

The crucial step forward taken by Dirac was to introduce several wavefunction components into the theory and to find a system of first order differential equations in both space and time. Thus, he proposed the expression

$$H\psi = i\frac{\partial\psi}{\partial t} = -i\left(\alpha_1\frac{\partial}{\partial x^1} + \alpha_2\frac{\partial}{\partial x^2} + \alpha_3\frac{\partial}{\partial x^3}\right)\psi + \beta m\psi \tag{C.11}$$

where α and β are some constant $N \times N$ matrices to be determined, and ψ is a column vector with N components (as usual we do not write the unit matrix explicitly). The idea is that each of the components ψ_σ should satisfy the relativistic Klein Gordon equation $(\Box + m^2)\psi_\sigma = 0$. Multiplying (C.11) by $i\partial/\partial t$, and using the equation itself to replace the time derivative on ψ on the right-hand side, we find

$$-\frac{\partial^2\psi}{\partial t^2} = -M_{ij}\frac{\partial^2\psi}{\partial x^i \partial x^j} - iN_j m\frac{\partial\psi}{\partial x^j} + m^2\beta^2\psi \tag{C.12}$$

(remember that we use the summation convention), where

$$M_{ij} = \frac{\alpha_i\alpha_j + \alpha_j\alpha_i}{2} \tag{C.13}$$

and

$$N_i = \alpha_i\beta + \beta\alpha_i \tag{C.14}$$

We see that (C.12) becomes the diagonal Klein Gordon equation only if

$$M_{ij} = \delta_{ij} \tag{C.15}$$

and

$$N_i = 0 \tag{C.16}$$

thus, introducing the anticommutator $\{,\}$:

$$\{A, B\} \equiv AB + BA \tag{C.17}$$

$$\{\alpha_i, \alpha_j\} = 2\delta_{ij} \tag{C.18}$$

$$\{\alpha_i, \beta\} = 0 \tag{C.19}$$

In addition, we must have

$$\beta^2 = 1 \tag{C.20}$$

From (C.18) it also follows that

$$\alpha_i^2 = 1 \tag{C.21}$$

for all i.

Of course, we still have to show that there exist matrices fulfilling these relations. Also, we have to show that (C.12) is Lorentz invariant: that is, that it has the same form in all inertial frames. Some properties of the matrices are easy to derive. Since the square of any of them is the unit matrix, the eigenvalues have to be ± 1. From (C.19) we see that $\alpha_i = -\beta\alpha_i\beta$. Taking the trace of this equation and using $\text{Tr}(AB) = \text{Tr}(BA)$ it is seen that the trace of α_i has to be zero. In the same way, $\text{Tr}(\beta) = 0$ is proven. Since the trace is the sum of the eigenvalues, which are $+1$ or -1, we see that we must have an even number of dimensions. The Pauli matrices σ_i would almost fulfil the requirements. Since

$$\{\sigma_i, \sigma_j\} = 2\delta_{ij} \tag{C.22}$$

they could serve the role of α_i. However, the Pauli matrices span the space of 2×2 matrices together with the unit matrix $\sigma_0 = 1$. The unit matrix can not be used as β, however, since it commutes with all matrices in contradiction to the requirement (C.19). We must therefore consider 4×4 matrices. By trial-and-error, Dirac found a solution which uses the Pauli matrices as 2×2 blocks in the 4×4 matrices:

$$\alpha_i = \begin{pmatrix} 0 & \sigma_i \\ \sigma_i & 0 \end{pmatrix} \tag{C.23}$$

and

$$\beta = \begin{pmatrix} 1 & 0 \\ 0 & -1 \end{pmatrix} \tag{C.24}$$

The Dirac wave function Ψ can thus be considered as a column vector consisting of four components ψ_σ, $\sigma = 1, 2, 3, 4$. As usual, we define the hermitian conjugate wave function $\psi^\dagger\psi$ as a row vector with components $(\psi_1^*, \psi_2^*, \psi_3^*, \psi_4^*)$. Let us check that we can now form a positive-definite probability from $\psi^\dagger\psi$. We multiply (C.11) from the left by $\psi^\dagger$ and the hermitian conjugate of the equation from the right by ψ and subtract the two equations thus obtained. This gives (using the fact that both the α_i and β matrices are hermitian)

$$\frac{\partial\rho}{\partial t} + \nabla \cdot \mathbf{j} = 0 \tag{C.25}$$

where ρ is indeed the positive definite quantity $\psi^\dagger\psi$, and

$$\mathbf{j} = \psi^\dagger \alpha \psi \tag{C.26}$$

where we have assembled the 'component matrices' α_i to a 'vector of matrices' α.

To show Lorentz invariance, it is convenient to multiply (C.11) from the left by β, and rearrange the terms to obtain

$$i\left(\gamma^0 \frac{\partial}{\partial x^0} + \gamma^1 \frac{\partial}{\partial x^1} + \gamma^2 \frac{\partial}{\partial x^2} + \gamma^3 \frac{\partial}{\partial x^3}\right)\psi - m\psi = 0 \tag{C.27}$$

Here we have introduced the very convenient notation $\gamma^0 = \beta$ and $\gamma^i = \beta \alpha_i$. This allows us to write the anticommutation relations in the suggestive form

$$\{\gamma^\mu, \gamma^\nu\} = 2\eta^{\mu\nu} \tag{C.28}$$

In terms of block matrices,

$$\gamma^i = \begin{pmatrix} 0 & \sigma_i \\ -\sigma_i & 0 \end{pmatrix} \tag{C.29}$$

and

$$\gamma^0 = \begin{pmatrix} 1 & 0 \\ 0 & -1 \end{pmatrix} \tag{C.30}$$

The Dirac equation (C.27) can now be written in a form that looks manifestly Lorentz invariant:

$$(i\,\slashed{\partial} - m)\,\psi = 0 \tag{C.31}$$

where the 'slash' symbol will be used for an arbitrary contraction of a four-vector with the set of γ matrices, $\slashed{A} \equiv \gamma^\mu A_\mu$.

C.3 Plane-Wave Solutions

Let us consider solutions to the free Dirac equation (C.31) that can describe plane waves of positive energy and momentum $\mathbf{p}$, thus

$$\psi_{\mathbf{p}}(\mathbf{r}, t) = \frac{1}{\sqrt{2EV}} u(p) e^{-i(Et - \mathbf{p} \cdot \mathbf{r})} \tag{C.32}$$

where $u(p)$ is a four-column vector. Inserting this into (C.31) we find the Dirac equation in momentum space for $u(p)$:

$$(\slashed{p} - m)\,u(p) = 0 \tag{C.33}$$

In particular, we can ask for solutions at rest, $\mathbf{p} = 0$. Then only the $\gamma^0 \partial/\partial t$ term contributes, and we find with the representation (C.30) for the γ^0 matrix

$$-2m \begin{pmatrix} 0 & 0 & 0 & 0 \\ 0 & 0 & 0 & 0 \\ 0 & 0 & 1 & 0 \\ 0 & 0 & 0 & 1 \end{pmatrix} u(0) = 0 \tag{C.34}$$

which shows that we can take as basis states only

$$u^1(0) = \begin{pmatrix} 1 \\ 0 \\ 0 \\ 0 \end{pmatrix} \tag{C.35}$$

and

$$u^2(0) = \begin{pmatrix} 0 \\ 1 \\ 0 \\ 0 \end{pmatrix} \tag{C.36}$$

The two lowest components have to be zero to satisfy (C.34), but this leaves us with the problem that we only have a basis for the two-dimensional subspace spanned by the upper two components. Again, it is the negative energy states that have to be involved. If we try instead the plane-wave expression corresponding to negative energy (that is, we let the four-momentum $p \to -p$)

$$\psi_{\mathbf{p}}(\mathbf{r}, t) = \frac{1}{\sqrt{2EV}} v(-p) e^{+i(Et - \mathbf{p} \cdot \mathbf{r})} \tag{C.37}$$

then the Dirac equation at rest becomes

$$2m \begin{pmatrix} 1 & 0 & 0 & 0 \\ 0 & 1 & 0 & 0 \\ 0 & 0 & 0 & 0 \\ 0 & 0 & 0 & 0 \end{pmatrix} v(0) = 0 \tag{C.38}$$

which has solutions

$$v^1(0) = \begin{pmatrix} 0 \\ 0 \\ 1 \\ 0 \end{pmatrix} \tag{C.39}$$

and

$$v^2(0) = \begin{pmatrix} 0 \\ 0 \\ 0 \\ 1 \end{pmatrix} \tag{C.40}$$

(We will see later that the negative-energy solutions are related to antiparticles.) We can thus expand an arbitrary four-component Dirac-spinor at rest as

$$\psi(t) = e^{-iEt} \begin{pmatrix} \varphi \\ 0 \end{pmatrix} + e^{+iEt} \begin{pmatrix} 0 \\ \chi \end{pmatrix} \tag{C.41}$$

with φ and χ being two-component spinors.

Let us now check what happens when $\mathbf{p} \neq 0$. Then we can still write

$$u(p) = \begin{pmatrix} \varphi \\ \chi \end{pmatrix} \tag{C.42}$$

where φ and χ now will depend on $\mathbf{p}$. The Dirac equation was constructed to give $E = \pm p^0$ where we define p^0 to be the positive quantity $p^0 = +\sqrt{m^2 + \mathbf{p}^2}$. If we consider positive-energy solutions, inserting (C.42) into (C.33) implies

$$\chi = \frac{\sigma \cdot \mathbf{p}}{p^0 + m} \varphi \tag{C.43}$$

and therefore

$$u(p) = N \begin{pmatrix} \varphi \\ \frac{\sigma \cdot \mathbf{p}}{p^0+m} \varphi \end{pmatrix} \tag{C.44}$$

where N is a normalization constant. Similarly, the negative energy solutions $v(-p)$ are given by

$$v(-p) = N \begin{pmatrix} \frac{\sigma \cdot \mathbf{p}}{p^0+m} \chi \\ \chi \end{pmatrix} \tag{C.45}$$

The Dirac equation becomes, for these spinors

$$(\not{p} - m)\, u(p) = 0 \tag{C.46}$$

and

$$(\not{p} + m)\, v(-p) = 0 \tag{C.47}$$

respectively.

Let us define the conjugate Dirac spinor $\bar{\psi}$ by

$$\bar{\psi}(x) \equiv \psi^\dagger(x)\gamma^0 \tag{C.48}$$

and thus

$$\bar{u}(p) \equiv u(p)^\dagger \gamma^0 \tag{C.49}$$

and

$$\bar{v}(-p) \equiv v(-p)^\dagger \gamma^0 \tag{C.50}$$

Then using the identity

$$\gamma^0 \left(\gamma^\mu\right)^\dagger \gamma^0 = \gamma^\mu \tag{C.51}$$

which can be easily verified, we obtain for the conjugate spinors $\bar{u}$ and $\bar{v}$ the equations

$$\bar{u}(p)\, (\not{p} - m) = 0 \tag{C.52}$$

and

$$\bar{v}(-p)\, (\not{p} + m) = 0 \tag{C.53}$$

By multiplying (C.46) from the left by $\bar{u}\gamma^\mu$, (C.52) from the right by $\gamma^\mu u$, summing these equations and using the anticommutation relation

$$\{\gamma^\nu, \not{p}\} = 2p^\nu \tag{C.54}$$

we find

$$m\bar{u}(p)\gamma^\mu u(p) = p^\mu \bar{u}(p)u(p) \tag{C.55}$$

and

$$m\bar{v}(-p)\gamma^\mu v(-p) = -p^\mu \bar{v}(-p)v(-p) \tag{C.56}$$

Normalizing the two-spinors φ and χ to unity ($\varphi^\dagger \varphi = \chi^\dagger \chi = 1$), we choose to normalize the four-spinors not to unity but rather by the condition

$$\bar{u}(p)u(p) = 2m \tag{C.57}$$

$$\bar{v}(-p)v(-p) = -2m \tag{C.58}$$

which are relativistically covariant conditions. This means, for instance, that $u^\dagger u = 2E$. Then, according to (C.55) and (C.56)

$$\bar{u}(p)\gamma^\mu u(p) = 2p^\mu \tag{C.59}$$

and

$$\bar{v}(-p)\gamma^\mu v(-p) = 2p^\mu \tag{C.60}$$

This fixes the normalization constant N as $\sqrt{p^0 + m}$, so that

$$u^r(p) = \sqrt{p^0 + m}\left(\begin{array}{c} \varphi^r \\ \frac{\sigma \cdot \mathbf{p}}{p^0 + m}\varphi^r \end{array} \right) \tag{C.61}$$

where we choose the two independent two-spinor basis states to be

$$\varphi^1 = \begin{pmatrix} 1 \\ 0 \end{pmatrix} \tag{C.62}$$

and

$$\varphi^2 = \begin{pmatrix} 0 \\ 1 \end{pmatrix} \tag{C.63}$$

Similarly,

$$v^s(-p) = \sqrt{p^0 + m}\left(\begin{array}{c} \frac{\sigma \cdot \mathbf{p}}{p^0 + m}\chi^s \\ \chi^s \end{array} \right) \tag{C.64}$$

C.4 Coupling to Electromagnetism

The easiest way to introduce electromagnetism is as usual to make the minimal coupling substitution

$$p^\mu \to p^\mu - eA^\mu \tag{C.65}$$

where A^μ is the electromagnetic four-potential. Going back to the Dirac equation in Hamiltonian form (C.11), this gives

$$i\frac{\partial \psi}{\partial t} = [\alpha \cdot (\mathbf{p} - e\mathbf{A}) + \beta m + e\Phi]\,\psi \tag{C.66}$$

C.5 Lorentz Invariance

Although we have written the Dirac equation (C.31) in a way that looks Lorentz invariant:

$$(i\gamma^\mu \partial_\mu - m)\,\psi(x) = 0 \tag{C.67}$$

we have to demand that the Dirac matrices and the Dirac wave function transform in the required way. According to the relativity postulate, we have to find in a primed system

$$x^\mu \to x'^\mu = \Lambda^\mu{}_\nu x^\nu \tag{C.68}$$

(for example a Lorentz-boosted inertial frame) an equation of the same form:

$$\left(i\gamma'^\mu \partial'_\mu - m\right) \psi'(x') = 0 \tag{C.69}$$

($m' = m$ because the rest mass is itself an invariant quantity). Let us try to use the same constant Dirac matrices (that is, $\gamma'^\mu = \gamma^\mu$ – it can be proved that this is possible) but make a linear transformation of the Dirac spinor:

$$\psi'(x') = S\psi(x) \tag{C.70}$$

with inversion

$$\psi(x) = S^{-1}\psi'(x') \tag{C.71}$$

Then insertion into (C.31) gives

$$\left(i\gamma^\mu \Lambda^\nu{}_\mu \partial'_\nu - m\right) S^{-1}\psi'(x') = 0 \tag{C.72}$$

where we used (2.31). Multiplying from the left by S we find

$$\left(iS\gamma^\mu S^{-1}\Lambda^\nu{}_\mu \partial'_\nu - m\right)\psi'(x') = 0 \tag{C.73}$$

This is of the required form (C.69) if we can find an S such that

$$S\gamma^\mu S^{-1}\Lambda^\nu{}_\mu = \gamma^\nu \tag{C.74}$$

or

$$\Lambda^\mu{}_\nu \gamma^\nu = S^{-1}\gamma^\mu S \tag{C.75}$$

We now have to find S such that (C.75) is fulfilled. We first consider a Lorentz boost along the x^1 direction (see equation (2.22)), and write again for convenience

$$\beta = \tanh\zeta \tag{C.76}$$

Then

$$\Lambda(x^1;\zeta) = \begin{pmatrix} \cosh\zeta & -\sinh\zeta & 0 & 0 \\ -\sinh\zeta & \cosh\zeta & 0 & 0 \\ 0 & 0 & 1 & 0 \\ 0 & 0 & 0 & 1 \end{pmatrix} \tag{C.77}$$

and we see that for (C.75) to be fulfilled

$$S^{-1}\gamma^0 S = (\cosh\zeta)\gamma^0 - (\sinh\zeta)\gamma^1$$
$$S^{-1}\gamma^1 S = -(\sinh\zeta)\gamma^0 + (\cosh\zeta)\gamma^1$$
$$S^{-1}\gamma^2 S = \gamma^2 \tag{C.78}$$
$$S^{-1}\gamma^3 S = \gamma^3$$

$$\tag{C.79}$$

Using the anticommutation relations of the γ matrices, this can be summarized as (exercise: prove this!)

$$S^{-1}\gamma^\mu S = \left(\cosh\frac{\zeta}{2} + \gamma^0\gamma^1 \sinh\frac{\zeta}{2}\right)\gamma^\mu\left(\cosh\frac{\zeta}{2} - \gamma^0\gamma^1 \sinh\frac{\zeta}{2}\right) \tag{C.80}$$

which shows that

$$S = \cosh\frac{\zeta}{2} - \gamma^0\gamma^1 \sinh\frac{\zeta}{2} \tag{C.81}$$

The transformation matrices S for other Lorentz transformations can be found in a similar way, and can be summarized by

$$S = \cosh\frac{\zeta}{2} - \gamma^0\gamma^i \sinh\frac{\zeta}{2} \tag{C.82}$$

for a boost with boost parameter ζ along the x^i axis and

$$S = \cos\frac{\omega}{2} - \gamma^j\gamma^k \sin\frac{\omega}{2} \tag{C.83}$$

for a rotation of an angle ω around the x^i-axis ($i,j,k = 1,2,3$ cyclic).

In a similar way, it can be shown that the conjugate Dirac equation

$$i\partial_\mu\bar\psi(x)\gamma^\mu + m\bar\psi(x) = 0 \tag{C.84}$$

will be of the same form in a primed system provided that

$$\bar\psi'(x') = \bar\psi(x)S^{-1} \tag{C.85}$$

C.6 Bilinear Forms

To summarize, we have found the transformation matrix S that specifies how the four components of a Dirac spinor mix under a Lorentz transformation. Since S is not equal to Λ, the Dirac spinors are not four-vectors (in fact, the four spinor components live in an 'internal' space which is not space-time).

However, there is an interesting result from (C.75). Using the fact that ψ and $\bar\psi$ transform as (C.70) and (C.85) respectively, it is seen that

$$V^\mu(x) \equiv \bar\psi(x)\gamma^\mu\psi(x) \tag{C.86}$$

transforms as a four-vector:

$$V'^{\mu}(x') = \Lambda^{\mu}{}_{\nu}V^{\nu}(x) \tag{C.87}$$

Similarly, we see that with

$$s(x) \equiv \bar{\psi}(x)\psi(x) \tag{C.88}$$

$$s'(x') = s(x) \tag{C.89}$$

that is, $s(x)$ is a scalar quantity.

There exists an important 4×4 matrix, which is linearly independent from the set γ^{μ} of four matrices, but can be constructed from their product:

$$\gamma^5 \equiv i\gamma^0\gamma^1\gamma^2\gamma^3 \tag{C.90}$$

In our standard representation it becomes

$$\gamma^5 = \begin{pmatrix} 0 & 1 \\ 1 & 0 \end{pmatrix} \tag{C.91}$$

It satisfies

$$\left(\gamma^5\right)^2 = 1 \tag{C.92}$$

and

$$\{\gamma^5, \gamma^{\mu}\} = 0 \tag{C.93}$$

The γ^5 matrix plays an important role when discussing how Dirac spinors transform under parity. Let us see if we can find an operator P acting on a Dirac spinor that can represent the parity transformation of the space-time coordinates $t \to t$, $\mathbf{r} \to -\mathbf{r}$: that is, $x^{\mu} \to x'^{\mu} = (t, -\mathbf{r})$. The Lorentz transformation matrix which achieves this is

$$(\Lambda_P)^{\mu}{}_{\nu} = \begin{pmatrix} 1 & 0 & 0 & 0 \\ 0 & -1 & 0 & 0 \\ 0 & 0 & -1 & 0 \\ 0 & 0 & 0 & -1 \end{pmatrix} \tag{C.94}$$

which has $\det(\Lambda) = -1$ and is thus not a proper transformation (the latter is defined as one which can be continuously reached from the unit transformation). The requirement that the Dirac equation has the same form in the primed system, (C.75), is seen to be solved by

$$P = \eta_P\gamma^0 \tag{C.95}$$

where η_P is a phase factor.[1]

Since the γ^5 matrix anticommutes with γ^0,

$$P^{-1}\gamma^5 P = -\gamma^5 = \det(\Lambda_P)\gamma^5 \tag{C.96}$$

---

[1] A spin-1/2 particle can be shown to have the possible values ± 1 and $\pm i$. This means that in general it takes four reflections to bring the wave function back to its original value; similar in the case of rotations, where for fermions it takes a rotation through 4π radians.

For a general Lorentz transformation the γ^5 matrix thus transforms as

$$S^{-1}\gamma^5 S = \det(\Lambda)\gamma^5 \tag{C.97}$$

It therefore changes sign if we make a parity transformation $\mathbf{x} \to -\mathbf{x}$, (but $t \to t$) and remains unchanged for proper Lorentz transformations. It has the same properties as the box product $\mathbf{a} \cdot \mathbf{b} \times \mathbf{c}$ for three-vectors. The bilinear form

$$s^5(x) \equiv \bar{\psi}(x)\gamma^5\psi(x) \tag{C.98}$$

consequently behaves as a pseudoscalar quantity. The bilinear

$$A^\mu \equiv \bar{\psi}(x)\gamma^\mu\gamma^5\psi(x) \tag{C.99}$$

transforms under proper Lorentz transformations like a four-vector, but is a pseudovector under parity transformations. Thus,

$$A'^\mu(x') = \det(\Lambda)\,\Lambda^\mu{}_\nu A^\nu(x) \tag{C.100}$$

Let us compute how many 4×4 matrices we have now introduced. Each of the sets γ^μ and $\gamma^\mu\gamma^5$ contains four matrices. Together with γ^5 and the unit matrix we thus have 10 matrices. There should thus exist six more linearly independent matrices to span the 16-dimensional space of 4×4 matrices. A convenient choice of the remaining matrices is

$$\sigma^{\mu\nu} \equiv \frac{i}{2}[\gamma^\mu,\gamma^\nu] \tag{C.101}$$

Since this set is antisymmetric in μ and ν, it indeed contains six independent matrices. The bilinear constructed from $\sigma^{\mu\nu}$,

$$T^{\mu\nu}(x) \equiv \bar{\psi}(x)\sigma^{\mu\nu}\psi(x) \tag{C.102}$$

can be shown to transform like a rank-2 tensor (it is an antisymmetric tensor):

$$T'^{\mu\nu}(x') = \Lambda^\mu{}_\rho\Lambda^\nu{}_\sigma T^{\rho\sigma}(x) \tag{C.103}$$

C.7 Spin and Energy Projection Operators

Since the Dirac equation describes spin-1/2 particles, we should be able to find four-dimensional analogs to the 2×2 Pauli matrices σ^i. In fact, they are not difficult to find. If we define Σ^i by

$$\Sigma^i = \frac{1}{2}\epsilon^{ijk}\sigma_{jk} = \frac{i}{2}\epsilon^{ijk}\gamma_j\gamma_k \tag{C.104}$$

they have the form (in our usual representation of the γ matrices)

$$\Sigma^i = \begin{pmatrix} \sigma^i & 0 \\ 0 & \sigma^i \end{pmatrix} \tag{C.105}$$

The commutation relations for Σ^i are then the same as for σ^i

$$[\Sigma^i, \Sigma^j] = 2i\epsilon^{ijk}\Sigma_k \tag{C.106}$$

which means that $\Sigma^i/2$ satisfies the commutation relations of angular momentum. We also find trivially (since it is the same as for the σ matrices) that

$$\sum_i \left(\frac{\Sigma^i}{2}\right)^2 = \frac{3}{4} = \frac{1}{2}\left(\frac{1}{2}+1\right) \tag{C.107}$$

so that we are indeed dealing with a spin-1/2 particle.

In calculations it is sometimes helpful to use the formula

$$\Sigma^i = -\gamma^5\gamma^i\gamma^0 \tag{C.108}$$

which is easy to derive using the definition of γ^5. We have in (C.108) only defined the space-like part of the spin operator. To describe the spin operator in a covariant way we should find a four-vector that becomes $\Sigma^\mu = (0, \Sigma^1, \Sigma^2, \Sigma^3)$ in the rest frame. The four-dimensional generalization of ϵ^{ijk} is the four-dimensional Levi-Civita tensor $\epsilon^{\mu\nu\rho\sigma}$. Since the four-momentum vector is $p^\mu = (m, 0, 0, 0)$ in the rest frame we see that

$$\sigma^\mu = \frac{1}{2m}\epsilon^{\mu\nu\rho\sigma}p_\nu\sigma_{\rho\sigma} \tag{C.109}$$

is the operator we are looking for. To obtain the spin operator for a given direction $\hat{\mathbf{n}}$ we can similarly introduce a four-vector n^μ which has the value $n^\mu = (0, \hat{\mathbf{n}})$ in the rest frame. Then the covariant operator $n^\mu\Sigma_\mu$ will refer to this spin direction in any inertial frame. In particular, we can find the relativistic analog of the two-component spin-up and spin-down projection operators in the z-direction

$$P_\pm = \frac{1\pm\sigma_z}{2} \tag{C.110}$$

by introducing $n_z^\mu = (0, 0, 0, 1)$ and writing

$$P_\pm = \frac{1\pm\Sigma_\mu n_z^\mu}{2} = \frac{1\pm\gamma^5 \not{n}_z\gamma^0}{2} \tag{C.111}$$

Here we can write the γ^0 matrix as $\not{p}/m$, and use that $\not{p}$ acting on a free spinor produces $\pm m$. If we define the spinor $u^1(p)$ to have spin-up in the z-direction and $v^1(-p)$ to have spin-down, and vice versa for $u^2(p)$, $v^2(-p)$ (we will later see the motivation for this opposite assignment of spin), we can, for both types of spinor, use

$$P_\pm = \frac{1\pm\gamma^5 \not{n}_z}{2} \tag{C.112}$$

as spin projection operators.

We can also find projectors for positive and negative energies. The Dirac operator for positive-energy spinors $(\not{p}-m)$ gives zero on any spinor which has been multiplied by $(\not{p}+m)$ (see (C.46); use $p^2 = m^2$) and since the operator

$(-\not{p} + m)$ in a similar way gives negative-energy spinors (see C.47), we can expect them to be suitable projection operators, apart from normalization. Let us call the normalized projection operators Λ_- and Λ_+ respectively. Then we must have (as for all projection operators) $\Lambda_-^2 = \Lambda_-$, $\Lambda_+^2 = \Lambda_+$, $\Lambda_- + \Lambda_+ = 1$ and $\Lambda_- \Lambda_+ = \Lambda_+ \Lambda_- = 0$. We see that

$$\Lambda_+ = \frac{\not{p} + m}{2m} \tag{C.113}$$

and

$$\Lambda_- = \frac{-\not{p} + m}{2m} \tag{C.114}$$

have these required properties.

A related set of equations, often convenient to use in calculations, is

$$\sum_r u_\alpha^r(p)\bar{u}_\beta^r(p) = (\not{p} + m)_{\alpha\beta} \tag{C.115}$$

and

$$\sum_s v_\alpha^s(-p)\bar{v}_\beta^s(-p) = (-\not{p} + m)_{\alpha\beta} \tag{C.116}$$

Another set of projection operators which are important, for example, when describing the weak interactions of fermions, are the so-called chirality operators

$$P_L \equiv \frac{1 - \gamma^5}{2} \tag{C.117}$$

and

$$P_R \equiv \frac{1 + \gamma^5}{2} \tag{C.118}$$

They enable a Lorentz-invariant decomposition of an arbitrary Dirac spinor in left- and right-chirality components:

$$\psi_{L,R} \equiv P_{L,R}\psi \tag{C.119}$$

It can be shown that in the limit of zero mass, chirality is equal to helicity: that is, the projection of the spin onto the direction of motion. An eigenstate of P_L, for instance, has its spin oriented opposite to the direction of motion (and is said to describe a left-handed fermion). In the modern theory of elementary particles (quarks and leptons) it seems that the chiral states are the fundamental states of the theory. (For instance, the left- and right-chirality states may interact with different strength – for neutrinos it seems that the right-handed states do not interact at all!)

C.8 Non-Relativistic Limit

We now investigate the non-relativistic limit of the Dirac equation coupled to electromagnetism

$$i\frac{\partial\psi}{\partial t} = [\alpha\cdot(\mathbf{p} - e\mathbf{A}) + \beta m + e\Phi]\,\psi \tag{C.120}$$

If we start with a positive-energy solution the time-dependent Dirac spinor wave function is of the form

$$\psi = e^{-mt}\begin{bmatrix}\varphi\\\chi\end{bmatrix} \tag{C.121}$$

where we have taken out the large t dependence arising from the rest energy m, and where thus φ and χ should be slowly varying functions of t. Inserting this into (C.120) one finds

$$i\frac{\partial}{\partial t}\begin{bmatrix}\varphi\\\chi\end{bmatrix} = \sigma\cdot(\mathbf{p} - e\mathbf{A})\begin{bmatrix}\chi\\\varphi\end{bmatrix} + e\Phi\begin{bmatrix}\varphi\\\chi\end{bmatrix} - 2m\begin{bmatrix}0\\\chi\end{bmatrix} \tag{C.122}$$

Here we can approximate the solution to the lower equation by

$$\chi = \frac{\sigma\cdot(\mathbf{p} - e\mathbf{A})}{2m}\varphi \tag{C.123}$$

which, inserted into the upper equation, gives

$$i\frac{\partial\varphi}{\partial t} = \left(\frac{\sigma\cdot(\mathbf{p} - e\mathbf{A})\,\sigma\cdot(\mathbf{p} - e\mathbf{A})}{2m} + e\Phi\right)\varphi \tag{C.124}$$

which produces the correct Pauli term for a spin-1/2 particle. This is seen even more clearly by checking what the equation produces for the interaction with a weak magnetic field $\mathbf{B}$, in which case $\mathbf{A} = \frac{1}{2}\mathbf{B}\times\mathbf{r}$, expanding (C.124) to lowest order in $\mathbf{A}$:

$$i\frac{\partial\varphi}{\partial t} = \left(\frac{\mathbf{p}^2}{2m} - \frac{e}{2m}[\mathbf{L} + 2\mathbf{S}]\cdot\mathbf{B}\right)\varphi \tag{C.125}$$

Here we see that the total angular momentum $\mathbf{J}$ consists of two parts: the orbital angular momentum $\mathbf{L} = \mathbf{r}\times\mathbf{p}$ and a spin part $g\mathbf{S}$ with $S^i = \sigma^i/2$, the spin operator appropriate for a spin-1/2 particle, and the gyromagnetic ratio $g = 2$.

Thus, we see that the Dirac equation has the attractive property of being a relativistic equation with the correct non-relativistic limit to describe spin-1/2 particles like electrons.

If we want to compute the next-order corrections to (C.124), it is convenient to choose a slightly different basis for the Dirac matrix which makes the Hamiltonian diagonal to that order. A systematic scheme for carrying this out order by order was devised by Foldy and Wouthuysen in 1950. The idea is to transform

$$\psi \to \psi' = e^{iS}\psi \tag{C.126}$$

leading to

$$i\frac{\partial\psi'}{\partial t} = \left(e^{iS}\left[H - i\frac{\partial}{\partial t}\right]e^{-iS}\right)\psi' = H'\psi' \tag{C.127}$$

By using

$$H' = H + i[S, H] + \frac{(i)^2}{2!}[S, [S, H]] + \ldots \tag{C.128}$$

and choosing S so that the non-diagonal terms vanish, one finds after a straightforward though tedious calculation (see [7]) to second order:

$$H^{2nd} = \beta\left(m + \frac{(\mathbf{p} - e\mathbf{A})^2}{2m} - \frac{\mathbf{p}^4}{8m^3}\right) + e\Phi$$

$$- \frac{e}{2m}\beta\sigma \cdot \mathbf{B} - \frac{ie}{8m^2}\sigma \cdot (\nabla \times \mathbf{E})$$

$$- \frac{e}{4m^2}\sigma \cdot (\mathbf{E} \times \mathbf{p}) - \frac{e}{8m^2}\nabla \cdot \mathbf{E} \tag{C.129}$$

$$\tag{C.130}$$

In this formula one observes relativistic corrections to the kinetic energy and spin-orbit interactions. The last term (the so-called Darwin term) is specific to the Dirac equation and can be interpreted as a smearing of the electrostatic potential caused by the very fast vibrations ('zitter bewegung') of the electron at short distances.

C.9 Problems with the Dirac Equation

C.9.1 The Dirac Sea

We have seen that we can find a relativistic first-order linear wave equation which describes spin-1/2 particles such as electrons. The problem is that, as we have seen, it also allows negative-energy solutions. How shall we interpret them? Dirac found a suitable intermediate solution before quantum field theory for electrons was constructed. He postulated the existence of a 'sea' of electrons occupying all the negative-energy states. Since electrons obey the Pauli exclusion principle, there is no risk that positive-energy electrons will make transitions into this sea by radiating photons (gamma-rays) of huge energy $> 2m$, thus avoiding a catastrophic instability of the theory. However, the inverse process should be possible in this picture: by absorbing a gamma-ray of energy greater than $2m$, a negative-energy electron could be lifted to a positive-energy unoccupied state, leaving behind a *hole* in the Dirac sea (see Fig. C.1). If the sea electron had negative-energy $-E$, charge $-e$, and momentum $\mathbf{p}$, the new sea state has energy $+E$, charge $+e$ and momentum $-\mathbf{p}$ more than the original sea state. The hole will thus behave as a positive-energy particle with positive charge, with momentum $-\mathbf{p}$. It thus behaves as

a positron! The process where one negative-electron is lifted by a gamma-ray
to a positive-energy state can thus be interpreted as the creation of an e^+e^-
pair from the vacuum (allowed by the energy provided by the gamma-ray).
Indeed, if there is a hole already present, a positive-energy electron can fall
into that hole under emission of a gamma-ray. It is then equivalent to electron
positron annihilation.

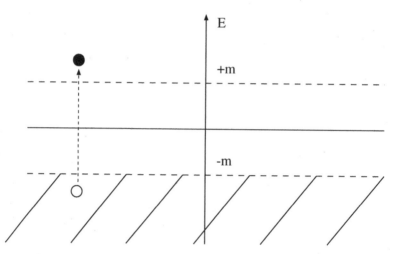

Fig. C.1. The energy spectrum of the free Dirac equation. Solutions with $E > m$
and $E < m$ are possible (there is a 'mass gap' of size $2m$). However, in the picture
of the Dirac sea, the negative-energy states are all occupied. By absorption of a
photon of energy greater than $2m$, one of the negative-energy electrons can be
lifted to an unoccupied positive-energy state, leaving a hole behind. The physical
interpretation of such a process is the production of an electron positron pair

How can we describe the presence of positrons (that is, the absence of
negative-energy electrons in the Dirac sea) by the Dirac equation? The key
point is that the charge is opposite. Therefore, since the negatively charged
electrons are described by

$$(i \, \partial\!\!\!/ - e \, A\!\!\!/ - m) \, \psi = 0 \tag{C.131}$$

we would like to find a 'charge conjugation' transformation on ψ such that
the transformed Dirac wave function ψ^c instead fulfils

$$(i \, \partial\!\!\!/ + e \, A\!\!\!/ - m) \, \psi^c = 0 \tag{C.132}$$

If we take the complex conjugate of (C.131) we find, using that A^μ is real-
valued,

$$\left[\left(i \frac{\partial}{\partial x^\mu} + e A^\mu \right) (\gamma^\mu)^* + m \right] \psi^* = 0 \tag{C.133}$$

This produces the possibility of obtaining (C.132) if we can find a matrix S_C with the property

$$S_C(\gamma^\mu)^* S_C^{-1} = -\gamma^\mu \tag{C.134}$$

It is a good exercise to check that the matrix

$$S_C = i\gamma_2 \tag{C.135}$$

fulfils this equation. Consequently, the charge conjugate Dirac field ψ^c can be represented by

$$\psi^c = i\gamma_2 \psi^* \tag{C.136}$$

If we introduce C by $S_C = C\gamma^0$, we see that we can also write

$$\psi^c = C\bar{\psi}^T \tag{C.137}$$

where we recall the definition (C.48) for $\bar\psi$.

Let us check how the charge-conjugation operator acts on our basis states. Consider a free negative-energy electron at rest with spin-down: that is

$$\psi \sim \begin{pmatrix} 0 \\ 0 \\ 0 \\ 1 \end{pmatrix} e^{+imt} \tag{C.138}$$

The charge-conjugated state will have the wave function

$$\psi^c = C\bar{\psi}^T = \begin{pmatrix} 1 \\ 0 \\ 0 \\ 0 \end{pmatrix} e^{-imt} \tag{C.139}$$

We see that ψ^c are the appropriate wave functions to use for the hole states, and we obtain the reasonable result that the absence of a spin-down negative-energy electron is equivalent to the presence of a spin-up positive-energy positron.

C.10 Central Potentials

For a central potential, the angular momentum operator

$$\mathbf{L} = \mathbf{r} \times \mathbf{p} \tag{C.140}$$

commutes with the Hamiltonian and can therefore be simultaneously diagonalized with it. For a given l we can couple the spin of an electron to a given $j = l \pm 1/2$. Let us see if we can do the same in the Dirac theory. With

$$H = \alpha \cdot \mathbf{p} + \beta m + V(r) \tag{C.141}$$

where $V(r)$ is a spherically symmetric potential, we find that

$$[H, \mathbf{L}] = -i\alpha \times \mathbf{p} \tag{C.142}$$

so $\mathbf{L}$ does in fact not commute with H. However, if we compute the commutator of H with the spin operator $\mathbf{\Sigma}$ (see (C.105)) we find

$$[H, \mathbf{\Sigma}] = 2i\alpha \times \mathbf{p} \tag{C.143}$$

so that the total angular momentum operator

$$\mathbf{J} = \mathbf{L} + \frac{1}{2}\mathbf{\Sigma} \tag{C.144}$$

commutes with the Hamiltonian. This is a general feature in quantum mechanics: rotational invariance of the potential is related to the conservation of total angular momentum.

C.11 Coulomb Scattering

We now compute the differential cross-section for the scattering of an electron on a heavy nucleus. If the electron energy is low compared to the mass of the nucleus, we can consider the latter as stationary. The effect of the nucleus can therefore be represented by a static Coulomb potential

$$V(r) = \frac{-Z\alpha}{r} \tag{C.145}$$

(we now drop the subscript on α_{em}).

To use the formalism of time-dependent perturbation theory, we should write the Hamiltonian as $H = H_0 + H_I$, where H_0 is the free Dirac Hamiltonian, and (see (C.141)) $H_I = V(r)$. The transition matrix element between initial momentum and spin p_i, r and final momentum and spin p_f, s caused by H_I is thus

$$M_{fi} = \langle p_f; s|H_I|p_i; r \rangle = \int d^3 r \psi_f^\dagger(\mathbf{r}) V(r) \psi_i(\mathbf{r}) \tag{C.146}$$

Here we insert our unperturbed plane-wave Dirac spinors:

$$\psi_i(\mathbf{r}) = \frac{1}{\sqrt{2E_i V}} u^{(r)}(p_i) e^{-(E_i t - \mathbf{p_i} \cdot \mathbf{r})} \tag{C.147}$$

$$\psi_f(\mathbf{r}) = \frac{1}{\sqrt{2E_f V}} u^{(s)}(p_f) e^{-(E_f t - \mathbf{p_f} \cdot \mathbf{r})} \tag{C.148}$$

where we have quantized in a large volume V, and where the four-spinor u is normalized such that $u^\dagger u = 2E$.

When inserting these expressions into (C.146) we find the integration over $d^3 r$ to give essentially the Fourier transform of the potential $V(r)$ (this is, of course, the Born approximation):

$$\int \frac{d^3 r}{r} e^{-i\mathbf{q} \cdot \mathbf{r}} = \frac{4\pi}{\mathbf{q}^2} \tag{C.149}$$

where $\mathbf{q} = \mathbf{p_f} - \mathbf{p_i}$. According to Fermi's golden rule, we should now sum over the possible final states of the same energy. In the momentum interval $d^3 p_f$ there are $V d^3 p_f/(2\pi)^3$ such states. Using the energy δ function $\delta(E_f - E_i)$, which follows from the time independence of the Coulomb potential, using $p_f dp_f = E_f dE_f$ (which follows from the energy equation $E^2 = p^2 + m^2$), we find after inserting the various factors in the golden rule (exercise: fill in the intermediate steps)

$$\frac{d\sigma}{d\Omega} = \frac{(Z\alpha)^2}{q^4} \left| u^{(s)\dagger}(p_f) u^{(r)}(p_i) \right|^2 \tag{C.150}$$

For given initial and final momenta and spins, we can compute the value of $|u^\dagger u|^2$ to obtain the differential cross-section. Often, however, we are interested in the unpolarized cross-section, which means that we take the average over the two possible initial spins and sum over the two possible final spins. Thus, we want to compute

$$\frac{d\sigma^{\text{unpol}}}{d\Omega} = \frac{(Z\alpha)^2}{q^4} \frac{1}{2} \sum_{r,s} \left| u^{(s)\dagger}(p_f) u^{(r)}(p_i) \right|^2 \tag{C.151}$$

We can do this by inserting all the four combinations of spins r, s and summing them separately. We will, however, introduce a technique which has also proven to be very useful in more complicated cases (and which is easy to implement using computer algebra). First, we note that (since $\bar{u} = u^\dagger \gamma^0$ and $(\gamma^0)^2 = 1$)

$$u^{(s)\dagger}(p_f) u^{(r)}(p_i) = \bar{u}^{(s)}(p_f) \gamma^0 u^{(r)}(p_i) \tag{C.152}$$

Also, for any 4×4 matrix Γ, we can write

$$\left| \bar{u}^{(s)}(p_f) \Gamma u^{(r)}(p_i) \right|^2 = \left(\bar{u}^{(s)}(p_f) \Gamma u^{(r)}(p_i) \right)^* \left(\bar{u}^{(s)}(p_f) \Gamma u^{(r)}(p_i) \right) \tag{C.153}$$

However,

$$\left(\bar{u}^{(s)}(p_f) \Gamma u^{(r)}(p_i) \right)^* = \left(u^{(r)\dagger}(p_i) \Gamma^\dagger \left[\bar{u}^{(s)}(p_f) \right]^\dagger \right)$$

$$= \left(\bar{u}^{(r)}(p_i) \tilde{\Gamma} u^{(s)}(p_f) \right) \tag{C.154}$$

with

$$\tilde{\Gamma} = \gamma^0 \Gamma^\dagger \gamma^0 \tag{C.155}$$

We now use

$$\sum_s u^{(s)}(p_f) \bar{u}^{(s)}(p_f) = \not{p}_f + m \tag{C.156}$$

so that (writing out the Dirac indices α and β – remember that we use the summation convention for them also)

$$\sum_{r,s} \left| u^{(s)\dagger}(p_f) u^{(r)}(p_i) \right|^2 = \sum_r \bar{u}_\alpha^{(r)}(p_i) \left(\gamma^0 \left[\slashed{p}_f + m \right] \gamma^0 \right)_{\alpha\beta} u_\beta^{(r)}(p_i) \quad \text{(C.157)}$$

Here we can move $\bar{u}_\alpha^{(r)}(p_i)$ to the far right, and using

$$\sum_r u^{(r)}(p_i) \bar{u}^{(r)}(p_i) = \slashed{p}_i + m \quad \text{(C.158)}$$

we see that the unpolarized spin sum can be written as

$$\sum_{r,s} \left| u^{(s)\dagger}(p_f) u^{(r)}(p_i) \right|^2 = \text{Tr} \left[\gamma^0 \left(\slashed{p}_f + m \right) \gamma^0 \left(\slashed{p}_i + m \right) \right] \quad \text{(C.159)}$$

C.12 Trace Formulae

We see that according to (C.159) an unpolarized cross-section can be written in terms of a trace of a product of matrices acting in the four-dimensional Dirac spinor space. Since this is a very general method to compute cross-sections, it is convenient to summarize some of the properties of such traces. We first note that

$$\text{Tr}(I) = 4 \quad \text{(C.160)}$$

where I is the 4×4 unit matrix. Also, we saw before that due to the anti-commutation relations

$$\text{Tr}(\gamma^\mu) = 0 \quad \text{(C.161)}$$

for all the four γ matrices. Also

$$\text{Tr}(\gamma^5) = 0 \quad \text{(C.162)}$$

as is obvious from the representation (C.91). We can use the fact that $(\gamma^5)^2 = I$ and that γ^5 anticommutes with all γ^μ to prove that the trace of any product of an odd number of γ matrices vanishes (here we do not write which index μ each γ matrix has):

$$\text{Tr}(\gamma_1 \gamma_2 \ldots \gamma_{2k+1}) = \text{Tr}(\gamma^5 \gamma^5 \gamma_1 \gamma_2 \ldots \gamma_{2k+1}) =$$
$$\text{Tr}(\gamma^5 \gamma_1 \gamma_2 \ldots \gamma_{2k+1} \gamma^5) = (-1)^{2k+1} \text{Tr}(\gamma_1 \gamma_2 \ldots \gamma_{2k+1}) = 0 \quad \text{(C.163)}$$

where in the second step we used the cyclic property of the trace, and in the final step we anticommuted one of the γ^5 matrices through all the $2k+1$ γ matrices, each step giving a factor of (-1).

For the product of two γ matrices, the result is easy to compute:

$$\text{Tr}(\gamma^\mu \gamma^\nu) = \text{Tr}(\gamma^\nu \gamma^\mu) = \frac{1}{2} \text{Tr}(\gamma^\mu \gamma^\nu + \gamma^\nu \gamma^\mu) = \eta^{\mu\nu} \text{Tr}(I) = 4\eta^{\mu\nu} \quad \text{(C.164)}$$

If a four-vector p_n is contracted with each γ matrix this formula can be written

$$\mathrm{Tr}\left(\not{p}_1 \not{p}_2\right) = 4\left(p_1 . p_2\right) \tag{C.165}$$

For an even number $n > 2$ we can use a recursion formula:

$$\mathrm{Tr}\left(\not{p}_1 \not{p}_2 \ldots \not{p}_n\right) = p_1 . p_2 \mathrm{Tr}\left(\not{p}_3 \not{p}_4 \ldots \not{p}_n\right)$$
$$-p_1 . p_3 \mathrm{Tr}\left(\not{p}_2 \not{p}_4 \ldots \not{p}_n\right) + \ldots + p_1 . p_n \mathrm{Tr}\left(\not{p}_2 \not{p}_3 \ldots \not{p}_{n-1}\right) \tag{C.166}$$

This formula is proven by using

$$\not{p}_1 \not{p}_2 = - \not{p}_2 \not{p}_1 + 2p_1 . p_2 \tag{C.167}$$

to move p_1 successively one step at a time to the right until it is last in the product. When that is the case we use the cyclic property to move it back to the first position, and then the formula follows.

An important example is for $n = 4$, which gives

$$\mathrm{Tr}\left(\not{p}_1 \not{p}_2 \not{p}_3 \not{p}_4\right) = 4\left(p_1 . p_2 p_3 . p_4 - p_1 . p_3 p_2 . p_4 + p_1 . p_4 p_2 . p_3\right) \tag{C.168}$$

It is easy to show that (exercise: do this, using the definition of γ^5)

$$\mathrm{Tr}\left(\gamma^5 \not{p}_1 \not{p}_2\right) = 0 \tag{C.169}$$

The trace of a product of γ^5 with four γ matrices is, however, non-zero:

$$\mathrm{Tr}\left(\gamma^5 \not{p}_1 \not{p}_2 \not{p}_3 \not{p}_4\right) = i\epsilon_{\mu\nu\rho\sigma} p_1^\mu p_2^\nu p_3^\rho p_4^\sigma \tag{C.170}$$

a rule that can be verified for one particular order of gamma matrices (for example, $\gamma^0\gamma^1\gamma^2\gamma^3$), and using the antisymmetry property of the result.

With this detour, we are now ready to compute the result (C.151), (C.159). It becomes

$$\frac{d\sigma^{\mathrm{unpol}}}{d\Omega} = \frac{(Z\alpha)^2}{2q^4} \mathrm{Tr}\left[\gamma^0 \not{p}_f \gamma^0 \not{p}_i + m^2 \left(\gamma^0\right)^2\right] \tag{C.171}$$

Using our rules for computing traces, this can now be evaluated:

$$\frac{d\sigma^{\mathrm{unpol}}}{d\Omega} = \frac{(Z\alpha)^2}{2q^4} \left[8E_f E_i - 4p_f . p_i + 4m^2\right] \tag{C.172}$$

Introducing spherical coordinates with the polar axis along the incident direction, one finds, with $\beta = v/c = p/E$

$$\frac{d\sigma^{\mathrm{unpol}}}{d\Omega} = \frac{(Z\alpha)^2}{4p^2\beta^2 \sin^4\left(\frac{\theta}{2}\right)} \left[1 - \beta^2 \sin^2\left(\frac{\theta}{2}\right)\right] \tag{C.173}$$

This is called the Mott cross-section, and reduces to the Rutherford formula in the non-relativistic limit $\beta \to 0$.

C.13 Quantization of the Dirac Field

So far, we have treated the Dirac field ψ as a classical field, without quantizing it. The first guess would be that to quantize the Dirac field we can use the canonical formalism and interpret the z coefficients of the plane-wave expansion as creation and annihilation operators. Since we have a complex field with four components ψ_α, we would need to introduce a_α and $b_\alpha^\dagger$ similarly to how we treated the complex Klein Gordon field in (B.48). However, this does not really work for the Dirac field, which as we have seen describes spin-1/2 particles. This is due to the requirement of Fermi statistics, which means that the Pauli principle has to be obeyed. For instance, a two-particle state must be antisymmetric in the exchange of the two particles. Also, we cannot have more than one particle in any state of given spin and momentum.

We have seen that in our description of scalar field quanta in terms of elementary excitations obtained by acting on the vacuum with creation operators $a_\mathbf{k}^\dagger$, we can obtain a state with n quanta of the same energy and momentum by acting n times with $a_\mathbf{k}^\dagger$. This is perfectly acceptable for bosons, but of course violates the Pauli principle for fermions. Moreover, if we describe a state with two particles of different momenta $\mathbf{k}$ and $\mathbf{k}'$ by acting on the vacuum state:

$$|0\ldots 1_\mathbf{k}\ldots 1_{\mathbf{k}'}\ldots\rangle = a_\mathbf{k}^\dagger a_{\mathbf{k}'}^\dagger |0\rangle \tag{C.174}$$

we see that it is symmetric when we exchange the particle labels:

$$a_\mathbf{k}^\dagger a_{\mathbf{k}'}^\dagger |0\rangle = a_{\mathbf{k}'}^\dagger a_\mathbf{k}^\dagger |0\rangle \tag{C.175}$$

because of the commutation relation

$$\left[a_\mathbf{k}^\dagger, a_{\mathbf{k}'}^\dagger \right] = 0 \tag{C.176}$$

If we want to describe fermions with creation operators $c_{r,\mathbf{k}}^\dagger$ and $c_{s,\mathbf{k}'}^\dagger$, we would instead expect antisymmetry:

$$c_{r,\mathbf{k}}^\dagger c_{s,\mathbf{k}'}^\dagger |0\rangle = -c_{s,\mathbf{k}'}^\dagger c_{r,\mathbf{k}}^\dagger |0\rangle \tag{C.177}$$

Furthermore, we should have

$$\left(c_{r,\mathbf{k}}^\dagger \right)^2 = 0 \tag{C.178}$$

for all r and $\mathbf{k}$, since we can only put one particle in each state. Both of these requirements can be met if we postulate that the creation and annihilation operators for fermions obey *anticommutation relations* instead of commutation relations:

$$\left\{ c_{r,\mathbf{k}}^\dagger, c_{s,\mathbf{k}'}^\dagger \right\} = \{ c_{r,\mathbf{k}}, c_{s,\mathbf{k}'} \} = 0 \tag{C.179}$$

Instead of

$$\left[a_{r,\mathbf{k}}, a_{s,\mathbf{k}'}^\dagger \right] = \delta_{rs}\delta_{\mathbf{k},\mathbf{k}'} \tag{C.180}$$

it is then also reasonable to postulate

$$\left\{ c_{r,\mathbf{k}}, c^{\dagger}_{s,\mathbf{k}'} \right\} = \delta_{rs}\delta_{\mathbf{k},\mathbf{k}'} \tag{C.181}$$

Indeed, it turns out that this way of quantizing the Dirac field is the correct one. To describe the quantized complex Dirac we thus write

$$\psi(\mathbf{r},t) = \sum_{r,\mathbf{p}} \frac{1}{\sqrt{2EV}} \left[c_{r,\mathbf{p}} u^{(r)}(p) e^{-ip_{\mu}x^{\mu}} + d^{\dagger}_{r,\mathbf{p}} v^{(r)}(-p) e^{ip_{\mu}x^{\mu}} \right] \tag{C.182}$$

where

$$\left\{ c^{\dagger}_{r,\mathbf{p}}, c^{\dagger}_{s,\mathbf{p}'} \right\} = \left\{ c_{r,\mathbf{p}}c_{s,\mathbf{p}'} \right\} = \left\{ d^{\dagger}_{r,\mathbf{p}}, d^{\dagger}_{s,\mathbf{p}'} \right\} = \left\{ d_{r,\mathbf{p}}d_{s,\mathbf{p}'} \right\} = 0 \tag{C.183}$$

and

$$\left\{ c_{r,\mathbf{p}}, c^{\dagger}_{s,\mathbf{p}'} \right\} = \left\{ d_{r,\mathbf{p}}, d^{\dagger}_{s,\mathbf{p}'} \right\} = \delta_{rs}\delta_{\mathbf{p},\mathbf{p}'} \tag{C.184}$$

C.14 Majorana Particles

We have assumed that the Dirac field is complex, which is indeed necessary to describe charged particles, as we saw also for the scalar Klein Gordon field. However, there are other spin-1/2 particles in nature which do not carry electric charge. Examples are neutrons and neutrinos. Are they their own antiparticles, and can they in that case be described by a real Dirac field? In the case of the neutron, the answer is 'no'. There is another type of charge, called the baryon number, which seems to be conserved to accuracy in nature, and which distinguishes a neutron from an antineutron.

In the case of neutrinos, the question is still open. In the Standard Model of particle physics, neutrinos carry a lepton number which also makes a difference between particles and antiparticles. However, in extensions of the model, neutrinos could be their own antiparticles and should then be described by a real field, called a Majorana field. Actually, whether the field is real depends on the choice of the representation of the γ matrices. There exists one representation: the Majorana representation, where the γ matrices are in fact real, and then the Dirac Majorana field in that representation is real. If we stick to our standard representation of γ matrices, the requirement (the so-called Majorana condition) is that the field is equal to its charge-conjugate: that is, (see (C.137))

$$\psi = \psi^c = C\bar{\psi}^T \tag{C.185}$$

with $C = i\gamma^0\gamma^2$. Since (check this using explicit expressions for u and v!)

$$u^{(s)}(p) = C\bar{v}^{(s)T}(-p) \tag{C.186}$$

and

$$v^{(s)}(-p) = C\bar{u}^{(s)T}(p) \tag{C.187}$$

it follows that the appropriate expansion of a Majorana field ψ_M is

$$\psi_M\left(\mathbf{r}, t\right) = \sum_{r,\mathbf{p}} \frac{1}{\sqrt{2EV}} \left[c_{r,\mathbf{p}} u^{(r)}\left(p\right) e^{-ip_\mu x^\mu} + c_{r,\mathbf{p}}^\dagger v^{(r)}\left(-p\right) e^{ip_\mu x^\mu} \right] \text{(C.188)}$$

Whether or not the neutrino is a Majorana particle can have observable consequences. Since a Majorana particle can have no conserved additive quantum number (like electric charge or lepton number), exotic decays of nuclei are possible which violate lepton number conservation. One example is neutrinoless double β decay, where a nucleus changes its charge by two units upon emission of two β particles (electrons). So far, none has been observed.

In so-called supersymmetric theories, Majorana fermions play a crucial role. In particular, the supersymmetric partner of the photon, which can be stable and perhaps make up the dark matter in the Universe, is a Majorana particle.

C.15 Lagrangian Formulation

You now have all the basic tools for computing quantities in quantum field theory. What you will learn in courses on that subject is how to carry out perturbation theory (and sometimes even estimate non-perturbative effects) in a Lorentz invariant way. It has turned out to be very convenient to use the Lagrangian formulation of field theory, since that makes Lorentz invariance manifest (and also various other symmetries that may exist in the theory).

It is not difficult to find a Lagrangian density which has the Dirac equation as its Euler Lagrange equations of motion. If we consider the complex Dirac field, we should treat ψ_α and ψ_α^* as independent variables, just as we treated φ and φ^* as independent variables for the complex scalar field. Since ψ^* and $\bar{\psi}$ are the same degrees of freedom (linearly related by the matrix γ^0), and $\bar{\psi}$ is more useful when constructing Lorentz covariant bilinears, we prefer to use ψ_α and $\bar{\psi}_\alpha$ as independent variables. You should check that the Lagrangian density

$$\mathcal{L} = \bar{\psi}\left(x\right) \left[i\, \partial\!\!\!/ - m\right] \psi\left(x\right) \tag{C.189}$$

produces the Dirac equation. From this we can also verify that the Hamiltonian has the form that we expect from (C.11). The canonically conjugate momenta are

$$\pi_\alpha = \frac{\partial \mathcal{L}}{\partial \dot{\psi}_\alpha} = i\psi_\alpha^\dagger. \tag{C.190}$$

and

$$\bar{\pi}_\alpha = \frac{\partial \mathcal{L}}{\partial \dot{\bar{\psi}}_\alpha} \equiv 0 \tag{C.191}$$

The Hamiltonian is then

$$H = \int d^3r \bar{\psi}(x) \left[-i\gamma^j \frac{\partial}{\partial x^j} + m \right] \psi(x) \tag{C.192}$$

Inserting the quantized Dirac field (C.182) and performing the integration, one finds with the use of the anticommutation relations (excluding the zero-mode contribution)

$$H = \sum_{r,\mathbf{p}} E(\mathbf{p}) \left[c_{r,\mathbf{p}}^\dagger c_{r,\mathbf{p}} + d_{r,\mathbf{p}}^\dagger d_{r,\mathbf{p}} \right] \tag{C.193}$$

and the total electric charge

$$Q = \int d^3r j^0 = \int d^3r \bar{\psi}\gamma^0\psi = -e \sum_{r,\mathbf{p}} \left[c_{r,\mathbf{p}}^\dagger c_{r,\mathbf{p}} - d_{r,\mathbf{p}}^\dagger d_{r,\mathbf{p}} \right] \tag{C.194}$$

These relations show that although negative energy states appear in the solutions to the Dirac equation, the total energy associated with the quantized complex Dirac field is positive definite. Also, although the energy contributed by the 'd-particles' is positive, the charge has the opposite sign to that of the 'c-particles'. Thus, the natural interpretation is that the theory contains positive energy electrons and positive energy positrons.

Since the Dirac equation is first-order in the space-time derivative, there are a couple of differences when determining the Feynman rules for spinor QED as opposed to scalar QED (Section B.3). First, since this means that after minimal coupling the interaction Lagrangian is also linear in the electro-magnetic field A^μ, there are no contact terms in the Feynman rules. Second, the propagator obtained from inverting the Green's function of the Dirac equation in Fourier space: that is, by solving

$$(i\gamma^\mu \partial_\mu - m)\psi(x) = i\delta^4(x) \tag{C.195}$$

is proportional to $1/p$ for large p instead of $1/p^2$ which is the case for bosons. Formally, we find from (C.195) for the Dirac propagator

$$P(p) = \frac{i}{\not{p} - m + i\epsilon} \tag{C.196}$$

where this expression should be interpreted as

$$P(p) = \frac{i(\not{p} + m)}{p^2 - m^2 + i\epsilon} \tag{C.197}$$

C.16   Summary

- The free Dirac equation for a spin $1/2$ particle of mass m reads

$$i\left(\gamma^0 \frac{\partial}{\partial x^0} + \gamma^1 \frac{\partial}{\partial x^1} + \gamma^2 \frac{\partial}{\partial x^2} + \gamma^3 \frac{\partial}{\partial x^3} \right) \psi - m\psi = 0$$

- The 4×4 matrices $\{\gamma^{\mu}\}$ fulfil the anticommutation relations

$$\{\gamma^{\mu}, \gamma^{\nu}\} = 2\eta^{\mu\nu}$$

- Plane-wave solutions are of the form

$$\psi_{\mathbf{p}}(\mathbf{r}, t) = \frac{1}{\sqrt{2EV}} u(p) e^{-i(Et - \mathbf{p} \cdot \mathbf{r})}$$

which have positive energy, or

$$\psi_{\mathbf{p}}(\mathbf{r}, t) = \frac{1}{\sqrt{2EV}} v(-p) e^{+i(Et - \mathbf{p} \cdot \mathbf{r})}$$

of negative energy.

- The four-spinors u and v fulfil the equations

$$(\not{p} - m) u(p) = 0$$

and

$$(\not{p} + m) v(-p) = 0$$

- Explicit forms of the spinor functions are

$$u^r(p) = \sqrt{p^0 + m} \begin{pmatrix} \varphi^r \\ \frac{\sigma \cdot \mathbf{p}}{p^0 + m} \varphi^r, \end{pmatrix}$$

and

$$v^s(-p) = \sqrt{p^0 + m} \begin{pmatrix} \frac{\sigma \cdot \mathbf{p}}{p^0 + m} \chi^s \\ \chi^s \end{pmatrix}$$

Here ϕ^r and χ^s are two-spinors of the Pauli type, and r and s are spin indices.

- Coupling to electromagnetism is performed through the minimal coupling prescription

$$p^{\mu} \rightarrow p^{\mu} - eA^{\mu}$$

- Quantization of the Dirac field is performed through anticommutation instead of commutation relations between the field variable and its canonical momentum.

- By imposing the so-called Majorana condition

$$\psi = \psi^c = C\bar{\psi}^T$$

on the solutions of the Dirac equation, half of the degrees of freedom are eliminated. What remains is a self-conjugate field which can describe Majorana particles: neutral spin-1/2 particles which are their own antiparticles.

- A Lagrangian density which has the Dirac equation as its equation of motion is

$$\mathcal{L} = \bar{\psi}(x) [i \not{\partial} - m] \psi(x)$$

D Cross-Section Calculations

D.1 Definition of the Cross-Section

In a quantum mechanical system, transitions between different quantum states are described by transition matrix elements, the squares of which define the probability for the transitions to occur. Formally, for an initial state i and a final state f

$$P_{i \to f} = |w_{fi}|^2 = |\langle f | i \rangle|^2 \tag{D.1}$$

The main complication when we deal with free particles in the initial and/or final state is that due to symmetries of the problem (for example, rotational and translational invariance) there is a large degree of degeneracy involved. This means, for instance, that the transition amplitude for scattering is not determined unless we specify boundary conditions for the incoming and outgoing states. The usual assumption is that the initial state is part of a Hilbert space $\mathcal{H}_{in}$, which is characterized by the fact that we have at time $t \to -\infty$ plane waves which converge to a region where scattering takes place, and the outgoing states belong to an $\mathcal{H}_{out}$ with plane waves at $t \to +\infty$ which diverge from the region of scattering. However, these two Hilbert spaces should really mean the same: there should exist a unitary matrix transferring between the two. The S matrix is defined by

$$w_{fi} = \langle f, \, out | i, \, in \rangle = \langle f, \, in | S | i, \, in \rangle = S_{fi} \tag{D.2}$$

Translational invariance means that the overall four-momentum is conserved. Thus, (D.2) should be non-zero only if $p_i = p_f$, meaning that we should be able to extract a δ-function $\delta^4(p_f - p_i)$. Also, one possibility is that the incoming plane waves just pass each other and no scattering occurs. This uninteresting part is just a unit matrix. We can thus define the transition matrix, or T matrix, by the expression

$$S_{fi} = \delta_{fi} + i \, (2\pi)^4 \, \delta^4 \, (p_f - p_i) \, T_{fi} \tag{D.3}$$

To each Feynman diagram there is a set of rules, the Feynman rules, from which the T-matrix elements can be obtained. We use Lorentz-invariant normalization of the plane waves, which means

$$\langle \mathbf{p} | \mathbf{p}' \rangle = 2E \, (2\pi)^3 \, \delta^3 \, (\mathbf{p} - \mathbf{p}') \tag{D.4}$$

or $2E$ particles per volume element in $\mathbf{r}$-space. Corresponding to this normalization of states there is a Lorentz invariant measure in momentum space

$$d\mu = \frac{d^3p}{2E\,(2\pi)^3} \tag{D.5}$$

When we insert (D.2) and (D.3) into (D.1) we run into the problem of 'squaring the δ function'. This is related to the use of plane waves and can be solved by performing the integral which defines the δ function

$$(2\pi)^4\,\delta^4\,(p) = \int dt d^3r e^{-i(p_0 t - \mathbf{p}\cdot\mathbf{r})} \tag{D.6}$$

only over a finite (but large) time T and volume V. We see by inserting $p = 0$ in this equation that formally $\delta^4(0) = VT$. Of course, when doing this, we should make sure that our final results do not depend on the auxiliary quantities V and T.

Let us treat the most common process of practical application, namely the collision of two particles giving n particles in the final state, $a + b \rightarrow c_1 + c_2 + \ldots + c_n$.

For simplicity, assume that the particles of type b are at rest, and the velocity of the particles a is $v = |\mathbf{p}_a|/E_a$. The number of particles of type b per target volume is, due to our normalization, $2E_b = 2m_b$ (since b is at rest, it only has rest energy). The incident flux is the velocity of the particles a times their number density $2E_a$: that is $2|\mathbf{p}_a|$. If the reaction volume is V and the reactions take place during the time T, we should obtain the transition rate per unit time and unit volume as the target density times the incident flux times the cross-section σ (this defines σ): that is, $2m_b \cdot 2|\mathbf{p}_a| \cdot \sigma$. However, it should also be given by $|w_{fi}|^2/(VT)$. Equating the two, and summing over all available momenta for the final state using the invariant measure $d\mu$, we thus obtain the 'master formula' for computing cross-sections

$$\sigma(a + b \rightarrow c_1 + \ldots + c_n) =$$

$$\frac{1}{4m_b|\mathbf{p}_a|} \int d\mu_1 \ldots d\mu_n\,(2\pi)^4\,\delta^4\,(p_a + p_b - p_1 - \ldots - p_n)\,|\widetilde{T}|^2 \tag{D.7}$$

Here we usually have to take into account that the colliding particles are unpolarized, and also that we are not interested in the final particle polarizations; so we have defined an effective square of the T-matrix element by

$$|\widetilde{T}|^2 = \frac{1}{S}\frac{1}{(2s_a + 1)\,(2s_b + 1)} \sum_{\text{final spins}} |\langle f|T|i\rangle|^2 \tag{D.8}$$

Here s_a and s_b are the spins of the initial state particles.[1] We have also introduced a symmetry factor S, which is needed due to the impossibility in quantum mechanics of distinguishing between two final states where the

---

[1] The photon has spin 1, but since it is massless it has only two polarization states, not three.

only difference is the exchange of identical particles. There is thus a risk of overcounting the number of available final states. In the cases we will treat we will have at most two identical particles in the final state. For that case $S = 2$. In general, if there are k groups of n_i ($i = 1, 2, \ldots, k$) identical particles in the final state, $S = n_1! n_2! \ldots n_k!$

The formula (D.7) still has the drawback that it is not manifestly Lorentz invariant. However, it can be made so by writing (Problem 6.4)

$$m_b |\mathbf{p}_a| = \sqrt{(p_a \cdot p_b)^2 - m_a^2 m_b^2} \tag{D.9}$$

An additional advantage with this form is that it can be also used for $m_b = 0$, which is the case for photons. Notice that in our system of units the cross-section has dimensions of $(\text{mass})^{-2}$, a fact which is very useful when checking results of computations.

For decay, $a \rightarrow c_1 + c_2 + \ldots + c_n$, the same formulae apply with the only exception being $4m_b |\mathbf{p}_a| \rightarrow 2m_a$ (in this case, the natural frame is the rest frame), and $(2s_a + 1)(2s_b + 1) \rightarrow (2s_a + 1)$. The quantity thus obtained is called the partial decay rate $\Gamma_{a \rightarrow n}$. Summing the partial decay rates over all possible decay channels one obtains the total decay rate Γ, which is related to the average lifetime τ by

$$\tau = \frac{1}{\Gamma} \tag{D.10}$$

Thus, an ensemble of N_0 particles at rest at time $t = 0$ will decay according to the law

$$N(t) = N_0 e^{-\Gamma t} \tag{D.11}$$

Sometimes the term branching ratio for a certain decay channel n is used. This is simply defined as

$$B.R. (a \rightarrow n) = \frac{\Gamma_n}{\Gamma} \tag{D.12}$$

and can be interpreted as the probability that a given particle of type a decays into the final state n. We now proceed to compute the cross-section of some interesting processes.

D.2 The Process $e^+e^- \rightarrow \mu^+\mu^-$

As a first example, we compute the cross-section for $e^+e^- \rightarrow \mu^+\mu^-$, which we estimated on dimensional grounds in Section 6.10.1. Let us call the initial four-momenta p_-, p_+ and the final state momenta k_-, k_+.

We need the Feynman rules for the process seen in Fig. 6.6. It breaks down into the vertices of the type fermion photon fermion (where the fermion is e in the first vertex and μ in the second vertex), and a photon propagator. With the minimal coupling prescription and the Lagrangian density for a

Dirac field given in Appendix C, it is seen that the coupling at a vertex is of
the form

$$V = ie \left(\bar{\psi} \gamma_\mu \psi \right) A^\mu \tag{D.13}$$

Here e is the elementary charge, fulfilling

$$e^2 = 4\pi\alpha \tag{D.14}$$

in our units, with $\alpha \simeq 1/137$ the fine-structure constant. The Feynman rule
can more or less be read off from this, giving $l_\mu = e\bar{v}(-p_+)i\gamma_\mu u(p_-)$, where
the appearance of $v(-p_+)$ can be shown using methods of quantum field
theory, but can also be inferred from the Dirac hole theory (a positron of
momentum p_+ is equivalent to a negative energy spinor of momentum $-p_+$).
If we instead had considered scattering of an electron off a muon, the Feynman
rule would have given $\bar{u}(p'_-)i\gamma_\mu u(p_-)$.

For the muon photon muon vertex, we similarly obtain

$$L_\nu = \bar{u}(k_-)i\gamma_\nu v(-k_+) \tag{D.15}$$

The photon propagator is a tensor function $P^{\mu\nu}(s)$ which connects the two
vertices. Its appearance depends on the gauge chosen for the computation. A
very convenient choice is the Feynman gauge, in which

$$P^{\mu\nu}(s) = \frac{-ig^{\mu\nu}}{s} \tag{D.16}$$

Here s is the usual Mandelstam variable, $s = (p_- + p_+)^2 = (k_- + k_+)^2$.

We can thus write the T-matrix element

$$T(e^+e^- \rightarrow \mu^+\mu^-) = -i\frac{l_\mu L^\mu}{s} \tag{D.17}$$

Since the electron and muon spinors act in different spinor spaces, we see
that the matrix element is factorized. We now have to take the initial spin
average, and sum over the final state spins. As is explained in the section
on trace formulae in Appendix C, such a spin sum becomes a trace in the
respective spinor space,

$$|\tilde{T}|^2 = \frac{1}{4} l_{\mu\nu} L^{\mu\nu} \tag{D.18}$$

where $1/4$ is from the initial spin average, and

$$l_{\mu\nu} = Tr \left(\gamma_\mu (\not{p}_- + m_e)\gamma_\nu (\not{p}_+ - m_e) \right) \tag{D.19}$$

and a similar expression for $L_{\mu\nu}$, with obvious substitutions of momenta and
masses. Using the trace formulae, it is not difficult to compute

$$l^{\mu\nu} = 4 \left[p_-^\mu p_+^\nu + p_-^\nu p_+^\mu - \left(p_- . p_+ + m_e^2 \right) g^{\mu\nu} \right] \tag{D.20}$$

and a corresponding expression for $L_{\mu\nu}$.

The contraction of $l_{\mu\nu}$ with $L^{\mu\nu}$ will generate Lorentz invariant scalar
products which can be expressed in terms of s, t and u (see Section 2.4.1).

For a two-to-two process and unpolarized particles only two of these are independent. Since the usual situation is that the initial energy, and therefore s, is given, there is only one independent variable, which we take to be t. To obtain the differential cross-section $d\sigma/dt$ from our master formula we need only introduce a δ function $\delta(t - (p_- - k_-)^2)$ in the phase-space integral. This makes it possible to perform this integral, and by using the formulae of Section 2.4.3,

$$\frac{d\sigma}{dt} = \frac{|\widetilde{T}|^2}{16\pi\lambda\left(s, m_a^2, m_b^2\right)} \tag{D.21}$$

(in our case, $m_a = m_b = m_e$). The integration limits for the t variable were given in (2.60). At this point, it is an excellent exercise for the dedicated student to assemble the results obtained so far and perform the t-integration to obtain

$$\sigma(e^+ e^- \to \mu^+ \mu^-) = \frac{2\pi\alpha^2}{s} v \left(1 - \frac{v^2}{3}\right) \tag{D.22}$$

where the only approximation made is to neglect m_e (this is allowed, since $m_e^2/m_\mu^2 \ll 1$). Here v is the velocity of one of the out-going muons in the centre-of-momentum frame, $v = \sqrt{1 - 4m_\mu^2/s}$.

D.3 The Process $\bar{\nu}_e e^- \to \bar{\nu}_\mu \mu^-$

Now we have the machinery to compute the cross-section $\bar{\nu}_e e^- \to \bar{\nu}_\mu \mu^-$ with small modifications. One change is that antineutrinos always have positive helicity (spin projection on the direction of motion), so there is no average over initial spin for the antineutrino. On the other hand, the fact that only right-handed antineutrinos take part in the interactions means that there is a helicity projection $P_L = (1 - \gamma_5)/2$ in the vertex (again, we can regard the positive energy right-helicity antineutrino as a negative energy negative-helicity hole). The W boson propagator, which replaces the photon propagator can, for $s \ll m_W^2$, be approximated by

$$P_W^{\mu\nu} \sim \frac{ig^{\mu\nu}}{m_W^2} \tag{D.23}$$

and the electromagnetic coupling e is replaced by $g_{weak} = e/\sin\theta_W$, see (6.21). Again neglecting the fermion masses, a very similar calculation as for $e^+ e^- \to \mu^+ \mu^-$ gives

$$\sigma\left(\bar{\nu}_e e^- \to \bar{\nu}_\mu \mu^-\right)_{m_\mu^2 \ll s \ll m_W^2} = \frac{g_{weak}^4 s}{96\pi m_W^4} \tag{D.24}$$

D.4 The Processes $ee\gamma\gamma$

We now consider the electromagnetic processes, which are perhaps the most important processes in astrophysics. They are $\gamma + e^{\pm} \to \gamma + e^{\pm}$, $e^{+} + e^{-} \to \gamma\gamma$ and $\gamma + \gamma \to e^{+} + e^{-}$. They are all $2 \to 2$ processes which differ only in the combination of initial and final state particles. Therefore, the calculation of the T matrix elements is essentially one and the same for all processes. (In fact, there are systematic so-called crossing rules for how to go from one amplitude to another by just changing the direction of four-momenta in the expressions for the amplitude.)

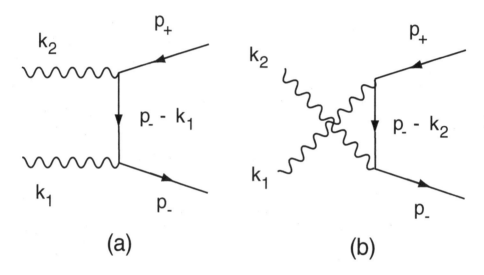

Fig. D.1. Feynman diagrams for the process $\gamma + \gamma \to e^{+}e^{-}$. The only difference between (a) and (b) is the order in which the photons are attached to the fermion vertices.

Let us indicate how the calculation is carried out for $\gamma + \gamma \to e^{+}e^{-}$. A new feature is that to conserve charge in the process, and using only the fermion photon fermion vertex (which is the only fundamental interaction in QED), we have to have a fermion propagator in the process (see Fig. D.1 (a)). There is also an ambiguity in which order to attach the two photons in the graph, and according to the basic principles of quantum mechanics we must add the one in Fig. D.1 (b) coherently to the one in Fig. D.1 (a). For the external spinors we use, as before, $\bar{u}(p_{-})$ and $v(p_{+})$, and the wave function for a free plane-wave photon of four-momentum k is just one of the constant polarization vectors $\epsilon_{1,2}^{\mu}(k)$ encountered in Section 2.6. (They are transverse: that is, $k_{\mu}\epsilon_{1,2}^{\mu} = 0$.) The propagator for a fermion of mass m and four-momentum p is given by

$$S_F(p; m) = i\frac{\not{p} + m}{p^2 - m^2} \tag{D.25}$$

This means that we can write the amplitude as

$$T = \bar{u}(p_-)\mathcal{F}_{\mu\nu}v(-p_+)\epsilon_1^\mu \epsilon_2^\nu \tag{D.26}$$

with

$$\mathcal{F}_{\mu\nu} = -e^2 \left[\gamma_\mu S_F(p_- - k_1; m_e)\gamma_\nu + \gamma_\nu S_F(p_- - k_2; m_e)\gamma_\mu \right] \tag{D.27}$$

The sum over initial photon polarizations is very simple in the Feynman gauge, which gives

$$\sum_{i=1,2} \epsilon_i^\mu \epsilon_i^\nu = -g^{\mu\nu} \tag{D.28}$$

The sum over final state electron and positron polarizations again produces a trace. This time there are quite a few terms of products of up to six γ matrices, so it is advisable to perform this calculation using a symbolic manipulation program.[2] However, once this is done the result may again be expressed as a function of s and t ($s > 4m_e^2$ is of course required for kinematical reasons), and inserted into (D.21). The result is

$$\sigma\left(\gamma\gamma \to e^+e^-\right) = \frac{\pi\alpha^2}{2m_e^2}\left(1 - v^2\right) \times$$
$$\left[\left(3 - v^4\right) \ln\left(\frac{1+v}{1-v}\right) + 2v\left(v^2 - 2\right) \right] \tag{D.29}$$

Here $v = \sqrt{1 - 4m_e^2/s}$ is the velocity of one of the out-going fermions in the centre of momentum system.

The computation of the reverse process $e^+e^- \to \gamma\gamma$ is very similar, and gives

$$\sigma\left(e^+e^- \to \gamma\gamma\right) = \frac{\pi\alpha^2\left(1 - v^2\right)}{2vm_e^2}\left[\frac{3 - v^4}{2v}\ln\left(\frac{1+v}{1-v}\right) - 2 + v^2\right] \tag{D.30}$$

Note the similarity with (D.29), especially after taking into account that for $e^+e^- \to \gamma\gamma$, a symmetry factor $S = 2$ has to be divided out because of the presence of two identical particles in the final state.

As the final example, we consider Compton scattering $\gamma+e^- \to \gamma+e^-$ (see Fig. D.2 (a) and (b)). Usually, there is an incoming photon beam of energy ω which hits electrons at rest. For scattering by an angle θ with respect to the incident beam, the out-going photon energy ω' is given by energy-momentum conservation to be

$$\omega' = \frac{m_e\omega}{m_e + \omega\left(1 - \cos\theta\right)} \tag{D.31}$$

---

[2] The first full calculation of this type was performed by Klein and Nishina in 1928. Klein later described how they spent several summer weeks doing the calculation by hand, comparing at the end of each day the intermediate results.

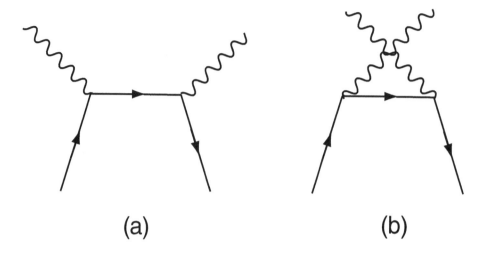

Fig. D.2. Feynman diagrams for the process $\gamma + e^- \rightarrow \gamma + e^-$. The only difference between (a) and (b) is the order in which the photons are attached to the fermion vertices.

In this frame, the unpolarized differential cross-section, as first computed by Klein and Nishina, is

$$\frac{d\sigma}{d\Omega} = \frac{\alpha^2}{2m_e^2} \left(\frac{\omega'}{\omega}\right)^2 \left[\frac{\omega'}{\omega} + \frac{\omega}{\omega'} - \sin^2\theta\right] \tag{D.32}$$

We can integrate this over the scattering angle to obtain

$$\sigma(\gamma + e \rightarrow \gamma + e) =$$
$$\frac{\pi\alpha^2 (1 - v)}{m_e^2 v^3} \left[\frac{4v}{1 + v} + \left(v^2 + 2v - 2\right) \ln\left(\frac{1 + v}{1 - v}\right) - \frac{2v^3 (1 + 2v)}{(1 + v)^2}\right] \tag{D.33}$$

where v is now the incoming electron velocity in the centre of momentum frame, $v = (s - m_e^2)/(s + m_e^2)$.

E Quantum Fluctuations of the Inflaton

E.1 Quantum Fields in General Relativity

We shall now investigate how a single scalar field, the inflaton field in the simplest models for inflation, evolves during a cosmological inflationary epoch of rapid expansion. It may be instructive to first read Appendix B, where quantization of a scalar field is carried out for a static Minkowski background. We now want to generalize this to a general space-time metric $g_{\mu\nu}(x)$, where $x = x^\mu$ are the four space-time coordinates. Defining

$$g \equiv \det g_{\mu\nu}(x), \tag{E.1}$$

we can make a Lagrangian that is consistent with general relativity and which has the correct flat-space limit by replacing (B.28) by

$$\mathcal{L}(x) = \frac{1}{2}\sqrt{-g}\left(g^{\mu\nu}(x)\partial_\mu\varphi\partial_\nu\varphi(x) - [m^2 + \xi R(x)]\varphi^2(x)\right). \tag{E.2}$$

Here we have multiplied by $\sqrt{-g}$ which is a Jacobian factor which makes $\mathcal{L}(x)$ a scalar density and thus the action

$$S = \int \mathcal{L}(x)d^4x \tag{E.3}$$

a scalar. For a scalar field, the general relativistic covariant derivative is just the ordinary partial derivative, and the only difference with (B.28) is that we use the metric $g^{\mu\nu}(x)$ to contract the derivatives. The mass of the field is m, but there is an additional quadratic term possible $\xi R(x)\varphi^2$, which vanishes in Minkowski space (where the Ricci scalar $R(x) = 0$), but which may exist here. If $\xi = 0$ we say that the scalar field is minimally coupled to gravity, if $\xi = 1/6$ it turns out that the Lagrangian is invariant under so-called conformal rescalings of the metric,

$$g_{\mu\nu}(x) \to \Omega^2(x)g_{\mu\nu}(x), \tag{E.4}$$

if $m = 0$ and φ is chosen to transform as $\varphi(x) \to \Omega^{-2}(x)\varphi(x)$. Without losing generality for our application, we shall use the minimally coupled case here. The potential terms for the inflaton, have been taken into account through their influence on the scale factor, which enters the metric. (Therefore this analysis also applies to any other kind of scalar field during inflation.) We will later see the changes caused by the slow roll of φ. To get inflation, we need a patch of the Universe where the average value of the field is homogenous, so the metric is of the FLRW type

$$g_{\mu\nu}(x) = \text{diag}(1, -a^2(t), -a^2(t), -a^2(t)), \tag{E.5}$$

so that

$$\sqrt{-g} = a^3(t). \tag{E.6}$$

Now we can use the general relativistic version of the Euler Lagrange equations (B.27)

$$\frac{\partial}{\partial x^\mu}\left[\frac{\partial\sqrt{-g}\mathcal{L}}{\partial(\partial\phi/\partial x^\mu)}\right] - \frac{\partial\sqrt{-g}\mathcal{L}}{\partial\phi} = 0 \tag{E.7}$$

to derive the equation of motion for the field φ,

$$\left(\Box_g + m^2\right)\varphi(x) = 0, \tag{E.8}$$

where

$$\Box_g\varphi(x) = g^{\mu\nu}(x)D_\mu D_\nu\varphi(x) = \frac{1}{\sqrt{-g}}\partial_\mu[\sqrt{-g}g^{\mu\nu}(x)\partial_\nu\varphi(x)]. \tag{E.9}$$

Equation (E.8) is the general relativistic Klein Gordon equation. In the diagonal metric (E.5) we obtain

$$\ddot{\varphi}(x) + 3H\dot{\varphi}(x) - \frac{\nabla^2\varphi(x)}{a^2} + m^2\varphi(x) = 0. \tag{E.10}$$

E.2 Evolution in de Sitter Space-Time

We can now evaluate how the field evolves for different behaviour of the scale factor $a(t)$. As we saw in Chapter 10, the generic inflation involves a nearly constant Hubble parameter, and an exponential scale factor

$$a(t) = a_0 e^{H(t-t_0)}, \tag{E.11}$$

where $H = \dot{a}/a \sim const$ during inflation. This means that the de Sitter space-time is a good approximation. We can simplify the analysis somewhat by using instead of the cosmic time t the conformal time η (see (4.34)), $dt = ad\eta$ so that the FLRW metric is the same as the Minkowski metric apart from a conformal factor

$$ds^2 = a^2(\eta)[d\eta^2 - d\mathbf{r}^2]. \tag{E.12}$$

In our case, we can derive how the scale factor depends on conformal time by computing

$$\eta = \int^t \frac{dt}{a(t)} = -\frac{1}{a_0 H e^{H(t-t_0)}} + c = -\frac{1}{a(\eta)H} + c \tag{E.13}$$

Choosing the constant $c = 0$, we find

$$a(\eta) = -\frac{1}{\eta H}, \tag{E.14}$$

with $0 < a < \infty$ obtained in the interval $-\infty < \eta < 0$.

To quantize the field, we have to find its canonical momentum (B.31) (from now on, in this section the dot means derivative with respect to conformal time)

$$\pi = \frac{\partial \mathcal{L}}{\partial \dot{\varphi}} = a^2(\eta)\dot{\varphi} \tag{E.15}$$

and impose the canonical quantization condition at equal conformal time

$$[\varphi, \pi]_\eta = a^2(\eta)[\varphi(\eta, \mathbf{r}), \dot{\varphi}(\eta, \mathbf{r}')] = i\delta^3(\mathbf{r} - \mathbf{r}') \tag{E.16}$$

(there should really be a factor $\hbar$ on the right hand side, but as usual we are putting it to unity).

Now we insert a mode expansion for the field, similar to (B.35)

$$\varphi(\eta, \mathbf{r}) = \int \frac{d^3k}{(2\pi)^{\frac{3}{2}}} \left[a_{\mathbf{k}} \phi_{\mathbf{k}}(\eta) e^{i\mathbf{k}\cdot\mathbf{r}} + a_{\mathbf{k}}^\dagger \phi_{\mathbf{k}}^*(\eta) \right] e^{-i\mathbf{k}\cdot\mathbf{r}}. \tag{E.17}$$

To fulfil the quantization condition (E.16), we then have to demand

$$a^2(\eta) \left(\phi_{\mathbf{k}} \dot{\phi}_{\mathbf{k}}^* - \phi_{\mathbf{k}}^* \dot{\phi}_{\mathbf{k}} \right) = i. \tag{E.18}$$

All the unknown physics which comes from the expansion of the Universe now has been put into the functions $\phi_{\mathbf{k}}(\eta)$. In the Minkowski case, we would have followed a similar procedure, writing $\varphi(t, \mathbf{r}) = \phi_{\mathbf{k}}(t) e^{\pm i\mathbf{k}\cdot\mathbf{r}}$, and finding by using the equations of motion

$$\ddot{\phi}_{\mathbf{k}}(t) + [k^2 + m^2]\phi_{\mathbf{k}}(t) = 0, \tag{E.19}$$

which has the well-known solutions

$$\phi_{\mathbf{k}}(t) = c_+\phi_{\mathbf{k}}^+(t) + c_-\phi_{\mathbf{k}}^-(t) = c_+ e^{-i\omega_{\mathbf{k}} t} + c_- e^{i\omega_{\mathbf{k}} t}, \tag{E.20}$$

where $\omega_{\mathbf{k}} = \sqrt{k^2 + m^2}$. Here the term proportional to c_+ has positive energy, and we expand the field as

$$\varphi(t, \mathbf{r}) = \int \frac{d^3k}{(2\pi)^{\frac{3}{2}}} \left[a_{\mathbf{k}} \phi_{\mathbf{k}}^+(t) e^{i\mathbf{k}\cdot\mathbf{r}} + a_{\mathbf{k}}^\dagger \phi_{\mathbf{k}}^-(t) e^{-i\mathbf{k}\cdot\mathbf{r}} \right]. \tag{E.21}$$

E.3 The Vacuum State

A vacuum state of the theory $|0\rangle$ is obtained by demanding $a_{\mathbf{k}}|0\rangle = 0$ for all values of $\mathbf{k}$, which means that no positive energy particles are present in any of the Fourier modes. Since the Minkowski space is time independent, the vacuum state, which if Lorentz invariant, remains the same empty state forever.

In an expanding Universe, however, the situation is different. Through the expansion there is a dependence on the scale factor in all our expressions. Even if we demand that all the annihilation operators annihilate a state at one time, the time evolution means that at another time this is no longer the case. Particles may be spontaneously produced in an expanding Universe! Also, as we will see, the evolution of a particular vacuum state, called the adiabatic or Bunch Davies vacuum, is such that after the inflationary period quantum fluctuations have survived in the form of nearly scale-invariant adiabatic density perturbations, which is just what is observed in the microwave background.

Let us go back to (E.17) and write $\phi_{\mathbf{k}}(\eta) = \chi_{\mathbf{k}}(\eta)/a(\eta)$. The equation of motion (E.8) for our de Sitter space background then becomes

$$\ddot{\chi}_{\mathbf{k}} + \left(k^2 + \frac{m^2}{H^2\eta^2} - \frac{\ddot{a}}{a} \right) \chi_{\mathbf{k}} = 0, \tag{E.22}$$

where in the de Sitter case,

$$\frac{\ddot{a}}{a} = \frac{2}{\eta^2} \tag{E.23}$$

The equation (E.22) then has well-known solutions in terms of Hankel functions,

$$\chi_{\mathbf{k}}(\eta) = \sqrt{-\eta} \left(c_1 H_\nu^{(1)}(-k\eta) + c_2 H_\nu^{(2)}(-k\eta) \right), \tag{E.24}$$

where $H_\nu^{(1)*} = H_\nu^{(2)}$ and $\nu^2 = 9/4 - m^2/H^2$. How shall we choose the vacuum state? We want to identify the positive energy states $\chi_{\mathbf{k}}^+(\eta)$ and chose the 'in' vacuum as one which is annihilated by all positive energy (or positive frequency) annihilation operators. Using the property of the Hankel functions

$$H_\nu^{(1)}(x \gg 1) \to \sqrt{\frac{2}{\pi x}} e^{i(x - \nu\frac{\pi}{2} - \frac{\pi}{4})} \tag{E.25}$$

we see that the asymptotic behaviour for $k\eta \to -\infty$, that is an 'in state' corresponding to very early times, or large k which means small distances, is

$$\chi_{\mathbf{k}}(k\eta \to -\infty) \to \sqrt{\frac{2}{\pi k}} \left(c_1 e^{-ik\eta} + c_2 e^{ik\eta} \right). \tag{E.26}$$

The 'adiabatic' or Bunch Davies vacuum is now obtained by choosing as positive frequency states those with $c_2 = 0$ and $c_1 = \sqrt{\pi}/2$ so that

$$\chi_{\mathbf{k}}^+(\eta \to -\infty) \to \frac{1}{\sqrt{2k}} e^{-ik\eta}, \tag{E.27}$$

and the vacuum state is the one which is annihilated by all annihilation operators of these positive frequency plane waves.[1]

The inflaton field can then finally be expanded as

$$\varphi = \frac{\sqrt{-\pi\eta}}{2a(\eta)} \int \frac{d^3k}{(2\pi)^{\frac{3}{2}}} \left[a_{\mathbf{k}} H_\nu^{(1)}(-k\eta)e^{i\mathbf{k}\cdot\mathbf{r}} + a_{\mathbf{k}}^\dagger H_\nu^{(2)}(-k\eta)e^{-i\mathbf{k}\cdot\mathbf{r}} \right], \quad \text{(E.28)}$$

where the normalization is such that (E.16) is fulfilled.

E.4 Connection to Observations

To make a connection between the fluctuations in a scalar field and the density changes they cause is a delicate subject with many pitfalls having to do with gauge invariance. Here we only give a flavour of the subject, directing the interested reader to specialized treatises [28, 32, 40].

The simplest gauge to choose is the so-called conformal Newtonian gauge, where the gauge freedom (i.e. invariance under reparametrization of the coordinates) can be used to put the metric in the form

$$ds^2 = a^2(\eta) \left[(1 + 2\Phi)d\eta^2 - (1 - 2\Psi)\delta_{ij}dx^i dx^j \right], \quad \text{(E.29)}$$

where in addition $\Phi = \Psi$ from the variation of the $i \neq j$ part of the Einstein equations, $\delta G_{ij} = 0$ in the absence of anisotropic stress. (We can also write $\Psi = \phi_1$ connecting to (11.40).) From the other components,

$$\delta G_j^0 = 8\pi G \delta T_j^0 \quad \text{(E.30)}$$

one obtains [40], using $T_{\mu\nu}$ for a single scalar field,

$$T_{\mu\nu} = \partial_\mu\varphi\partial_\nu\varphi - g_{\mu\nu} \left[\frac{1}{2}g^{\rho\sigma}\partial_\rho\varphi\partial_\sigma\varphi - V(\varphi) \right], \quad \text{(E.31)}$$

$$\dot{\Psi} + H\Psi = 4\pi G\dot{\varphi}\delta\varphi = \varepsilon H^2 \frac{\delta\varphi}{\dot{\varphi}}, \quad \text{(E.32)}$$

where ε is the slow-roll parameter introduced in (10.25). From analysing the perturbation δG_{ij} one can see that $\dot{\Psi}$ is very small (proportional to slow-roll parameters) on super-horizon scales, so that

$$\Psi \sim \varepsilon H \frac{\delta\varphi}{\dot{\varphi}}. \quad \text{(E.33)}$$

---

[1] In principle, there is a loophole in this argument. For a given conformal time η, the smallest distance we can speak about in conventional theory is the Planck length. It is not excluded that the field enters from the Planck length in some other state through violent, unknown physics. This state is then evolved by the inflationary expansion. This may give rise to 'transplanckian' effects that may be searched for in the microwave background, for example.

This fluctuation in Ψ can be regarded as a curvature perturbation, since we may replace the lowest-order Friedmann equation for a flat Universe

$$H^2 = \frac{8\pi G}{3}\rho_0,$$
(E.34)

by a perturbed one (*cf.* (4.16))

$$H^2 = \frac{8\pi G}{3}(\rho_0 + \rho_1) - \frac{\delta k}{a^2},$$
(E.35)

where δk is a perturbation in the parameter which defines the curvature in (4.16). Forming the difference between the two at epochs when the two values of H are equal, we find

$$\frac{\delta\rho}{\rho} = \frac{\rho_1}{\rho_0} = \frac{\delta k}{a^2 H^2}$$
(E.36)

This shows that variations in the total energy density, such as in the inflaton field which governs the expansion during inflation, has a geometric nature. The perturbation δk is sometimes written

$$\frac{\delta k}{a^2} = \frac{2}{3}\nabla^2 \mathcal{R},$$
(E.37)

which defines what is usually called the curvature perturbation, $\mathcal{R}$. For a perfect fluid, with stress-energy tensor given by (3.54) we can deduce the general relativistic versions of the continuity equation from the vanishing divergence of the energy-momentum tensor,

$$\partial_t \rho = -3H(\rho + p)$$
(E.38)

and similarly the general relativistic Euler equation

$$a_i = -\frac{\partial_i p}{p + \rho},$$
(E.39)

where **a** is the acceleration. (This shows that in general relativity it is the sum of p and ρ that acts as source for acceleration, not just ρ as in Newtonian mechanics.) In the presence of a pressure perturbation δp, the equation for the acceleration, (4.28), sometimes called the *Raychaudhuri equation* takes the form

$$\frac{\ddot{a}}{a} = \frac{-4\pi G}{3}(\rho + 3p) - \frac{1}{3}\frac{\nabla^2 \delta p}{p + \rho}.$$
(E.40)

We can now use the time derivative of (E.35) together with (E.38), (E.40) and (11.40) to obtain

$$\dot{\mathcal{R}}_{\mathbf{k}} = -H\frac{\delta p_{\mathbf{k}}}{p + \rho}.$$
(E.41)

It can be shown [28] that the right hand side of this equation is very small compared to $H\mathcal{R}_{\mathbf{k}}$ on super-horizon scales, so that $\mathcal{R}_{\mathbf{k}}$ stays essentially constant. That is in fact the great virtue of the curvature perturbation, which makes it very useful when analysing cosmological perturbations. In addition, a simple relation between $\mathcal{R}_\mathbf{k}$ and the Newtonian potential perturbation $\Phi_{1\mathbf{k}}$ is found:

$$\mathcal{R}_\mathbf{k} = -\frac{5 + 3\alpha}{3 + 3\alpha} \Phi_{1\mathbf{k}} \tag{E.42}$$

where as usual we have used the equation of state parameter α to write $p = \alpha\rho$. In the slow-roll case, connecting to (E.33), one finds

$$\mathcal{R}_\mathbf{k} = \Psi_\mathbf{k} + H\frac{\delta\varphi_\mathbf{k}}{\dot\varphi} = (1 + \varepsilon)\frac{\delta\varphi_\mathbf{k} H}{\dot\varphi} \approx \frac{\delta\varphi_\mathbf{k} H}{\dot\varphi}, \tag{E.43}$$

and the power spectrum

$$\mathcal{P}_\mathcal{R}(k) \equiv \frac{k^3}{2\pi^2}|\mathcal{R}_\mathbf{k}|^2 = \frac{k^3}{2\pi^2}\frac{H^2}{\dot\varphi^2}\Delta_\varphi^2(k), \tag{E.44}$$

where the field fluctuation is $\Delta_\varphi^2(k) \equiv |\delta\varphi_\mathbf{k}|^2$.

We now return to the computation of this vacuum fluctuation of the inflaton field which we compute as

$$\langle|\delta\varphi|^2\rangle = \langle 0_{in}|\varphi^\dagger(\eta, \mathbf{r})\varphi(\eta, \mathbf{r})|0_{in}\rangle =$$
$$\int_0^\infty \frac{k^2 dk}{2\pi^2}|\phi_\mathbf{k}(\eta)|^2 = \frac{-\eta}{8\pi a^2(\eta)}\int_0^\infty k^2 dk |H_\nu^{(1)}(-k\eta)|^2. \tag{E.45}$$

The fluctuation per logarithmic k range, $d\ln k = dk/k$ then is

$$\Delta_\varphi^2(k) = \frac{k^3}{2\pi^2}|\phi_\mathbf{k}(\eta)|^2 = \frac{H}{8\pi}(-k\eta)^3 |H_\nu^{(1)}(-k\eta)|^2. \tag{E.46}$$

During inflation, $-k\eta = k/(aH) \to 0$ exponentially fast. That means that the physical linear size of the fluctuation a/k becomes super-horizon, i.e., larger than $1/H$. By using the small argument expansion

$$H_\nu^{(1)}(x << 1) \sim -i\frac{\Gamma(\nu)}{\pi}\left(\frac{x}{2}\right)^{-\nu}, \tag{E.47}$$

we find for the super-horizon scale fluctuations

$$\Delta_\varphi^2(k) = \left(\frac{H}{2\pi}\right)^2 2^{2\nu-3}\left(\frac{\Gamma(\nu)}{\Gamma(\frac{3}{2})}\right)^2\left(\frac{k}{aH}\right)^{3-2\nu}. \tag{E.48}$$

When inflation ends, these fluctuations again come inside the horizon and will act as primordial seeds for structure formation. If we remember that $\nu^2 = 9/4 - m^2/H^2$, we see that the massless limit is particularly simple,

$$\Delta_\varphi^2(k) = \left(\frac{H}{2\pi}\right)^2. \tag{E.49}$$

This does not depend on k, so it is a scale-invariant fluctuation. Since its origin is in a set of uncoupled harmonic oscillators representing the vacuum fluctuation of the inflaton, these fluctuations are also gaussian and independent for each mode $\mathbf{k}$. They contribute to the curvature perturbation which means that they are adiabatic.

In the slow-roll approximation, we can insert in (E.22)

$$\frac{\ddot{a}}{a} = a^2(2 - \varepsilon)H^2 \tag{E.50}$$

and $a(\eta) = -1/(H\eta[1 - \varepsilon])$ instead of (E.23). Using also $m^2/H^2 = 3\eta$ with η the second slow-roll parameter (10.26), one finds

$$\nu \simeq \frac{3}{2} + \varepsilon - \eta, \tag{E.51}$$

and inserted into (E.44) the result is

$$\mathcal{P}_\mathcal{R}(k) = \frac{G}{2\varepsilon} \left(\frac{H}{2\pi} \right)^2 \left(\frac{k}{aH} \right)^{n_\mathcal{R}-1}, \tag{E.52}$$

where we find the spectral index to be

$$n_\mathcal{R} = 1 - 6\epsilon + 2\eta. \tag{E.53}$$

We see that the deviations from scale invariance are small and a generic prediction of inflation is thus an almost scale-invariant, gaussian, adiabatic spectrum of primordial fluctuations. This prediction has recently been vindicated by observations of the microwave background, as explained in Chapter 11.

F Suggestions for Further Reading

- Bahcall, J. N., *Neutrino Astrophysics*, Cambridge Univ. Press, 1989.
- Berezinsky, V.S., Bulanov, S.V, Dogiel, V.A., Ginzburg, V.L., and Ptuskin, V.S., *Astrophysics of Cosmic Rays*, Elsevier, 1990.
- Börner, G., *The Early Universe: facts and fiction*, Springer-Verlag, 1993.
- Cheng, T.-P. and Li L.-F., *Gauge theory and elementary particle physics*, Clarendon Press, 1984.
- Close, F. E., *An introduction to quarks and leptons*, Academic Press, 1979.
- Dolgov, A. D., Sazhin, M. V. and Zeldovich, Ya. B., *Basics of Modern Cosmology*, Editions Frontieres, 1990. Translation of a Russian book published in 1988.
- Gaisser, T. K., *Cosmic Rays and Particle Physics* Cambridge Univ. Press, 1990.
- Kenyon, I. R., *General Relativity*, Oxford Science Publications, 1990.
- Kolb, E. W. and Turner, M. S., *The Early Universe*, Addison-Wesley (Frontiers in Physics, 69), 1990.
- Linde, A., *Particle Physics and Inflationary Cosmology*, Harwood Academic (Contemporary Concepts in Physics, v. 5), 1990. Translation of a Russian book.
- Liddle, A. R, *An Introduction to Modern Cosmology*, Wiley, 1999.
- Liddle, A. R. and Lyth, D. H., *Cosmological Inflation and Large-Scale Structure*, Cambridge University Press, 2000.
- Longair, M. S., *High Energy Astrophysics, 2nd ed.*, Cambridge Univ. Press, 1992.
- Mandl, F. and Shaw, G., *Quantum Field Theory*, Wiley, 1984.
- Narayan, R. and Bartelmann, M., *Lectures on Gravitational Lensing, 13th Jerusalem Winter School in Theoretical Physics: Formation of Structure in the Universe*, Jerusalem, Israel, arXiv:astro-ph/9606001, 1996.
- Narlikar, J. V., *Introduction to Cosmology*, 2nd ed. Cambridge Univ. Press, 1993.
- Peacock, J. A., *Cosmological Physics*, Cambridge Univ. Press, 1999.
- Peebles, P .J .E., *Principles of Physical Cosmology*, Princeton Univ. Press (Princeton Series in Physics), 1993.

- Raffelt, G. G., *Stars as Laboratories for Fundamental Physics: The Astrophysics of neutrinos, axions, and other weakly interacting particles*, Chicago Univ. Press, 1996.
- Rees, M., *Perspectives in Astrophysical Cosmology* Cambridge Univ. Press, 1995.
- Rindler, W., *Introduction to Special Relativity*, Oxford University Press, 1982.
- Roos, M., *Introduction to Cosmology*, Wiley, 1994.
- Roulet, E. and Mollerach, S., *Microlensing*, Physics Reports, **279**, 67:118, 1997.
- Rowan-Robinson, M., *Cosmology, 2nd ed.*, Clarendon Press, 1981.
- Ryder, L. H., *Quantum Field Theory, 2nd ed.*, Cambridge University Press, 1996.
- Sakurai, J. J., *Advanced Quantum Mechanics*, Addison-Wesley, 1967.
- Schneider, P., Ehelers, J. and Falco, E. E., *Gravitational lenses*, Springer-Verlag (Astronomy and Astrophysics Library), 1994.
- Schutz, B. F., *A first course in general relativity*, Cambridge University Press, 1985.
- Taylor, E., F. and Wheeler, J. A., *Spacetime Physics*, Freeman, 1992.
- Weinberg, S., *Gravitation and Cosmology: principles and applications of the general theory of relativity*, Wiley, 1972.

More specialized articles by researchers can be found on the web at location
http://arxiv.org/astro-ph (and hep-ph). Also, Proceedings from *The Texas Conference on Relativistic Astrophysics* (every even year) and *TAUP – Topics in Astroparticle and Underground Physics* (every odd year) contain, among other things, reviews of many of the topics covered in this book. The level is, however, quite advanced.

References

1. Alcock, C., *et al.* (The MACHO collaboration), *Ap. J.*, **471**, 774, 1996.
2. Alcock, C., *et al.* (The MACHO collaboration), *Nature*, **365**, 621, 1993.
3. AMS home page at http://mitlns.mit.edu/~elsye/ams.html.
4. Bahcall, J. N., Neutrino Astrophysics, Cambridge University Press, 1989. Recent updates and figures are available at http://www.sns.ias.edu/~jnb.
5. Bennett, D.L., et al., WMAP Collaboration, arXiv:astro-ph/0302207 (2003).
6. Berezinsky, V. S., Bulanov, S. V., Dogiel, V. A., Ginzburg, V. L., and Ptuskin, V. S., *Astrophysics of Cosmic Rays*, Elsevier, 1990.
7. Bjorken, J. D. and Drell, S. D., *Relativistic Quantum Mechanics*, McGraw-Hill, 1964.
8. Blandford, R. and Eichler, D., *Phys. Rep.*, **154**,1, 1987.
9. Bratton, C. B.,*et al.*, (The IMB collaboration), *Phys. Rev. D*, **37**, 3361, 1988.
10. Börner, G., *The Early Universe: facts and fiction*, Springer-Verlag, 1993.
11. Copi, C. J, Schramm, D. N. and Turner, M. S., *Science*, **267**, 192-199, 1995.
12. Damour, T., *Class. Quant. Grav.*, **13**, A33-A42, 1996.
13. Elgarøy, Ø., et al., *Phys. Rev. Lett.*, **89**, 061301, 2002.
14. Fixen, D. J., *et al.*, *Ap.J.*, **473**, 576, 1996.
15. GLAST home page at http://www-glast.stanford.edu.
16. Goldenfeld, N., *Lectures on Phase Transitions and the Renormalization Group*, Addison-Wesley, 1992.
17. Goldstein, H.,*Classical Mechanics*, Addison-Wesley, 1980.
18. Hayashida, N., *et al.*, *Phys. Rev. Lett.*, **73**, 3491, 1994.
19. Hamuy, M., *et al.* 1996, *Ap. J.*, **112**, 2391, 1996.
20. Hawking, S. W., *Nature*, **248**, 1974. *Commun. Math. Phys.* **43**, 1975.
21. Hirata, K. S.,*et al.* (The Kamiokande collaboration), *Phys. Rev. D*, **38**, 448, 1988.
22. Itzykson, C. and Zuber, J. B., *Quantum Field Theory*, McGraw-Hill, 1980.
23. Jackson, J. D., *Classical Electrodynamics*, Wiley, 1975.
24. Kenyon, I. R., *General Relativity*, Oxford University Press, 1990.
25. Knop, R. *et al.*, *Ap. J.* in press, 2003.
26. Kolb, E. W. and Turner, M. S., *The Early Universe*, Addison-Wesley, 1990.
27. Learned, J. G., and Pakvasa, S., *Astropart. Phys.*, **3**, 267, 1995.
28. Liddle, A. R. and Lyth, D. H., *Cosmological Inflation and Large-Scale Structure*, Cambridge University Press, 2000.
29. Linde, A., *Phys. Lett.* **B129**, 177, 1983.
30. Longair, M. S., *High Energy Astrophysics*, 2nd ed., Cambridge University Press, 1992.
31. Mandl, F. and G. Shaw, G., *Quantum Field Theory*, Wiley, 1984.

32. Mukhanov, V.F, Feldman, H.A. and Brandenberger, R.H., *Physics Reports* **215**, 203, 1992.

33. Peebles, P. J. E., *Principles of Physical Cosmology*, Princeton University Press, 1993.

34. Peebles, P.J.E., *Ap. J.* **153**, 1, 1968.

35. Perlmutter, S. *et al.*, *Ap. J.*, **517**, 565, 1999.

36. Press, W. and Thorne, K. S., *Rev. Astron. Astrophys.*, **10**, 335, 1972.

37. Raffelt, G. G., *Stars as Laboratories for Fundamental Physics*, University of Chicago Press, 1996.

38. Riess, A. G., *et al.*, *AJ*, **117**, 707, 1999.

39. Rindler, W., *Introduction to Special Relativity*, Oxford University Press, 1982; Taylor, E., F. and Wheeler, J. A., *Spacetime Physics*, Freeman, 1992.

40. Riotto, A., Lectures delivered at the *ICTP Summer School on Astroparticle Physics and Cosmology*, Trieste, 17 June - 5 July 2002, arXiv:hep-ph/0210162.

41. Roos, M., *Introduction to Cosmology*, Wiley, 1994.

42. Sakurai, J. J., *Modern quantum mechanics*, Addison-Wesley, 1987.

43. Schutz, B. F., *A First Course in General Relativity*, Cambridge University Press, 1985.

44. Schramm, D. N. and Turner, M. S., *Rev. Mod. Phys.*, **70** (1998) 303.

45. Sofue, Y., *Ap.J.*, **458**, 120, 1996; Sofue, Y., *PASJ*, **49**, 17, 1997; Sofue, Y., Tutui, Y., Honma, M., and Tomita, A., *A.J.*, **114**, 2428, 1997; Sofue, Y., Tomita, A., Tutui, Y., Honma, M., and Takeda, Y., *PASJ*, submitted 1998.

46. Stecker, F. W., and De Jager, O. C., *Ap.J.*, **473**, L75, 1996; Berezinsky, V. S., Bergström, L., and Rubinstein, H. R., *Phys. Lett.*, **B407**, 53, 1997.

47. Swordy, S., private communication. The points are from published results of the LEAP, Proton, Akeno, AGASA, Fly's Eye, Haverah park and Yakutsk experiments.

48. Szabo, A. P., and Protheroe, R., J., in *Proc. High Energy Neutrino Astrophysics*, eds. Stenger, V. J., Learned, J. G., Pakvasa, S. and Tata, X., World Scientific, 1992.

49. Taylor, J. H., *Class. Quant. Grav.*, **10**, S167, 1993. (Supplement 1993).

50. Thorne, K. S., in *Black Holes and Relativistic Stars*, Proceedings of a Conference in Memory of S. Chandrasekhar, ed. R. M. Wald (University of Chicago Press, Chicago, 1998).

51. Weinberg, S., *Gravitation and Cosmology*, Wiley, 1972.

Index

Plate 1. With the most modern optical telescopes such as the Hubble Space Telescopes (HST) we can see objects over 2 billion times fainter than with the unaided eye. This picture, containing hundreds of distant galaxies, shows a part of the Hubble Deep Field, a small region on the sky in the direction of Ursa Major which was in the focus of HST for more than 10 consecutive days. Since the finite speed of light means that the more distant objects are seen as they appeared at much earlier time than the present, pictures such as this can be used to learn about how galaxies formed in the early Universe and how they have since evolved. Credit: R. Williams, The HDF Team (STScI), NASA.

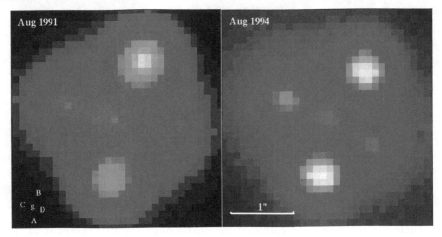

Plate 2.CCD imaging of the the Einstein Cross gravitational system seen over a three year period. A distant QSO is split into four images as its light passes through a foreground galaxy at $z = 0.04$. The brightness variations on the individual images are thought to be caused by passing stars in the foreground galaxy, thereby microlensing the QSO. Images from G. Lewis and M. Irwin at the William Herschel Telescope.

Plate 3. Gravitational lensing by the cluster of galaxies CL0024+1654. The cluster produces four or possibly five separate images of the conspicuous blue galaxy which happens to be located behind the cluster. Analyses of images like this shows that the mass of galaxy clusters is dominated by dark matter. HST image, credit: W.N. Colley and E. Turner (Princeton), J.A. Tyson (Lucent Technologies), HST, NASA.

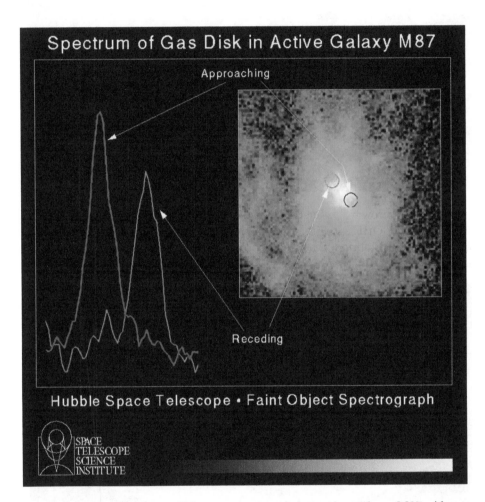

Plate 4. Hubble Space Telescope image of the active galaxy M87. About 60 light years from the centre of the galaxy the gas is moving at 550 km/s towards the Earth on one side and in the opposite direction on the other side. Courtesy of STScI/NASA.

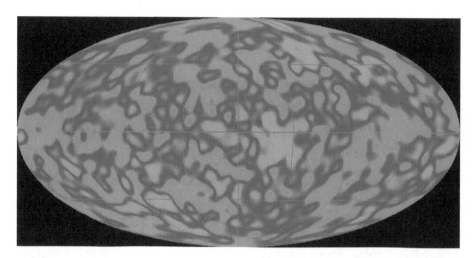

Plate 5. Microwave image of the entire sky as seen by the DMR instrument on board the COBE satellite, after removal of foreground objects and effects due to the local motion. The red 'spots' show regions with higher measured temperature. Credit: The COBE/DMR Team, NASA.

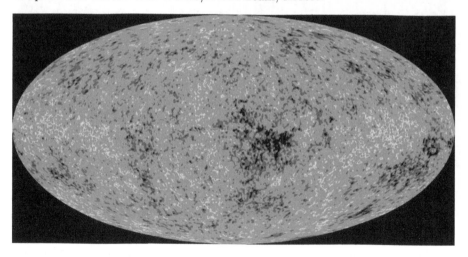

Plate 6. The all-sky map of the cosmic microwave background as measured by the Wilkinson Microwave Anisotropy Probe (WMAP). Notice the dramatic increase in angular resolution compared with the COBE data in Plate 6. Credit: The WMAP Science Team, NASA.